Curriculum and Environmental Education

This collection traces the development and findings of curriculum studies of environmental education since the mid-1970s. Based on a virtual special issue of the *Journal of Curriculum Studies*, the volume identifies a series of curriculum challenges *for* and *from* environmental education. These return us to key questions in curriculum politics, planning and implementation, including which educative experiences should a curriculum foster and why; what the scope of a worthwhile curriculum should be and how it should be decided, organised and reworked; why distinctive curricula are provided to different groups of students; and how might curriculum best be enacted and evaluated?

The editor and contributors call for renewed attention to the possibilities for future directions in research, in light of previously published work and innovations in scholarship. They also offer critical commentary on curriculum, critique and crisis in environmental education, through new material and previous studies from the journal, that addresses three key themes: perspectives on curriculum and environment education; accounting for curriculum in environmental education; and changes in curriculum for environmental education.

Alan Reid is an Associate Editor of the *Journal of Curriculum Studies* and Editor of *Environmental Education Research*. He conducts a wide range of studies focused on teachers' thinking and practice in environmental and sustainability education, and associated traditions, capacities and issues in theory, research and practice. His recent work considers the history and possible futures of the field.

Curriculum and Environmental Education

Perspectives, Priorities and Challenges

Edited by
Alan Reid

LONDON AND NEW YORK

First published 2018
by Routledge

2 Park Square, Milton Park, Abingdon, Oxfordshire OX14 4RN
52 Vanderbilt Avenue, New York, NY 10017

Routledge is an imprint of the Taylor & Francis Group, an informa business

First issued in paperback 2020

British Library Cataloguing in Publication Data
A catalogue record for this book is available from the British Library

ISBN 13: 978-1-138-50183-6 (hbk)
ISBN 13: 978-0-367-58982-0 (pbk)

Typeset in Plantin
by RefineCatch Limited, Bungay, Suffolk

Publisher's Note
The publisher accepts responsibility for any inconsistencies that may have
arisen during the conversion of this book from journal articles to book chapters,
namely the possible inclusion of journal terminology.

Disclaimer
Every effort has been made to contact copyright holders for their permission to
reprint material in this book. The publishers would be grateful to hear from any
copyright holder who is not here acknowledged and will undertake to rectify
any errors or omissions in future editions of this book.

Contents

CONTENTS

Citation Information

The following chapters were originally published in the *Journal of Curriculum Studies*. When citing this material, please use the original page numbering for each article, as follows:

Chapter 4
Environmental education and the issue of nature
Michael Bonnett
Journal of Curriculum Studies, volume 39, issue 6 (2007), pp. 707–721

Chapter 5
*'Littered with literacy': an ecopedagogical reflection on whole language,
pedocentrism and the necessity of refusal*
David W. Jardine
Journal of Curriculum Studies, volume 26, issue 5 (1994), pp. 509–524

Chapter 6
From epistemology to ecopolitics: renewing a paradigm for curriculum
Noel Gough
Journal of Curriculum Studies, volume 21, issue 3 (1989), pp. 225–241

Chapter 7
Sustainability and the learning virtues
John Foster
Journal of Curriculum Studies, volume 43, issue 3 (2011), pp. 383–402

Chapter 8
*Ideology, political education and teacher education: matching paradigms and
models*
John Fien
Journal of Curriculum Studies, volume 23, issue 3 (1991), pp. 239–256

Chapter 9
Ecological consciousness and curriculum
Marla Morris
Journal of Curriculum Studies, volume 34, issue 5 (2002), pp. 571–587

Chapter 19

Towards a socially critical environmental education: water quality studies in a coastal school

Annette Greenall Gough and Ian Robottom

Journal of Curriculum Studies, volume 25, issue 4 (1993), pp. 301–316

Chapter 20

Teacher receptivity to curriculum change in the implementation stage: the case of environmental education in Hong Kong

John Chi-Kin Lee

Journal of Curriculum Studies, volume 32, issue 1 (2000), pp. 95–115

Chapter 21

Complementary curriculum: the work of ecologically minded teachers

Christy M. Moroye

Journal of Curriculum Studies, volume 41, issue 6 (2009), pp. 789–811

For any permission-related enquiries please visit:
http://www.tandfonline.com/page/help/permissions

Preface

Curriculum is found in human eyes, in rivers, in animals, in the language of music, poetry, art, science, history, anthropology, in what is public, intimate, beloved.

(Dunlop, 2009, p. 16)

This edited collection reworks 'Curriculum Challenges *for* and *from* Environmental Education', the first Virtual Special Issue prepared for the *Journal of Curriculum Studies*. At its core, the volume reproduces eighteen studies from the Journal, topping and tailing these with a suite of original commentaries in the following ways.

First, the studies of curriculum and environmental education are arranged into three themes with six chapters in each section:

Part 1. Perspectives on Curriculum and Environment Education
Part 2. Accounting for Curriculum in Environmental Education
Part 3. Changes in Curriculum for Environmental Education.

Second, additional material in the Virtual Special Issue has been redeveloped into two online sections (at http://explore.tandfonline.com/content/ed/jcs-vsi-ee), focusing on:

Part 4. Questioning the Curriculum, from the Mainstream to the Margins
Part 5. For Wisdom, Justice and Action in Curriculum?

Third, after a general introduction to *environmental education and curriculum*, new and updated material from the Virtual Special Issue offers this volume's readers the following:

- a **non-technical introduction** to the theme of 'curriculum challenges *for* and *from* environmental education'
- an **essay** on 'how to understand curriculum challenges *for* and *from* environmental education'
- a **reflection** on 'curriculum, critique and crisis in environmental education'.

In brief, drawing inspiration from the source for this collection's epigram (Dunlop, 2009), the general and non-technical introductions offer a primer of sorts for key considerations in Parts 1–3. The introductions are also largely written for those who might regard themselves more as outsiders than insiders to the field of environmental education, if not somewhere in between or relative newcomers to the scene. In common, they sketch various aspects of the background to studies from the Journal as a platform for advancing critical understandings of curriculum and environmental education. They also tap into distinct but key aspects of the wider international literatures on the volume's key themes of curriculum perspectives, accountings and changes, rather than rehearse key debates about curriculum *per se*.

Next, in the essay on 'How to understand curriculum challenges *for* and *from* environmental education', we dig into some of the key arguments and complexities arising from the curriculum challenges raised in five decades of studies from the Journal. On the one hand, readers may recognise that some of the authors of these studies have set out to reiterate or reinforce particular aspects of curriculum thinking and scholarship about environmental education, while others have sought to reimagine or refine these by offering what amount to a series of 'deviations' from what might be otherwise treated as its standard motifs (for these and other possibilities, see for example, Hungerford et al., 1980; Gough, 1994; Luke, 2001; Vare & Scott, 2007; Kopnina, 2014). In other words, as a form of commentary, the essay introduces broader questions looping through the volume as a whole: what comes—and has come—to count as timely critique of the curriculum field of environmental education since the 1970s, and how to evaluate the (continued?) relevance of various curriculum challenges in the field as either historic, current or anticipated areas of debate (cf. Covert, 1969; Krall, 1978; Brennan, 1991; Smyth, 1998; Stevenson, 2006; Payne, 2006; Jickling & Wals, 2012; Hilbur et al., 2016).

Another key feature of the essay is to suggest that the making of curriculum in and for environmental education must address three key areas of consideration, including how interconnected or not these are in scholarship when establishing and debating the 'languages and grammars' of:

(i)   the curriculum field of environmental education,
(ii) that of education more generally conceived, and
(iii) broader questions of how to pursue and evaluate human life and flourishing on the planet in education.

Finally, after the three sets of studies, the collection concludes with a reflective essay. As but one way of wrapping up such a volume, the reflection is designed to reconnect readers with questions of the timeliness and relative power of curriculum critiques, noting too there are associated sets of studies to consider online for this collection, and studies that continue to be published in the Journal and elsewhere on these themes. More specifically, the reflection invites readers to probe how the key challenges identified in those studies might inform and rework the activity of three spheres of action that have variously established, developed or evaluated curriculum in environmental education, be that in the past through to today's concerns. These spheres, as introduced in the general and non-technical introductions, are typified by scholarship addressing:

- international events and initiatives in this field (such as the work of the United Nations or the World Environmental Education Congress—see, for example, Grün, 1996; Lotz-Sisitka, 2005; Sauvé et al., 2007),
- the field of environmental education and closely related areas known as 'adjectival educations' (see Gayford, 1991; Ferreira, 2000; Sterling, 2014), and
- environmental education and curriculum studies more broadly (cf. Le Grange, 2004; Blenkinsop & Egan, 2009; Schinkel, 2009).

In other words, as befits studies published in the Journal, by drawing together a series of examples and reflections on a range of empirical, philosophical, sociological or policy-related investigations, as well as a range of key arguments about curriculum and environmental education presented in the articles, we hope this collection helps readers identify curriculum challenges both *from* and *for* environmental education.

On behalf of the *Journal of Curriculum Studies*, I thank the scholars who have contributed to five decades of curriculum studies about environmental education, and trust you will find this a thoughtful and stimulating collection that informs further curriculum making, studies and critique in this and related areas.

References

Blenkinsop, Sean, & Egan, Kieran (2009) Three 'big ideas' and environmental education. In: Marcia McKenzie, Paul Hart, Heesoon Bai & Bob Jickling (Eds.), *Fields of Green: Restorying Culture, Environment, and Education* (pp. 85–93) (Cresskill: Hampton Press).

Brennan, Andrew (1991) Environmental awareness and liberal education. *British Journal of Educational Studies*, 39:3, 279–296.

Covert, Douglas C. (1969) Toward a curriculum in environmental education. *Environmental Education*, 1:1, 11–12.

Dunlop, Rishma (2009) Primer: alphabet for the new republic. In: Marcia McKenzie, Paul Hart, Heesoon Bai, and Bob Jickling (Eds.), *Fields of Green: Restorying Culture, Environment, and Education* (pp. 11–63) (Cresskill: Hampton Press).

Ferreira, Jo-Anne (2000) Learning to govern oneself: environmental education pedagogy and the formation of an ethical subject. *Australian Journal of Environmental Education*, 16, 31–36.

Gayford, Chris (1991) Environmental education: a question of emphasis in the school curriculum, *Cambridge Journal of Education*, 21:1, 73–79.

Gough, Noel (1994) Playing at catastrophe: ecopolitical education after poststructuralism, *Educational Theory*, 44:2, 189–210.

Grün, Mauro (1996) An analysis of the discursive production of environmental education: terrorism, archaism and transcendentalism, *Curriculum Studies*, 4:3, 329–347.

Hillbur, Per, Ideland, Malin, & Malmberg, Claus (2016) Response and responsibility: fabrication of the eco-certified citizen in Swedish curricula 1962–2011, *Journal of Curriculum Studies*, 48:3, 409–426.

Hungerford, Harold R., Peyton, Ben, & Wilke, Richard J. (1980) Goals for curriculum development in environmental education, *The Journal of Environmental Education*, 11:3, 42–47.

Jickling, Bob, & Wals, Arjen E.J. (2012) Debating education for sustainable development 20 years after Rio: a conversation between Bob Jickling and Arjen Wals. *Journal of Education for Sustainable Development*, 6:1, 49–57.

Kopnina, Helen (2014) Future scenarios and environmental education, *The Journal of Environmental Education*, 45:4, 217–231.

Krall, Florence R. (1979) Living metaphors: the real curriculum in environmental education. *Journal of Curriculum Theorizing*, 1:1, 180–185.

Le Grange, Lesley (2004) Embodiment, social praxis and environmental education: some thoughts, *Environmental Education Research*, 10:3, 387–399.

Lotz-Sisitka, Heila (2005) Reflections on the '3rd World Environmental Education Congress: educational pathways towards sustainability', Italy, 2005, *Southern African Journal of Environmental Education*, 22, 172–177.

Luke, Tim W. (2001) Education, environment and sustainability: what are the issues, where to intervene, what must be done? *Educational Philosophy and Theory*, 33:2, 187–202.

Payne, Phillip G. (2006) Environmental education and curriculum theory, *The Journal of Environmental Education*, 37:2, 25–35.

Reid, Alan, & Dillon, Justin (Eds.) (2017) *Environmental Education: Critical Concepts in the Environment*. London: Routledge.

Sauvé, Lucie, Berryman, Tom, & Brunelle, Renée (2007) *Three decades of international guidelines for environment-related education*: A critical hermeneutic of the United Nations discourse, *Canadian Journal of Environmental Education*, 12:1, 33–54.

Schinkel, Anders (2009) Justifying compulsory environmental education in liberal democracies, *Journal of Philosophy of Education*, 43:4, 507–526.

Smyth, John (1998) Environmental education—the beginning of the end or the end of the beginning. *Environmental Communicator*, 28:4, 14–15.

Sterling, Stephen (2014) Separate tracks or real synergy? Achieving a closer relationship between Education and SD, post-2015, *Journal of Education for Sustainable Development*, 8:2, 89–112.

Stevenson, Robert B. (2006) Tensions and transitions in policy discourse: recontextualizing a decontextualized EE/ESD debate, *Environmental Education Research*, 12:3–4, 277–290.

Vare, Paul, & Scott, William (2007) Learning for a change: exploring the relationship between education and sustainable development, *Journal of Education for Sustainable Development*, 1:2, 191–198.

Curriculum and environmental education

Perspectives, priorities and challenges

ALAN REID

Introduction

To coincide with the 8th meeting of the World Environmental Education Congress in Göteborg, Sweden (WEEC, 29 June – 2 July, 2015), the *Journal of Curriculum Studies* created its first Virtual Special Issue on 'Curriculum Challenges *for* and *from* Environmental Education'. Drawing on five decades of studies previously published in the Journal, we wanted the collection to provide both a broad-based and 'long view' of curriculum scholarship on these topics, while in so doing, surface longstanding to fresh debates on curriculum and environmental education.

The Virtual Special Issue was made available for two years at http://explore.tandfonline.com/content/ed/jcs-vsi-2015, and the core of that collection is presented here in book form. For the Virtual Special Issue, the themes of articles included were organised in ways that related, first, to Congress themes and deliberations at the event in Sweden, and second, those that would broach wider considerations in the literature, grounded and exemplified through careful selections of high quality scholarship. As with the Journal's overarching mandate and editorial focus, our continuing hope is that by bringing a wide range of work together online and in book form, the collection can foster further scholarship of various stripes that analyses 'the ways in which the social and institutional conditions of education and schooling contribute to shaping curriculum, including political, social and cultural studies; education policy; school reform and leadership; teaching; teacher education; curriculum development; and assessment and accountability'.

That the interest in curriculum challenges *for* and *from* environmental education has been at such a level to merit a book version of the collection is (to our mind, at least) testament to the ongoing significance of its core themes and concerns about perspectives, accountings and changes in curriculum for this field. In fact, the ongoing relevance of these matters to scholars, policy makers and practitioners is easily recognised around the world, particularly now that the 9th Congress has met in Vancouver in September, 2017. Over the last two years, for example, further questions and inquiries have focused on preparations for and the announcement of the UN's Sustainable Development Goals (SDGs) in September 2015 to form the '2030 Agenda' (readers may recall Agenda 21 included Chapter 36 on education and environmental matters), alongside the launch of a UNESCO Global Action Programme (GAP) on Education for Sustainable Development

(ESD) that followed a United Nations Decade on ESD (2005–14) (see http://en.unesco.org/gap). These and other such markers of activity have met with various pragmatic, political and academic responses both within and beyond the pages of the Journal. Most notably, they have focused on various Decade reports and evaluations (Wals, 2012; UNESCO, 2014b) as well as offered critically-informed investigations of the claims and focus of analysis taking place in relation to curriculum initiatives, policy and evaluation (e.g. Gough, 2005; Sauvé et al., 2005; Ferreira, 2009; Wals, 2009; Feinstein et al., 2013; Van Poeck, et al. 2013; Madsen, 2013; Huckle & Wals, 2015; Simovska & Prøsch, 2016; Hillbur et al., 2016; Waldron et al., 2016).

To illustrate some of the most recent foci, we note that the SDGs contain 7 specific targets on quality education (under Goal 4), as well as emphasise climate change education and educating for sustainable lifestyles (see https://sustainabledevelopment.un.org/sdg4). As examples of related but additional foci for education using an 'adjectival' form (Sterling, 2010, pp. 215–217), we will return to this theme throughout the commentary in this volume, but for now, we note that under the heading, *I. Vision, rationale and principles*, the United Nations underlines the central role of education in the realisation of all 17 SDGs:

> 4. Education is at the heart of the 2030 Agenda for Sustainable Development and essential for the success of all SDGs. Recognizing the important role of education, the 2030 Agenda for Sustainable Development highlights education as a stand-alone goal (SDG 4) and also includes targets on education under several other SDGs, notably those on health; growth and employment; sustainable consumption and production; and climate change. In fact, education can accelerate progress towards the achievement of all of the SDGs and therefore should be part of the strategies to achieve each of them. The renewed education agenda encapsulated in Goal 4 is comprehensive, holistic, ambitious, aspirational and universal, and inspired by a vision of education that transforms the lives of individuals, communities and societies, leaving no one behind.

The targets and means of implementation of SDG4 are as follows:

> 4.1 By 2030, ensure that all girls and boys complete free, equitable and quality primary and secondary education leading to relevant and effective learning outcomes
>
> 4.2 By 2030, ensure that all girls and boys have access to quality early childhood development, care and pre-primary education so that they are ready for primary education
>
> 4.3 By 2030, ensure equal access for all women and men to affordable and quality technical, vocational and tertiary education, including university
>
> 4.4 By 2030, substantially increase the number of youth and adults who have relevant skills, including technical and vocational skills, for employment, decent jobs and entrepreneurship
>
> 4.5 By 2030, eliminate gender disparities in education and ensure equal access to all levels of education and vocational training for the vulnerable, including persons with disabilities, indigenous peoples and children in vulnerable situations
>
> 4.6 By 2030, ensure that all youth and a substantial proportion of adults, both men and women, achieve literacy and numeracy

4.7 By 2030, ensure that all learners acquire the knowledge and skills needed to promote sustainable development, including, among others, through education for sustainable development and sustainable lifestyles, human rights, gender equality, promotion of a culture of peace and non-violence, global citizenship and appreciation of cultural diversity and of culture's contribution to sustainable development

4.a Build and upgrade education facilities that are child, disability and gender sensitive and provide safe, non-violent, inclusive and effective learning environments for all

4.b By 2020, substantially expand globally the number of scholarships available to developing countries, in particular least developed countries, small island developing States and African countries, for enrolment in higher education, including vocational training and information and communications technology, technical, engineering and scientific programmes, in developed countries and other developing countries

4.c By 2030, substantially increase the supply of qualified teachers, including through international cooperation for teacher training in developing countries, especially least developed countries and small island developing States.

Meanwhile, the Global Action Programme aims 'to generate and scale up action in all levels and areas of education and learning to accelerate progress towards sustainable development', while Education for Sustainable Development has been repositioned via the GAP to be the primary mechanism in education that contributes substantially to the 2030 Agenda and the SDGs (see also, UNESCO, 2014a). Pulling this all together, as UNESCO (2017) puts it, 'Quality education for all at the heart of the Sustainable Development Goals'.

These movements of thought and their related foci for action have not gone unnoticed in the field of curriculum studies. Directly and indirectly, they both underscore and implicate a re-visioning and re-purposing of curriculum in environmental education and education more broadly (see, for example, UNESCO, 2012, 2016; McKenzie, 2012; Bengtsson & Östman, 2013; Jickling & Sterling, 2017). The stakes are high too: particularly if the GAP's twin objectives are to be realised (if not contextualised, contested or critiqued):

- reorienting education and learning so that everyone has the opportunity to acquire the knowledge, skills, values and attitudes that empower them to contribute to a sustainable future,
- strengthening education and learning in all agendas, programmes and activities that promote sustainable development.

Turning to the Congress itself, this is a biennial international event that is designed to share and discuss thinking and practice about the world of 'education for environment and sustainable development' (http://www.environmental-education.org/). As a glance at the online programmes for any of the Congresses will show, since the early days there have been longstanding interests in debating trends and innovation in core aspects of education, including priorities for curriculum foci and arrangements, and the role of research and evaluation in developing curriculum and pedagogy.

Thus in bringing these congress-related considerations into conversation with specific and wider curriculum developments and thinking, our hope is that the collection helps surface key challenges that may also apply to these and other such initiatives as the SDGs and the GAP, by illustrating a context for and the content of what constitutes the taking of a 'long view' on key issues and challenges for curriculum and environmental education.

Minding the gaps

All the contributions to the collection were selected and organised to address the general theme of: 'curriculum challenges *for* and *from* environmental education', with the assistance of the editorial board and publishers. The subthemes that were used to group the studies persist across the online and hard copy versions, and are as follows:

Part 1. Perspectives on Curriculum and Environment Education
Part 2. Accounting for Curriculum in Environmental Education
Part 3. Changes in Curriculum for Environmental Education
Part 4. Questioning the Curriculum, from the Mainstream to the Margins
Part 5. For Wisdom, Justice and Action in Curriculum?

In preparing a physical version of the collection, we draw on the first three Parts to form the main content of the book, while Parts 4 and 5 have been revised and are available online, via http://explore.tandfonline.com/content/ed/jcs-vsi-ee.

For this general introduction, we note that a key benefit arising from the way the collection emerged is the opportunity to combine examples of historic to recent scholarship drawn from the Journal between the covers of one volume, that starts from the early 1970s. The timeline is important because studies in the Journal go back to not long after the first attempts to establish the field as a matter of concern in education if not interest to academic scholarship (see, for example, Covert, 1969; Swan, 1969; IUCN, 1970; then from the Journal, Marsden, 1971). Thus in being able to afford such a long view chronologically, the horizon for understanding curriculum and environmental education via the Journal's studies can be pushed back much further than that offered at the World Environmental Education Congress, which first met in Espinho in May 2003.

After Espinho, subsequent congresses took place in Rio de Janeiro, Torino, Durban, Montreal, Brisbane, Marrakech, Göteborg, and Vancouver. According to its official website, the Congress provides a forum for discussions about key issues in environmental education, through the exchange of thoughts, experiences and proposals, and the creation of a worldwide community of research and practice in environmental education and sustainability. Thus cognizant of the likely interests of both the audiences of the Göteborg Congress and the Journal on the curriculum aspects of these discussions, key challenges identified within the Virtual Special Issue asked critical questions of:

(i) 'mainstream' versus 'alternative' curriculum orientations and movements;
(ii) the place and critique of 'instrumentalism' in curriculum policy and practice; and

(iii) how various options for attending to a particular form of 'extensionism'—in essence, those attempts to stretch the curriculum to include another theme without removing or compromising attention to others—have come to shape key debates on curriculum priorities and politics regarding environmental and related areas.

Transitioning the online collection into a traditional format, the content, questions and arguments have been redeveloped so they might engage a broader audience interested in education, curriculum, environment and sustainability, such as those coming to grips with the UN's 2030 Agenda. We have also updated the collection with new material in the form of four fresh contributions: first, to provide this general introduction; then, to provide a non-technical introduction to curriculum challenges *for* and *from* environmental education; and finally, to explore matters of curriculum, critique and crisis in environmental education, through an essay on how we might understand these topics, and a final reflection on the studies reproduced in this collection.

As noted in the Preface, the papers that contribute to the collection are drawn from all five decades of studies published on curriculum and environmental education in the *Journal of Curriculum Studies*. Thus they return readers to the very early days of the field as much as they also illustrate a range of markers, trends and shifts since that formative period (see also the various volumes of Reid & Dillon, 2017, for an in-depth analysis and further illustration of curriculum themes and debates from other journals and academic publications, especially Volume IV, Part 8, on *Curriculum and Pedagogy in Environmental Education – Liberal, Critical and Place-Based Considerations*). To illustrate, the earliest study of curriculum and environmental education in this collection (Marsden, 1971) predates an important international watershed for the field: the United Nations Conference on the Human Environment (UNCHE), held in Stockholm, 1972.

From Stockholm to Tbilisi and beyond

The Stockholm event is one of many emblematic moments from what is now half a century of national and international concerted activity in this area (for a timeline, see www.thegeep.org, including events held at the University of Keele in 1965, as well as Carson City (Nevada), Belgrade, Tbilisi, Moscow, Thessaloniki, Ahmedabad, Bonn, Aichi-Nagoya). Contemporary (if not younger) scholars and students of curriculum and environmental education may be forgiven for not automatically appreciating the novelty factor of, say, the *Earthrise* images taken by William Anders during the December 1968 lunar explorations, or recognising the titles of some of the ground-breaking publications on key themes from that period: *Silent Spring* by Rachel Carson (1962), *Only One Earth* by Barbara Ward and Rene Dubos (1972), the Club of Rome's *Limits to Growth* report (1972), and *Small Is Beautiful* by Fritz Schumacher (1973). But as 'signs of those times', whether it was the fragilities they identified or the faith they expressed in the power of education to address the environmental crisis, in the long run these were often more significant symbolically than in terms of the perspicacity or longevity of their prognostications or prescriptions. Put simply, taken together they helped crystallise and represent an emerging argument for a broader audience that would address a wide range of pressing environmental concerns into the future, and in hopeful ways (see Hicks, 1998; Dahlbeck, 2014). In short, as UNESCO and UNEP have both repeatedly argued, the audience required must engage all citizens of the world, not just scientists, economists or members of special interest groups, and thus the cultural institutions of education, and especially the activities of curriculum making and reform, should be expected to play a key part in achieving this goal.

But consciousness of what, exactly?

A key debate in those early days continues to the present. It is whether environmental education is inevitably characterised by a 'pedagogy of despair' or something other (Hart, 2003). For example, how would citizens around the world learn not to be disheartened by the dire predictions, warnings and trends documented in various '*State of the World*'-type reports (e.g. Clark, 1975; Sessions, 1983; NGO Forum, 1992; Foster, 2008)? In curriculum programmes, commentaries and studies, this challenge often translates into various calls for critical and creative action in and through educational systems, institutions, thinking and reform movements.

In this, scholars, activists, students and educators reasoned (albeit in diverse ways) that the key issue is what constitutes the centre of education, and how this might be recentred through critique and reform. Their goal, in effect, is to bring environmental education out of the peripheries of curriculum thinking and practice, mainstreaming considerations of environmental predicaments that can be addressed in education to make the world 'a better place' for all (see, for example, Hungerford, 1975; Vidart, 1978; Döbler, 1995; Stevenson, 1987; Bowers, 1990; Gough, 1991; van Rossen, 1995; Postma, 2002; Weston, 2004; UNESCO, 2007; Bonnett, 2010; Government of Georgia, 2012).

In the late 1960s and throughout the 1970s, much of the analysis of these predicaments tended to prioritise questions of pollution near and far, land use and reform, and the role of active citizens in curriculum initiatives (see, for example, UNESCO-UNEP, 1976). In particular, an invocation of a *Stockholm*, *Silent Spring*, or *Limits to Growth* would often serve to anchor the first flush of (hopefully) worldwide action by a now global audience of concerned groups looking to harness educational means to address environmental problems (UNESCO-UNEP, 1978).

Other policy tools and interventions are, of course, available such as institutional development, legislation and regulation. But in foregrounding the role of education, was it as simple as 'capture the imagination' of policy makers and the general public through teaching and learning, because imagination links to passion and passion to action? As this collection shows, it is neither that simple nor simplistic. Nor would it necessarily be carried out in that order or addressed in timely ways. (Cf. Linke, 1976; Kauckak et al.,

1978; Stevenson, 1987; Robottom, 1989; Grün, 1996; White et al., 2007; Schinkel, 2009; for various perspectives on the vexed question of charges of indoctrination.) Thus despite the efforts of the United Nations and its Environment Programme in the 1970s (UNEP, see below), arguably it wasn't until particular curriculum texts came onto the scene—as exemplified in the late 1980s by social justice education priorities such as 'Teaching Geography for a Better World' (Fien & Gerber, 1988) in Australia—or until nations saw the development of cross-curricular themes on environmental education as a key component to establishing a 'National Curriculum' (or 'standards') (such as in England and Wales in the early 1990s), that we see the mainstream of curriculum development and debate productively tapping into interest in environmental issues beyond analysis of its niche programs and initiatives (see also Greenall, 1986; Nixon, 1991; and Scott & Reid, 1998).

In the case of *Teaching Geography for a Better World*, a particular feature of that work was the attempt to tie curriculum about the environment to a more socially critical approach to teaching and learning, including educational reform more broadly conceived (Morgan, 2002; see also Fien, 1993; Gruenewald, 2004; Sandell et al., 2005; Sauvé et al., 2007; Chapman, 2011; cf. van Rossen, 1995; Walker, 1997). In fact, while some syllabi and textbooks for school geography, biology and civics had already embedded key curriculum questions on how the world and its regions and peoples were changing and why, in line with UNESCO's earlier recommendations and principles set out for environmental education at Belgrade (1975) and Tbilisi (1977) (see Fensham, 1976, 1978; and then Gough, 1989, for another Australian perspective), a 'new wave' of concern and expectation wasn't far behind. This, its advocates suggested, required inflecting a much more politically-linked orientation to action-taking in the curriculum among students and teachers through promoting socially critical forms of environmental education in existing subject areas such as secondary school geography (e.g. Huckle, 1983; Stowell & Bentley, 1988; Morgan, 2002).

To understand the underpinning of such initiatives, we can ask Dewey-type questions about which particular 'ends-in-view' curriculum fosters when the status quo is shown to be clearly inadequate in addressing environmental predicaments, as well as whether reformist to radical changes are required, if not what are the possibilities for other routes and alternatives to those ends-in-view (Jardine, 1994; Weston, 1996; Reid & Taylor, 2003)? To illustrate some of the curriculum dilemmas surfaced in the 1980s and 1990s, which resonate through to today, when does it help the local to wider cause, for example, if 'education for advanced capitalism' is pitted against 'education for ecological sustainability', mandated teaching is contrasted with free-choice learning, or expert-driven curriculum development trumps practitioner-led action research on curriculum making?

While such contrasts serve to represent some of the poles to contemporary and ongoing debates about curriculum and curriculum development (see Huckle, 1983; Maher, 1986; Naidoo et al., 1990; Orr, 1992; Posch, 1997; Smith, 1998; Heimlich & Falk, 2009; Nelson & Cassell, 2012), clearly whatever one's political, professional or pedagogical allegiances, a recurring point of contention centres on an aspect we've briefly skipped over above.

That is, how to respond to the charge of 'indoctrination' as much as 'sloganizing' or 'rhetorical excess' (. . . education *for survival*, education *for peace*, education *for global citizenship* . . .? see Scheffler, 1960; Huebner, 1967; Hare, 1986; Reichenbach, 2016). In fact, while some curriculum scholarship has added fuel to the fire (e.g. Linke, 1976; Jickling, 1992) other work has tried to de-escalate these concerns, showing for example, that the charges can't be levelled at each and every variant of environmental education curriculum emerging from that period or in place now, as if there was no diversity of view about or forms of environmental education (e.g. Kauchak et al., 1978; Marsden, 1997).

In fact, arguments and counter arguments have also shown how educators, including environmental educators, might respond to any heavy-handedness in how assumed normative dimensions to such education and curriculum are operationalised or critiqued (see van Rossen, 1995; Jickling & Spork, 1998; Ferreira, 2009). Equally, others have probed the contention that the field was actually inadequately resourced philosophically and practically, be that in terms of its educational or environmental terms of reference, as well as in relation to the infrastructures and capacities to address its own problems, leading to the question of whether environmental education required another 'fix' entirely—financially, philosophically, ontologically, epistemologically, technically, ethically, culturally . . . (e.g. Payne, 2006; Blumstein & Saylan, 2007; Cole, 2007).

In short, emerging from such debates is an existential challenge for the field: does environmental education imperil itself with its own extinction by becoming unrecognisable to the concerns and discourses of the mainstream of education by separating itself off (e.g. in its language and grammar) from that self-same mainstream? More pointedly, is this risk heightened by following the agendas of certain scholars or activists as to what the silences are in education? While as some went on to argue, could a critical measure of success be ultimately counter-intuitive for the movement: on the one hand, that it actually makes itself redundant by achieving its aims, or on the other, it celebrates a certain redundancy and its difference by refusing to play the games of mainstream education (see Gruenewald, 2004)?

While there is some degree of irony in any field that imagines its own demise, it must be said that these challenges are often engaged to avoid more talk of what is being shoehorned into the field, or how environmental education is becoming surpassed by other concerns and priorities for curriculum making. Arguments on this might typically concern a focus on sustainable development or climate change, if not the denial of their relevance to the 'core business' of education (e.g. Sessions, 1983; Orr, 1992; Stables & Scott, 2001; cf. McKeown & Hopkins, 2003; and Jickling & Wals, 2008; Coughlan, 2017). But to return to how we might understand the context of the period of the early days of environmental education (e.g. from the late 1960s, through markers that include Covert, 1969; Stapp et al. 1969; Swan, 1969; IUCN, 1970; Fenwick, 1972), we single out Andrew Dobson's (2016) recent work, which exemplifies how contemporary scholarship prepared for a public audience traces the ongoing influence of environmental movements and politics, rather than solely that of educational and curriculum theory on the field.

Dobson's *Very Short Introduction to Environmental Politics* identifies a series of cultural, social and environmental factors (amongst others) that have come to shape not just environmental politics but environmental education too. In common, they raise the question of how a space for addressing the relative influence of materialist and post-materialist ideologies for people-environment relations is created and sustained in contemporary society, including in its cultural institutions such as education (see also O'Riordan, 1981; Pepper, 1987; Russell & Bell, 1996; Oulton & Scott, 2000; Zembylas, 2005; Blühdorn & Welsh, 2008; Kahn, 2008; Bai, 2009). Key amongst these factors, says Dobson, is the broad raft of environmentally-focused social movements and special interest groups that had already begun to surface in the West during the mid- to late 20th century. Many became networked but also differentiated themselves, for example, in relation to dominant and other alternative priorities for 'science and society' in the post-war period. Using the UK as an example, Dobson also highlights how leading bodies and voices sought to have their concerns taken up in a broader range of socio-cultural and politically charged spheres, especially through education, and what some of the barriers to progress were and continue to be (on education and sustainability specifically, see Martin et al., 2013, on international initiatives, funding, teacher education, recruitment and retention of curriculum experts alongside staff turnover, and educational reform).

For environmental education, these interacting spheres highlight the significance of the dynamic worlds, traditions and actors involved in curriculum making, educational assessment and the training (pre-service and inservice) of educators. They also reveal tensions and challenges when these run counter to prevailing governing logics of unfettered economic development and inevitable social progress (for an account from the early days of environmental education, see for example, Ginzbarg, 1971, or McInnis, 1975; or more recently, Lotz-Sisitka, 2008, or Sund & Öhman, 2014).

Yet it is with much more detail than Dobson that this collection shows it was actually at milestone events like UNCHE that the movements and groups—and most importantly, their initial arguments in relation to environment and education—first united *with curriculum in mind*. In particular, an explicit goal was advocated: that of generating a broader, global and deeper shift in people's 'consciousness' through education on those planetary conditions that circumscribe human wellbeing and endeavours, and for generations to come (see for example, Schumacher, 1973, on 'Education as the Greatest Resource').

'Education which fails to clarify our central convictions is mere training or indulgence' (Schumacher, 1973)

For those educationalists becoming increasingly interested in environmental matters (if not the equivalent for environmentalists in relation to education—see Skolimowski, 1991; Monroe, 2012), corresponding efforts and approaches to curriculum quickly moved to those of praxis. Key debates were often on what was required as interventions or changes to the goals, content and experience of curriculum, most often by arguing for their revision, reorientation, and possibly, transformation through a deliberate focus on the provision of environmental education (e.g. Winn, 1970; Simmons & Simmons, 1973; Nash, 1976; O'Riordan, 1981; Fox, 1986; Di Chiro, 1987; Greenall Gough, 1991; Tilbury, 1995; Marsden, 1997; Bonnett, 1999; Scott & Oulton, 1999; Saul, 2000; McKeown & Hopkins, 2003; Gruenewald & Manteaw, 2007).

As we also show in this collection though, it was how the prevailing scaffolds for curriculum design could be revisited, critiqued, supplanted and innovated, rather than the minutiae of syllabi and curricula that attracts serious attention. To use Biesta's (2012) terms, what is required is a rebalancing as well as reconceiving of how education might function to *qualify, socialise* and *subjectify* on these matters. For example, a key characteristic of many earlier attempts (e.g. prior to addressing sustainability as a goal within UNESCO's vision of 'environmental education for sustainable development') was the express intent of shifting and/or reinventing curriculum theory and practice so that it addressed a raft of 'eco-logics' (Gough, 1989) of various forms and configurations, such as

- to foster deeper levels of ecological consciousness through teaching and learning,
- to test whether sustainability worked as a well-conceived 'learning virtue', or
- to ascertain whether it was desirable and possible to make curriculum in general simply less anthropocentric, if not, fully ecocentric

(e.g. Bowers, 1990; Morris, 2002; Foster, 2011; see also Hart, Jickling & Kool, 1999; Gough et al., 2000; Sauvé et al., 2005; McKenzie, 2012).

As can be readily appreciated, such intentions and expectations represent a significant and complex set of endeavours for creating, revising or reforming curriculum and pedagogy when such ends are in view. They also come with a clear set of challenges, particularly if environmental educators are to be able to move beyond tinkering from the margins towards having their work taken up in the mainstream of curriculum, particularly given what else is prioritised or the default in efforts at training, education and public awareness. Campbell (2009), for example, highlights the role of the global media in shaping environmental consciousness not always towards the 'better' or 'common' good or in ways compatible with environmental education, i.e. 'education for uncritical consumption' continues apace. Yet as Irina Bokova, Director-General of UNESCO, in the Foreword to the Global Monitoring and Evaluation Report on its Decade of Education for Sustainable Development, claims: 'The Decade has activated hundreds of thousands of people to reorient education globally towards a central goal: to learn to live and work sustainably' (UNESCO, 2014, p. 3).

Working through such a maze of claims and counter claims then requires scholarship that addresses not only their empirical underpinnings or challenges but also engages whether assumptions about curriculum, including curriculum making, debates and action *on* or *as* an 'environmental education', inform public and scholarly discourses in revising and/or reforming curriculum and pedagogy. As we will show, these should offer grounds for both reasoned and reasonable ways forward, but as with Martin et al. (2013), help identify likely obstacles and blocks to movement and progress in these regards, if not also about their associated conditions and contexts, and how likely these might change too.

The work of the journal

Studies on these matters have drawn the attention of various editors and readers of the *Journal of Curriculum Studies* over five decades, and so its back catalogue of studies is well placed to test out how various local to international, as well as fleeting to enduring, challenges and responses have fared. Moreover, throughout this collection, we trust the studies also illustrate how compelling forms of empirically- and theoretically-based scholarship might advance the critique of curriculum in general if not environmental education *per se*, such as in relation to its anthropocentric biases, assumptions about people and environment (or nature-culture), and why 'education for a better world' isn't always automatically about having people at the centre (see UNESCO, 2014, and then Ideland & Malmberg, 2014; cf. Waldron et al., 2017).

Elsewhere, as Heimlich and Falk (2009) remind us, whether it is in the form of a formal excursion or the informal learning that may take place at a zoo, aquarium or museum, a 'free choice' environmental curriculum typically has a loose teaching-learning relationship. The experiences and composition of these 'learnings' begs a series of questions in relation to biases and inherited and emerging patterns of teacher-learner interaction, including who or what constitutes which in curriculum design, making and critique. Thus there are questions to be asked of both the deliberate and unintended aspects of any associated curriculum (built, intended, experienced, resisted) of environmental education.

To put such provocations to work, we might ask, is education now predominantly and unconsciously 'mono-species', i.e. fixated on *homo sapiens*? How many relationships, and of what forms, does the typical pre-school, school or university experience encourage as interactions with other-than-human species, including in immediate or distant environments? While as some have recently observed, is curriculum making increasingly that which doesn't encourage or press for much by way of interactions with other-than-human biota (Kahn, 2008) if not abiotic materiality (Duhn, 2012)? Consequently, is the range of ethical relationships in curriculum (still) predicated on principles of ecological mastery, dominance, dominion, death or functionality of our environments, even when environmental education is part of the mix (e.g. Lloro-Bidart, 2017; Russell, 2017; Spannring, 2017)?

UNthinking environmental education

To begin to answer these questions and identify some of the key threads that might be followed throughout the collection, since the early days of environmental education many of the expected transitions in direction and substance have been instigated through explicit reference to the work of UNESCO and UNEP, particularly during the late 1970s and into the 1980s. (This period might be regarded as the heyday of the field, before the rise of sustainable development and sustainability-related discourse— see, for example, Sauvé et al., 2007). UNESCO and UNEP sought to determine, establish and delimit the role, goals, objectives, guiding principles, characteristics and audiences for environmental education at events after Stockholm, including in Belgrade (1975) and Tbilisi (1977) (see UNESCO-UNEP, 1976, 1978), through to more recent versions focused on aligning and blending themes in environmental and sustainability education, as discussed in Ahmedabad (UNESCO, 2007), Göteborg (GMV, 2009) and Aichi-Nagoya (UNESCO, 2014), when creating the Global Action Programme on Education for Sustainable Development (GAP).

While this has been previously likened to the creation of a Möbius strip (Reid & Scott, 2006) between environmental and sustainability education, key lines of curriculum inquiry have primarily responded to questions of foundations and contradictions in themes arising from the early and ongoing work of UNESCO and UNEP. These include their assumptions, outworkings, instantiations and alternatives to these, but as this collection also illustrates, this can be brought into conversation with a recognition of a complex and shifting terrain of curriculum and wider environmental and educational debates, such as through dialogue with provocations from the philosophy of education, sociology of education, and anthropology of education. Put bluntly, some studies invite us to ask whether environmental education has lost its way, so to speak, or did it ever start from the right place, when it could have proceeded from somewhere else, or become something other by now? (For example, on ecocentric rather than anthropocentric framings, see Bonnett, 2007; Greenall Gough, 1991; Jardine, 1994; Jickling & Wals, 2008; Smyth, 1995; cf. Kopnina, 2012.)

Overview of the collection

We dig into these and other questions further in terms of the principles and politics of curriculum, as well as themes and challenges *for* and *from* environmental education in subsequent pages on the collection especially in the essay and reflection on curriculum, crisis and critique in the field of environmental education. But before we get to those, we mention a few other features of the collection to aid the reader in navigating the studies that it brings together.

First, the studies are organised thematically into three main parts, with each section containing six studies from the Journal's archives. Given the aforementioned concerns, these aren't arranged chronologically as may be the case with other collections on environmental education (e.g. Reid & Dillon, 2017). This is because there are both ebbs and flows to demonstrate internationally, rather than a linearity to the development of the field and its studies and debates. Also, given the deliberate form of a *virtual* special issue for the initial selection, we note that two supporting selections remain available elsewhere. These illustrate wider curriculum studies of relevance to the focus and breadth of the challenges, and provide other resources and ideas for developing thinking and practice about curriculum and environmental education from the *Journal of Curriculum Studies*, points we also return to in the essay and reflection.

We believe that this approach to organising the studies enables us to show that while the set of issues and challenges that have emerged from five decades of curriculum inquiry about environmental education has not remained stable, in another sense, certain concerns endure, and hence the value of creating thematic titles for each section. By flagging the importance of perspectives, accountabilities and changes (alongside questioning, and wisdom, justice and action), we trust that the collection will be of wider interest than to specialist audiences alone, particularly if it is to explore challenges *from* and *for* the field. Consequently, the five groupings and the studies for both parts of the physical and online collection are as follows:

Perspectives on curriculum and environment education

- Environmental education and the issue of nature, *Michael Bonnett* (2007)
- 'Littered with literacy': an ecopedagogical reflection on whole language, pedocentrism and the necessity of refusal, *David Jardine* (1994)
- From epistemology to ecopolitics: renewing a paradigm for curriculum, *Noel Gough* (1983)
- Sustainability and the learning virtues, *John Foster* (2011)
- Ideology, political education and teacher education: matching paradigms and models, *John Fien* (1991)
- Ecological consciousness and curriculum, *Marla Morris* (2002)

Accounting for curriculum in environmental education

- Environmental education and the secondary school curriculum, *Chris Gayford* (1986)
- Subjects for Study: Aspects of a Social History of Curriculum, *Ivor Goodson* (1983)
- Greening the future for education: changing curriculum content and school organization, *Annette Greenall Gough* (1991)
- Globalization and environmental education: looking beyond sustainable development, *Bob Jickling & Arjen Wals* (2008)
- Environmental Studies Courses in Colleges of Education, *Bill Marsden* (1971)
- Environment in the curriculum: representation and development in the Scottish physical and social sciences, *Hamish Ross* (2007)

Changes in curriculum for environmental education

- Environmental and health education viewed from an action-oriented perspective: a case from Denmark, *Bjarne Bruun Jensen* (2004)
- Implementing curriculum guidance on environmental education: the importance of teachers' beliefs, *Debby Cotton* (2006)
- Curriculum change and climate change: Inside outside pressures in higher education, *Shireen Fahey* (2012)
- Towards a socially critical environmental education: water quality studies in a coastal school, *Annette Greenall Gough & Ian Robottom* (1993)
- Teacher receptivity to curriculum change in the implementation stage: The case of environmental education in Hong Kong, *John Chi-Kin Lee* (2000)
- Complementary curriculum: the work of ecologically minded teachers, *Christy Moroye* (2009)

Questioning the curriculum, from the mainstream to the margins

- Subject departments and the 'implementation' of National Curriculum policy: an overview of the issues, *Stephen Ball & Richard Bowe* (1992)
- No child, no school, no state left behind: schooling in the age of accountability, *Stefan Hopmann* (2008)
- Teacher voice and ownership of curriculum change, *David Kirk & Doune MacDonald* (2001)
- Power and the knowledge produced by educationists, *Elizabeth H. McEneaney* (2002)
- Overcoming the crisis in curriculum theory: a knowledge-based approach, *Michael Young* (2013)
- Reading Schwab's the 'Practical' as an invitation to a curriculum enquiry, *Ian Westbury* (2013)

For wisdom, justice and action in curriculum?

- Becoming world-wise: an educational perspective on the rhetorical curriculum, *Gert Biesta* (2012)
- A new look at big history, *Kate Hawkey* (2014)
- Public understanding of science and science education for action, *Edgar Jenkins* (1994)
- Civic learning: moving from the apolitical to the socially just, *Kristina Llewellyn, Sharon Cook & Alison Molina* (2010)
- (Re)creating citizenship: Saskatchewan high school students' under standings of the 'good' citizen, *Jennifer Tupper & Michael Cappello* (2012)
- Response and responsibility: fabrication of the eco-certified citizen in Swedish curricula 1962–2011, *Per Hillbur, Malin Ideland & Claus Malmberg* (2016)

(The studies in Parts 4 and 5 can be accessed at the revised VSI site, or in libraries for readers wishing to read or review those.)

The beginning of the end, or the end of the beginning?

Alongside producing the iconic Earthrise image, the Apollo 8 mission included a reading from the first lines of a creation myth written into the book of *Genesis*, snippets of which were also evoked on a 1969 US Postal Service commemorative stamp (Scott #1371). As in the source for the epigram to this volume, the late Rishma Dunlop's (2009) 'Primer: Alphabet for the new republic', invites us to probe the poetic and mythic dimensions to any education addressing environments, terrestrial or otherwise. If we are to look at matters of beginnings and endings, to consider the familiar from afar, to surface existential and psychic shocks from an environmental perspective, and to consider how we might respond to them in education, . . . these, like Dunlop's, are key themes to this as much as they were to any broader collection on the 'restorying' of culture, environment and education.

Thus it goes without saying that others have tackled these themes too from a wide range of perspectives: see, for example, the pioneering work of Arthur Lucas (1980) on the question of 'disciplinary chauvinism' in science education in relation to environmental education, Annette Gough (2002) on reworking this through a notion of mutualism, and John Smyth (1995) on the Churchillian notions of this section's subtitle as these apply to environmental and sustainability education. (We return to these and other possibilities later in the collection.) However, sticking with Dunlop's sensibility for a while longer, we note that not long after the Belgrade Charter or the Tbilisi Recommendations on environmental education, Florence Krall (1979) wrote on the place of 'living metaphors' in the 'real curriculum' of environmental education. Thus in commending this collection to its readers, we ask you to consider who we listen to and see as the authors and audiences for 'ourselves living in the curriculum'. Will we continue to appraise whether environmental education can (ever?) achieve the equivalent of a 'Copernican revolution' in curriculum in empirical to theoretical ways? While does it become little more than a form of philately to consider what scholars identify as the challenges for curriculum in environmental education if curriculum making, review and reform don't also accompany this? In other words, how do we show we are 'for' wisdom, justice and action in curriculum as we question it, including in relation to the aims and scope of environmental education?

To bring this general introduction to a close, we note that the 2017 World Environmental Education Congress in Vancouver took place in Dunlop's country of citizenship, Canada, which 'celebrated' 150 years of its modern existence in 2017. The Congress was hosted by staff from Simon Fraser University, whose campuses are acknowledged to be located on the '*unceded traditional territories of the Coast Salish peoples of the Musqueam, Squamish, and Tsleil-Waututh Nations*' (sfu.ca). Not all universities recognise the traditional owners of the land, nor do the scholars who work within and from these particular places, educating and researching as they go. Thus our final hope is to trust that this collection will help readers grasp how various scholarly perspectives, priorities and challenges can be brought to bear on attempts to bring officially-sanctioned as much as incidentally-occurring environmental education into critical conversation with curriculum studies, and vice versa, including to inspire further work in this area. Mindful of the 2017 Congress theme, this might include on 'Culturenvironment: Weaving new connections', if not how, for example, a land education (McCoy et al., 2016) as another form of environmental education (and much else besides) can catalyse new developments in thinking and practice on these topics, such as by rethinking curriculum, sustainability and environment in critical conversation with 'pedagogies of place', and 'Indigenous, postcolonial, and decolonizing perspectives' (see also Gruenewald, 2003).

In sum, we trust that throughout this innovative collection, the historic to contemporary questions and challenges arising *for* and *from* environmental education also serve to invite a vibrant community of curriculum scholars to return to some of its core questions and foster further curriculum inquiry, debate and action on:

- which educative experiences should a curriculum foster and why?
- what should be the scope of a worthwhile curriculum and how should it be decided, organised and reworked?
- why is distinctive curricula provided to different groups of students? and
- how should (a) curriculum be best enacted and evaluated?

We look forward to receiving follow-up submissions to the Journal that help address such questions, and in particular scholarly pieces that help advance our collective understandings of the curriculum challenges *for* and *from* environmental education.

Acknowledgements

The following colleagues have all shared helpful and provocative thoughts with me in a range of venues on the themes and selection for this collection: Stefan Hopmann, Katrien Van Poeck, Jonas Lysgaard, Blanche Higgins, Justin Dillon, Annette Gough, David Greenwood, Leesa Fawcett, William Scott, Marcia McKenzie, Andrew Stables, Martha Monroe, Paul Hart, Connie Russell, Noel McInnis, Judy Braus, Kartikeya Sarabhai, Annette Gough, Lesley Le Grange, Heila Lotz-Sisitka, Hamish Ross, and Zane Ma Rhea. I thank them for their suggestions and steers, and remain mindful that all subsequent errors in judgement and omission, alongside the views expressed in the introductions, essay and reflection, are most safely treated as my own.

References

Bai, Heesoon (2009) Reanimating the universe: environmental education and philosophical animism. In: Marcia McKenzie, Paul Hart, Heesoon Bai & Bob Jickling (Eds.), *Fields of Green: Restorying Culture, Environment, and Education* (pp. 135–151) (Cresskill: Hampton Press).

Ball, Stephen J., & Bowe, Richard (1992) Subject departments and the 'implementation' of National Curriculum policy: an overview of the issues, *Journal of Curriculum Studies*, 24:2, 97–115.

Bengtsson, Stefan, & Östman, Leif (2013) Globalisation and education for sustainable development: emancipation from context and meaning, *Environmental Education Research*, 19:4, 477–498.

Biesta, Gert (2012) Becoming world-wise: an educational perspective on the rhetorical curriculum, *Journal of Curriculum Studies*, 44:6, 815–826.

Blühdorn, Ingolfur, & Welsh, Ian (Eds.) (2008) *The Politics of Unsustainability: Eco-Politics in the Post-Environmental Era* (London: Routledge).

Blumstein, Daniel T., & Saylan, Charlie (2007) The failure of environmental education (and how we can fix it), *PLoS Biology*, 5:5: e120. doi:10.1371/journal.pbio.0050120

Bonnett, Michael (1999) Education for sustainable development: a coherent philosophy for environmental education? *Cambridge Journal of Education*, 29:3, 313–324.

Bonnett, Michael (2007) Environmental education and the issue of nature, *Journal of Curriculum Studies*, 39:6, 707–721.

Bonnett, Michael (2010) 'Getting and spending, we lay waste our powers': environmental education and the culture of the school, *Forum*, 52:1, 87–92.

Bowers, Chet (1990) Educational computing and the ecological crisis: some questions about our curriculum priorities, *Journal of Curriculum Studies*, 22:1, 72–76.

Campbell, Elizabeth H. (2009) Corporate power: the role of the global media in shaping what we know about the environment. In: Kenneth A. Gould & Tammy L. Lewis (Eds.), *Twenty Lessons in Environmental Sociology* (pp. 68–84) (Oxford: Oxford University Press).

Chapman, David (2011) Environmental education and the politics of curriculum: a national case study, *The Journal of Environmental Education*, 42:3, 193–202.

Clark, Edward (1975) Good education is environmental, *The Journal of Environmental Education*, 6:4, 1–5.

Cole, Anna Gahl (2007) Expanding the field: revisiting environmental education principles through multidisciplinary frameworks, *The Journal of Environmental Education*, 38:2, 35–45.

Cotton, Debby (2006) Implementing curriculum guidance on environmental education: the importance of teachers' beliefs, *Journal of Curriculum Studies*, 38:1, 67–83.

Coughlan, Sean (2017) Did Michael Gove really try to stop teaching climate change? *BBC News*. http://www.bbc.com/news/education-40250214

Covert, Douglas C. (1969) Toward a curriculum in environmental education, *Environmental Education*, 1:1, 11–12.

Dahlbeck, Johan (2014) Hope and fear in education for sustainable development, *Critical Studies in Education*, 55:2, 154–169.

Di Chiro, Giovanna (1987) Applying a feminist critique to environmental education, *Australian Journal of Environmental Education*, 3, 10–17.

Döbler, Matthias (1995) Common values and value conflicts in environmental education, *History of European Ideas*, 21:1, 37–46.

Dobson, Andrew (2016) *Environmental Politics: A Very Short Introduction* (Oxford: Oxford University Press).

Duhn, Iris (2012) Making 'place' for ecological sustainability in early childhood education, *Environmental Education Research*, 18:1, 19–29.

Dunlop, Rishma (2009) Primer: alphabet for the new republic. In: Marcia McKenzie, Paul Hart, Heesoon Bai, and Bob Jickling (Eds.), *Fields of Green: Restorying Culture, Environment, and Education* (pp. 11–63) (Cresskill: Hampton Press).

Fahey, Shireen J. (2012) Curriculum change and climate change: inside outside pressures in higher education, *Journal of Curriculum Studies*, 44:5, 703–722.

Feinstein, Noah W., Jacobi, Pedro R., & Lotz-Sisitka, Heila (2013) When does a nation-level analysis make sense? ESD and educational governance in Brazil, South Africa, and the USA, *Environmental Education Research*, 19:2, 218–230.

Fensham, Peter (1976) A report on the Belgrade workshop on environmental education (Canberra: Curriculum Development Centre).

Fensham, Peter J. (1978) Stockholm to Tbilisi – the evolution of environmental education, *Prospects*, VIII, 446–455.

Fenwick, Walter P. (1972) Education and environment, *The Ecologist*, 2:8, 7–10.

Ferreira, Jo-Anne (2009) Unsettling orthodoxies: education for the environment/for sustainability, *Environmental Education Research*, 15:5, 607–620.

Fien, John (1991) Ideology, political education and teacher education: matching paradigms and models, *Journal of Curriculum Studies*, 23:3, 239–256.

Fien, John (1993) *Education for the Environment: Critical Curriculum Theorising and Environmental Education*. Geelong: Deakin University Press.

Fien, John, & Gerber, Rob (Eds.) (1988) *Teaching Geography for a Better World* (Edinburgh: Oliver & Boyd).

Foster, John (2008) Widening our options. In: *The Sustainability Mirage* (pp. 135–148) (London: Earthscan).

Foster, John (2011) Sustainability and the learning virtues, *Journal of Curriculum Studies*, 43:3, 383–402.

Fox, Warwick (1986) Ways of Thinking Environmentally (and Some Brief Comments on their Implications for Acting Educationally). In: Jim Wilson, Giovanna Di Chiro & Ian M. Robottom (Eds.), *Thinking Environmentally ... Acting Educationally: Proceedings of the Fourth National Conference of the Australian Association for Environmental Education* (pp. 21–29). Melbourne: Victorian Association for Environmental Education.

Gayford, Chris (1986) Environmental education and the secondary school curriculum, *Journal of Curriculum Studies*, 18:2, 147–157.

Ginzbarg, Hana M. (1971) Environmental education for political action, *The Journal of Environmental Education*, 3:1, 26–27.

GMV [Centre for Environment and Sustainability] (2009) The Gothenburg Recommendations on Education for Sustainable Development, http://uf.gu.se/digitalAssets/1271/1271420_1270373_Gothenburg_recommendations_on_ESD_Adopted_Nov_12_2008.pdf

González-Gaudiano, Edgar (2005) Education for sustainable development: configuration and meaning, *Policy Futures in Education*, 3:3, 243–250.

Goodson, Ivor (1983) Subjects for study: aspects of a social history of curriculum, *Journal of Curriculum Studies*, 15:4, 391–408.

Gough, Annette (2002) Mutualism: a different agenda for environmental and science education, *International Journal of Science Education*, 24:11, 1201–1215.

Gough, Annette (2005) Sustainable schools: renovating educational processes, *Applied Environmental Education & Communication*, 4:4, 339–351.

Gough, Noel (1989) From epistemology to ecopolitics: renewing a paradigm for curriculum, *Journal of Curriculum Studies*, 21:3, 225–241.

Gough, Noel (1991) A do-it-yourself guide to dismantling Trojan horses – a response to national curriculum initiatives in environmental education, *Australian Journal of Environmental Education*, 7, 112–115.

Gough, Noel (2009) Becoming transnational: rhizosemiosis, complicated conversion, and curriculum inquiry. In: Marcia McKenzie, Heesoon Bai, Paul Hart, and Bob Jickling (Eds.), *Fields of Green: Re-storying Culture, Environment, and Education*, pp. 67–83. Creskill, NJ: Hampton Press.

Gough, Stephen, Scott, William, & Stables, Andrew (2000). Beyond O'Riordan: balancing anthropocentrism and ecocentrism, *International Research in Geographical and Environmental Education*, 9:1, 36–47.

Government of Georgia (2012) The Tbilisi Communiqué – Educate Today for a Sustainable Future (Tbilisi: Ministry of the Environment/UNESCO/UNEP).

Greenall, Annette (1986) Search for a meaning: what is environmental education? *Geographical Education*, 5:2, 9–12.

Greenall Gough, Annette (1991) Greening the future for education: changing curriculum content and school organization, *Journal of Curriculum Studies*, 23:6, 559–571.

Greenall Gough, Annette, & Robottom, Ian (1993) Towards a socially critical environmental education: water quality studies in a coastal school, *Journal of Curriculum Studies*, 25:4, 301–316.

Greenwood (formerly Gruenewald), David A. (2008) A critical pedagogy of place: from gridlock to parallax, *Environmental Education Research*, 14:3, 336–348.

Gruenewald, David (2003) The best of both worlds: a critical pedagogy of place, *Educational Researcher*, 32:4, 3–12.

Gruenewald, David A. (2004) A Foucauldian analysis of environmental education: toward the socioecological challenge of the Earth Charter, *Curriculum Inquiry*, 34, 71–107.

Gruenewald, David A., & Manteaw, Bob O. (2007) Oil and water still: how No Child Left Behind limits and distorts environmental education in US schools, *Environmental Education Research*, 13:2, 171–188.

Grün, Mauro (1996) An analysis of the discursive production of environmental education: terrorism, archaism and transcendentalism, *Curriculum Studies*, 4:3, 329–347.

Hare, William (1986) Reflections on some contemporary educational slogans, *International Review of Education*, 32:1, 71–84.

Hart, Paul, Jickling, Bob, & Kool, Richard (1999) Starting points: questions of quality in environmental education, *Canadian Journal of Environmental Education*, 4, 104–124.

Hawkey, Kate (2014) A new look at big history, *Journal of Curriculum Studies*, 46:2, 163–179.

Heimlich, Joe E., & Falk, John H. (2009) Free-choice learning and the environment. In: John H. Falk, Joe E. Heimlich & Susan Foutz (Eds.), *Free-choice Learning and the Environment* (pp. 11–21) (Los Altos: AltaMira Press).

Hicks, David (1998) Stories of hope: a response to the 'psychology of despair', *Environmental Education Research*, 4:2, 165–176.

Hillbur, Per, Ideland, Malin, & Malmberg, Claus (2016) Response and responsibility: fabrication of the eco-certified citizen in Swedish curricula 1962–2011, *Journal of Curriculum Studies*, 48:3, 409–426.

Hopmann, Stefan (2008) No child, no school, no state left behind: schooling in the age of accountability, *Journal of Curriculum Studies*, 40:4, 417–456.

Huckle, John (1983) Environmental education. In: John Huckle (Ed.), *Geographical Education: Reflection and Action* (pp. 99–111) (Oxford: Oxford University Press).

Huckle, John, & Wals, Arjen E. J. (2015) The UN Decade of Education for Sustainable Development: business as usual in the end, *Environmental Education Research*, 21:3, 491–505.

Huebner, Dwayne (1967) Curriculum as concern for man's temporality, *Theory Into Practice*, 6(4), 172–179.

Hungerford, Harold R. (1975) Myths of environmental education, *The Journal of Environmental Education*, 7:2, 21–26.

Ideland, Malin, & Malmberg, Claes (2014) 'Our common world' belongs to 'Us': constructions of otherness in education for sustainable development, *Critical Studies in Education*, 55:3, 369–386.

International Union for the Conservation of Nature (IUCN) (1970) International working meeting on environmental education in the school curriculum. Final Report. IUCN/UNESCO.

Jardine, David W. (1994) 'Littered with literacy': an ecopedagogical reflection on whole language, pedocentrism and the necessity of refusal, *Journal of Curriculum Studies*, 26:5, 509–524.

Jenkins, Edgar W. (1994) Public understanding of science and science education for action, Journal of Curriculum Studies, 26:6, 601–611.

Jensen, Bjarne Bruun (2004) Environmental and health education viewed from an action-oriented perspective: a case from Denmark, *Journal of Curriculum Studies*, 36:4, 405–425.

Jickling, Bob (1992) Why I don't want my children to be educated for sustainable development, *The Journal of Environmental Education*, 23:4, 5–8.

Jickling, Bob, & Spork, Helen (1998) Education for the environment: a critique, *Environmental Education Research*, 4:3, 309–327.

Jickling, Bob, & Wals, Arjen (2008) Globalization and environmental education: looking beyond sustainable development, *Journal of Curriculum Studies*, 40:1, 1–21.

Jickling, B., & Sterling, S. (Eds.) (2017) *Post-sustainability and Environmental Education: Remaking Education for the Future*. (Palgrave Macmillan).

Kahn, Richard (2008) Towards ecopedagogy: weaving a broad-based pedagogy of liberation for animals, nature and the oppressed peoples of the Earth. In: A. Darder, R. Torres & M. Baltodano (Eds.), *The Critical Pedagogy Reader* (2nd ed.) (pp. 522–540) (New York: Routledge).

Kauchak, Don, Krall, Flo, & Heimsath, Kim (1978) The need for education, not indoctrination, *The Journal of Environmental Education*, 10:1, 19–22.

Kirk, David & Macdonald, Doune (2001) Teacher voice and ownership of curriculum change, *Journal of Curriculum Studies*, 33:5, 551–567.

Kopnina, Helen (2012) Education for sustainable development (ESD): the turn away from 'environment' in environmental education? *Environmental Education Research*, 18:5, 699–717.

Krall, Florence R. (1979) Living metaphors: the real curriculum in environmental education, *Journal of Curriculum Theorizing*, 1(1), 180–185.

Lee, John Chi-Kin (2000) Teacher receptivity to curriculum change in the implementation stage: The case of environmental education in Hong Kong, *Journal of Curriculum Studies*, 32:1, 95–115.

Linke, Russell D. (1976) A case for indoctrination in environmental education, *South Pacific Journal of Teacher Education*, 4, 125–129.

Lucas, Arthur M. (1980) Science and environmental education: pious hopes, self praise and disciplinary chauvinism, *Studies in Science Education*, 26, 1–26.

Llewellyn, Kristina R. Cook, Sharon A., & Molina, Alison (2010) Civic learning: moving from the apolitical to the socially just, *Journal of Curriculum Studies*, 42:6, 791–812.

Lloro-Bidart, Teresa (2017) A feminist posthumanist political ecology of education for theorizing human-animal relations/relationships, *Environmental Education Research*, 23:1, 111–130.

Lotz-Sisitka, Heila (2008) Reading Conference recommendations in a wider context of social change, *Southern African Journal of Environmental Education*, 25, 191–198.

Madsen, Katrine Dahl (2013) Unfolding education for sustainable development as didactic thinking and practice, *Sustainability*, 5:9, 3771–3782.

Maher, Mary (1986) Environmental education: what are we fighting for? *Geographical Education*, 5:2, 21–25.

Marsden, William E. (1971) Environmental studies courses in colleges of education, *Journal of Curriculum Studies*, 3:2, 163–178.

Marsden, William E. (1997) Environmental education: historical roots, comparative perspectives, and current issues in Britain and the United States, *Journal of Curriculum and Supervision*, 13(1), 6–29.

Martin, Stephen, Dillon, James, Higgins, Peter, Peters, Carl, & Scott, William (2013) Divergent evolution in education for sustainable development policy in the United Kingdom: current status, best practice, and opportunities for the future, *Sustainability*, 5:4, 1522–1544.

McCoy, Kate, Tuck, Eve, & McKenzie, Marcia (Eds.) (2016) *Land Education: Rethinking Pedagogies of Place from Indigenous, Postcolonial, and Decolonizing Perspectives* (London: Routledge).

McEneaney, Elizabeth H. (2002) Power and the knowledge produced by educationists, *Journal of Curriculum Studies*, 34:1, 103–115.

McInnis, Noel (1975) What makes environment educational? In: Noel McInnis & Don Albrecht (Eds.), *What Makes Education Environmental?* (pp. 21–29) (Kentucky: Data Courier).

McKenzie, Marcia (2012) Education for y'all: global neoliberalism and the case for politics of scale in sustainability education policy, *Policy Futures in Education*, 10:2, 165–177.

McKeown, Rosalyn, & Hopkins, Charles (2003) EE≠ESD: defusing the worry, *Environmental Education Research*, 9:1, 117–128.

Monroe, Martha C. (2012) The co-evolution of ESD and EE, *Journal of Education for Sustainable Development*, 6:1, 43–47.

Morgan, John (2002) 'Teaching geography for a better world'? The postmodern challenge and geography education, *International Research in Geographical and Environmental Education*, 11:1, 15–29.

Moroye, Christy (2009) Complementary curriculum: the work of ecologically minded teachers, *Journal of Curriculum Studies*, 41:6, 789–811.

Morris, Marla (2002) Ecological consciousness and curriculum, *Journal of Curriculum Studies*, 34:5, 571–587.

Naidoo, Prem, Kruger, Jackie, & Brookes, David (1990) Towards better education: environmental education's pivotal role in the transformation of education, *Southern African Journal of Environmental Education*, 11, 13–17.

Nash, Roderick (1976) Logs, universities, and the environmental education compromise, *The Journal of Environmental Education*, 8:2, 2–11.

Nelson, Thomas, & Cassell, John A. (2012) Pedagogy for survival: an educational response to the ecological crisis. In: Arjen E.J. Wals & Peter Blaze Corcoran (Eds.), *Learning for Sustainability in times of Accelerating Change* (pp. 63–75) (Wageningen: Wageningen Academic Publishers).

NGO Forum (1992) 'Treaty on Environmental Education for Sustainable Societies and Global Responsibility', Environmental Education for Sustainable Societies and Global Responsibility (Rio de Janeiro: International Council for Adult Education).

Nixon, Jon (1991) Reclaiming coherence: cross-curriculum provision and the National Curriculum, *Journal of Curriculum Studies*, 23:2, 187–192.

O'Riordan, Timothy (1981) Environmentalism and education, *Journal of Geography in Higher Education*, 5:1, 3–17.

Orr, David W. (1992) What is education for? In: *Ecological Literacy: Education and the Transition to a Postmodern World* (pp. 141–148) (Albany: SUNY Press).

Oulton, Christopher, & Scott, William (2000) Environmental education: a time for revisioning. In: Bob Moon, Miriam Ben-Peretz & Sally Brown (Eds.), *Routledge Companion to Education* (pp. 489–501) (London: Routledge).

Payne, Phillip G. (2006) Environmental education and curriculum theory, *The Journal of Environmental Education*, 37:2, 25–35.

Pepper, David (1987) The basis of a radical curriculum in environmental education. In: Colin Lacey & Roy Williams (Eds.), *Education, Ecology and Development: The Case for an Education Network* (pp. 65–79) (London: The World Wildlife Fund & Kogan Page).

Posch, Peter (1997) Social change and environmental education, *Australian Journal of Environmental Education*, 13, 1–6.

Postma, Dirk W. (2002) Taking the future seriously: on the inadequacies of the framework of liberalism for environmental education, *Journal of Philosophy of Education*, 36:1, 41–56.

Reichenbach, Roland (2016) The quest of educational slogans. In: Michael A. Peters (Ed.), *Encyclopedia of Educational Philosophy and Theory* (pp. 1–6) (Singapore: Springer Singapore). 10.1007/978-981-287-532-7_7-1

Reid, Alan, & Dillon, Justin (2017) Environmental education: the major work of ensuring quality and outcomes in connecting environment and education. In: Alan Reid & Justin Dillon (Eds.), *Environmental Education: Critical Concepts in the Environment* (pp. 23–47) (London: Routledge).

Reid, Alan, & Scott, William (2006) Researching education and the environment: retrospect and prospect, *Environmental Education Research*, 12:3–4, 571–588.

Reid, Herbert & Taylor, Betsy (2003) John Dewey's aesthetic ecology of public intelligence and the grounding of civic environmentalism, *Ethics & the Environment*, 8:1, 74–92.

Robottom, Ian (1989) Social critique or social control: some problems for evaluation in environmental education, *Journal of Research in Science Teaching*, 26:5, 435–443.

Ross, Hamish (2007) Environment in the curriculum: representation and development in the Scottish physical and social sciences, *Journal of Curriculum Studies*, 39:6, 659–677.

Russell, Constance, L., & Bell, Anne C. (1996) A politicized ethic of care: environmental education from an ecofeminist perspective. In: Karen Warren (Ed.), *Women's Voices in Experiential Education* (pp. 172–181) (Dubuque: Kendall/Hunt Publishing).

Russell, Joshua (2017) 'Everything has to die one day': children's explorations of the meanings of death in human-animal-nature relationships, *Environmental Education Research* 23:1, 75–90.

Sandell, Klas, Ohman, Johan, & Ostman, Leif (2005) Selective traditions within environmental education. In: *Education for Sustainable Development: Nature, School and Democracy* (pp.155–168) (Lund: Studentlitteratur AB).

Saul, Darin (2000) Expanding environmental education: thinking critically, thinking culturally, *The Journal of Environmental Education*, 31:2, 5–8.

Sauvé, Lucie, Brunelle, Renée, & Berryman, Tom (2005) Influence of the globalized and globalizing sustainable development framework on national policies related to environmental education, *Policy Futures in Education*, 3:3, 271–283.

Sauvé, Lucie, Berryman, Tom, & Brunelle, Renée (2007) Three decades of international guidelines for environment-related education: a critical hermeneutic of the United Nations discourse, *Canadian Journal of Environmental Education*, 12:1, 33–54.

Scheffler, Israel (1960) *The Language of Education* (Springfield: Charles C. Thomas).

Schinkel, Anders (2009) Justifying compulsory environmental education in liberal democracies, *Journal of Philosophy of Education*, 43:4, 507–526.

Schumacher, Ernst "Fritz" (1973) The greatest resource – education. In: *Small Is Beautiful: A Study of Economics As If People Mattered* (pp. 60–80) (New York: Harper & Row).

Scott, William, & Oulton, Christopher (1999) Environmental education: arguing the case for multiple approaches, *Educational Studies*, 25:1, 89–97.

Scott, William, & Reid, Alan (1998) The revisioning of environmental education: a critical analysis of recent policy shifts in England and Wales, *Educational Review*, 50:3, 213–223.

Sessions, George (1983) Ecophilosophy, utopias, and education, *The Journal of Environmental Education*, 15:1, 27–42.

Simmons, I. G., & Simmons, C. M. (1973) Environmentalism and education: a context for conservation, *School Science Review*, 54, 574–579.

Simovska, Venka & Prøsch, Åsa Kremer (2016) Global social issues in the curriculum: perspectives of school principals, *Journal of Curriculum Studies*, 48:5, 630–649.

Skolimowski, Henryk (1991) Ecology, education and the real world, *The Trumpeter*, 8(3), unpaginated.

Smith, Gregory A. (1998) Creating a public of environmentalists: the role of nonformal education. In: Greg Smith & Dilafruz Williams (Eds.), *Ecological Education in Action: On Weaving Education, Culture, and the Environment* (pp. 207–227) (Albany: SUNY Press).

Smyth, John C. (1995) Environment and Education: a view of a changing scene, *Environmental Education Research*, 1:1, 3–20.

Spannring, Reingard (2017) Animals in environmental education research, *Environmental Education Research*, 23:1, 63–74.

Stables, Andrew, & Scott, William (2001) Post-humanist liberal pragmatism? Environmental education out of modernity, *Journal of Philosophy of Education*, 35:2, 269–279.

Stapp, William B. et al. (1969) The concept of environmental education, *Environmental Education*, 1:1, 30–31.

Sterling, Stephen (2010) Living *in* the Earth: towards an education for our times, *Journal of Education for Sustainable Development*, 4:2, 213–218.

Stevenson, Robert B. (1987) Schooling and environmental education: contradictions in purpose and practice. In: Ian Robottom (Ed.), *Environmental Education: Practice and Possibility* (pp. 69–82) (Geelong: Deakin University Press).

Stowell, Rob, & Bentley, Lynn (ed.) (1988) *New-wave Geography*. Volumes 1 & 2. Geography Teachers Association of Victoria (Milton, Queensland: Jacaranda Press).

Sund, Louise, & Öhman, Johan (2014) On the need to repoliticise environmental and sustainability education: rethinking the postpolitical consensus, *Environmental Education Research*, 20:5, 639–659.

Swan, James A. (1969) The challenge of environmental education, *Phi Delta Kappan*, 51:1, 26–28.

Tilbury, Daniella (1995) Environmental education for sustainability: defining the new focus of environmental education in the 1990s, *Environmental Education Research*, 1:2, 195–212.

Tupper, Jennifer A., & Cappello, Michael P. (2012) (Re)creating citizenship: Saskatchewan high school students' understandings of the 'good' citizen, *Journal of Curriculum Studies*, 44:1, 37–59.

UNESCO (2007) The Ahmedabad Declaration 2007: A Call to Action – Education for Life: Life through Education. Report by the Director-General on the Follow-Up to Decisions and Resolutions Adopted by the Executive Board and the General Conference at their Previous Sessions. UNESCO Executive Board, Hundred and Seventy-Seventh Session (Paris: UNESCO).

UNESCO (2012) Shaping the Education of Tomorrow, Report on the UN Decade of Education for Sustainable Development, abridged (Paris: UNESCO). http://sustainabledevelopment.un.org/content/documents/919unesco1.pdf

UNESCO (2014a) Aïchi-Nagoya Declaration on Education for Sustainable Development, Paris: UNESCO. https://sustainabledevelopment.un.org/content/documents/5859Aichi-Nagoya_Declaration_EN.pdf

UNESCO (2014b) Shaping the Future We Want: UN Decade of Education for Sustainable Development (2005–2014) (Paris: UNESCO). https://sustainabledevelopment.un.org/content/documents/1682Shaping%20the%20future%20we%20want.pdf

UNESCO (2016) Education for People and Planet: Creating Sustainable Futures for All (Paris: UNESCO). http://unesdoc.unesco.org/images/0024/002457/245752e.pdf

UNESCO (2017) Quality education for all at the heart of the Sustainable Development Goals. Press release 27 June 2017. http://www.unesco.org/new/en/media-services/single-view/news/quality_education_for_all_at_the_heart_of_the_sustainable_de/

UNESCO-UNEP (1976) The Belgrade Charter: a global framework for environmental education, *Connect*, I:1, 1–9, 23.

UNESCO-UNEP (1978) The Tbilisi Declaration, *Connect*, III:1, 1–8.

Van Poeck, Katrien, Vandenabeele, Joke, & Bruyninckx, Hans (2013) Taking stock of the UN decade of education for sustainable development: the policy-making process in Flanders, *Environmental Education Research*, 20:5, 1–23.

van Rossen, Judy (1995) Conceptual analysis in environmental education: why I want my children to be educated for sustainable development, *Australian Journal of Environmental Education*, 11, 73–81.

Vidart, Daniel (1978) Environmental education—theory and practice, *Prospects*, VIII:4, 466–479.

Waldron, F., Ruane, B., Oberman, R., & Morris, S. (2016) Geographical process or global injustice? Contrasting educational perspectives on climate change, *Environmental Education Research*, 1–17. doi:10.1080/13504622.2016.1255876

Walker, Kim (1997) Challenging critical theory in environmental education, *Environmental Education Research*, 3:2, 155–162.

Wals, Arjen E. J. (2009) A Mid-DESD Review: Key findings and ways forward, *Journal of Education for Sustainable Development*, 3:2, 195–204.

Wals, Arjen E. J. (2012) Shaping the Education of Tomorrow: 2012 Full-length Report on the UN Decade of Education for Sustainable Development (Paris: UNESCO). http://unesdoc.unesco.org/images/0021/002164/216472e.pdf

Westbury, Ian (2013) Reading Schwab's the 'Practical' as an invitation to a curriculum enquiry, *Journal of Curriculum Studies*, 45:3, 640–651.

Weston, Anthony (1996) Instead of environmental education. In: Bob Jickling (Ed.) *Proceedings of the Yukon College Symposium on Ethics, Environment and Education* (pp. 148–157) (Whitehorse: Yukon College).

Weston, Anthony (2004) What if teaching went wild? *Canadian Journal of Environmental Education* 9:1, 31–46.

White, Damien, Rudy, Alan, & Wilbert, Chris (2007) Anti-environmentalism: Prometheans, contrarians and beyond. In: Jules Pretty, Alan Ball, Ted Benton, Julia Guivant, David Lee, David Orr, Max Pfeffer, & Hugh Ward (Eds.), *The SAGE Handbook of Environment and Society* (pp. 124–142) (London: SAGE Publications).

Winn, Ira (1970) The education in environmental education, *Environmental Education*, 1:4, 140–141.

Young, Michael (2013) Overcoming the crisis in curriculum theory: a knowledge-based approach, *Journal of Curriculum Studies*, 45:2, 101–118.

Zembylas, Michalinos (2005) Science education: for citizenship and/or for social justice?, *Journal of Curriculum Studies*, 37:6, 709–722.

A non-technical introduction to curriculum challenges *for* and *from* environmental education

ALAN REID

Introduction

In this non-technical introduction to curriculum challenges *for* and *from* environmental education, we broach some of the key themes, events and questions that have shaped environmental education curriculum, and provide examples of the stimuli for particular debates in this collection, from past to present. As we will show, the key concerns raised focus attention on:

- how a smudging and blurring of the boundaries between education and advocacy in environmental education might take place;
- whether meliorist rather than transformationist approaches are typically the order of the day with environmental education curriculum; and
- what might mark the 'end of the beginning' or 'the beginning of the end' for environmental education, two decades after those tropes were first used.

We start by revisiting the value of developing a Virtual Special Issue on these concerns from the archives of the *Journal of Curriculum Studies,* before discussing some of the key issues in the conceptual and linguistic freighting of mainstream to alternative notions of curriculum in environmental education. This allows us to cue up a range of further questions about the pliability and stretching of curriculum frameworks in general, and the 'opportunity costs' of pursuing disciplinary, whole-of-curriculum and pluralised curriculum models to achieve particular ends-in-view for education and environment. The chapter concludes by (re)commending the value of a 'long view' of those approaches to curriculum that 'adjectival educations' tend to form, and pursuing the development of critical historical perspectives on contemporary and ongoing debate about environment, education and curriculum, themes that are explored in more detail in the subsequent chapter.

Background

To begin, we start by exploring, *why was a Virtual Special Issue developed from the* Journal of Curriculum Studies *on environmental education?* While our general introduction has rehearsed some of the reasons, for the purposes of this non-technical introduction we offer three further lines of response.

First, its contributions highlight the value of bringing a critical perspective to what might be typically regarded as a specialist's or insider's curriculum discourse, no matter how important or significant its members and scholars consider the field to be. Second, the contributions reiterate the importance in curriculum studies of understanding how matters of public and global concern might be brought to consciousness in the history of the field in question, if not curriculum studies more widely. Third, the contributions raise the broader question of how curriculum traditions, debates, change and innovation are understood to develop and/or wither, with a particular focus in this instance, on how curriculum comes to address local to transnational environmental issues and priorities.

So to grasp why all this matters to the *Journal of Curriculum Studies*, as raised in the general introduction we must recognise that the Journal is one of the few non-specialist outlets that has provided extensive and prolonged scholarly elaboration and critique of environmental education in relation to curriculum. This has taken place since the field first emerged (albeit not quite fully-fledged) onto the education scene in the late 1960s, including during a 'toddling phase' in which curriculum practitioners and theorists sought to delineate what was distinctive about this newcomer in an already crowded curriculum space (see Covert, 1969; Stapp et al., 1969).

In more recent terms, interest has centred on such questions as the promise in environmental education of a 'less-than-anthropocentric' approach to curriculum than the prevailing 'deeply anthropocentric' shape to what teachers and students experience in the vast majority of their studies. Another take on the question of what is privileged is what is distinctive to this field, in comparison to other areas of education adopting an adjectival form (such as 'experiential education' or 'outdoor education'): is it the particular emphasis and reworking of the importance of experience and action-taking towards more markedly ecocentric concerns (Morris, 2002)? Or, is there actually something greater at stake that is possible and desirable here: questioning the very frame of reference for education itself, as Gruenewald (2004, p. 83) argues, when so much of environmental education:

> is legitimizing the practice of organizing education around standards in discrete subject areas, rather than constituting itself as a lens through which all content areas—and the purpose of education itself—might be viewed. This is why Martin (1996) wants to shed all "adjectival adjuncts" to education: to abolish EE completely may be the only way to save it from being coopted and weakened by the dominant discourse as merely another fragmented content area to be covered and assessed.

Establishing absence and presence

As noted in the general introduction, Gough (1989), Bonnett (2007) and Moroye (2009) among others are just some of the contributors to this collection who have explored such questions in the *Journal of Curriculum Studies*. But it must also be recognised that while the Journal has a rich archive of articles on environmental education, its track record of scrutinizing curriculum thinking and practice in environmental education stands in marked contrast to the wider output from other scholarly outlets available, including in the most highly regarded educational and curriculum journals.

Even a cursory scan of leading publications on teaching and learning would suggest a veritable silence towards environmental education and its questions. Perhaps, we might wonder, is this because of an indifference to such concerns? Or even a sense of disavowal of these matters stemming from, for example, an existential or environmental unease in reaction to what permeates the work of scholars and educators interested in these matters typically raise? (On these, see: Fen, 1961; Huebner, 1967; Clark, 1975; McInnis, 1975; Krall, 1979; Brennan, 1991; Sandall et al., 2005; Schinkel, 2009.)

We will return to this proto-theme of 'presence' (e.g. sustained and non-tokenistic) as well as 'absences' in educational scholarship later on, but before then, we remind readers that the Journal's earliest contributions to these debates (Marsden, 1971) predated the global impetus to establish curriculum for environmental education, the codification and development of which took place largely from the mid-1970s onwards (see Swan, 1969; IUCN, 1970; UNESCO-UNEP, 1976, 1978; Vidart, 1978; Hungerford, 1980; Knapp, 2000; Sauvé et al., 2007; Monroe, 2012). In other words, as the articles selected for this collection show, this particular segment of the community of curriculum scholars hasn't been one displaying a benign disinterest in environmental education or to that of the curriculum challenges its thinkers and practitioners have to wrestle with. Moreover, past and current editorial board members have actively sought to engage various challenges *for* and *from* environmental education through the pages of this Journal (e.g. Noel Gough (1989) and Andrew Stables (2006) as well as during their service on the editorial board, and Bowers, 1990). Recent contributions include a study by Angela Barthes (2017) on the construction of curricula in Education for Sustainable Development in France, and Per Hillbur et al.'s (2016) study in Sweden of how environmental education operates as hub for constructing desirable citizens. The latter, for example, recognises key challenges that are as complex as they are compelling: whether a focus on developing 'eco-certified citizens' with 'post-political' tendencies is sufficient for addressing sustainability challenges. For curriculum, this can create frictions, particularly within contemporary and historical frameworks for governmentality and the horizons and legacies of state education policies. Thus at a broader level, both studies raise the challenge of how curriculum might and should be constructed and reconstructed curriculum into the 21st century?

So, given this span of activity, historic to present, and the international reach of the Journal, it is perhaps unsurprising that the collection shows there are some key ideas in and for curriculum from a wide range of times and places that still seem to have traction in environmental education scholarship, if not beyond. These aren't always visible in a standard special issue as these typically draw on new or solicited papers. While to illustrate the value of a Virtual Special Issue format further, by digging into the archives of the Journal we can reveal something of the depth and difficulties in deliberations on the state and future of various content themes and approaches drawn from environmental education's 'feeder areas'. Historically, these have included rural studies and outdoor education, nature study and appreciation, school-level geography and biology, and civics and experiential learning (among many others[1]) (Gayford, 1986; Cotton, 2006; Ross, 2007). They also come back to the fore when we consider certain efforts to return the field to its putative 'roots', if not in considering how these debates and concerns might butt up against various attempts to move environmental education on to other priorities or adjectival forms for curriculum and pedagogy, such as 'ecological education', or education for 'sustainable development', 'sustainability', 'resilience' or 'ecojustice' (e.g. Jardine, 1994; Morris, 2002; Moroye, 2009).

So which articles have tracked these developments and shifts in the field? The complete list of studies for the collection is set out in the general introduction and Table of Contents. But in responding to this question, we should recognise that those which are included herein are drawn from a very wide range of Anglo- and non-Anglospheric settings and countries. They also analyse both major and minor trends and issues in a complex field, even if their core concerns are most typically those of curriculum content, orientation and ends-in-view.

Again, given the Journal's track record of publication and coupled with its pluralistic aims and scope, we note that such studies have largely been written with a non-specialist academic audience in mind. Thus throughout the papers selected for this collection, we find that many of the contributions help examine and advance scholarship on the linking of education and the environment, alongside the wider lessons that can be drawn from this work for curriculum studies in general from cases of the particular.

To illustrate this more readily, recall the chapters have been organised into sections on:

- Perspectives on Curriculum and Environment Education
- Accounting for Curriculum in Environmental Education
- Changes in Curriculum for Environmental Education
- Questioning the Curriculum, from the Mainstream to the Margins
- For Wisdom, Justice and Action in Curriculum?

A recurring theme across these sections is how to develop scholarship on specific 'barriers and enablers' to embedding and developing particular forms of education, most typically if environmental education is taken to be part of a broad and difficult family of 'adjectival educations'. Closely related to this is another set of questions: how this squares up to what is going on in studies of mainstream curriculum thinking and practice (e.g. on determining or questioning the relative status of knowledge and ways of knowing in the sciences, humanities and arts in schools); or how mainstream initiatives play out in a wide range of curriculum contexts, including for environmental education, recognising their associated and distinctive dynamics, priorities and challenges around the world, such as in relation to bottom-up versus top-down forms of curriculum change (see Goodson, 1983; Gayford, 1986; Gough, 1989; Lee, 2000; Bonnett, 2007; Jickling & Wals, 2008; see too the supporting material for the collection, e.g. Ball & Bowe, 1992; Young, 2013).

We also note that one of the consistently 'most read' and 'most cited' papers in the Journal in recent years (i.e. Jickling & Wals, 2008) has focused on appraising the international development and reception of prominent environment-related education initiatives around the world. These are typically sponsored by the United Nations and driven by the goal of using education to address pressing socio-ecological issues. In many instances, such initiatives are transnational but are expected to have local to national manifestations and uptake, e.g. on how curriculum does or doesn't address questions of 'environmental justice', or more recently, 'climate justice' (Fahey, 2012). While in the specific case of environmental education (and now more often than not, these days, some form of sustainability education), they typically orientate their arguments in relation to work initiated by the United Nations, UNESCO and IUCN to address public concerns about environmental issues, including via UNESCO-UNEP (United Nations Environment Programme).

This work can be traced back to the International Environmental Education Programme, and the Intergovernmental Conference on Environmental Education (Tbilisi), arising from the UNCHE Stockholm Conference, as discussed in the general introduction too. While the most recent contributions to the collection are those that bring this up to date, in that they have examined the remit, effects and evaluation of markers like the UN Decade of Education for Sustainable Development (2005–14) (Hawkey, 2014), a key theme to consider is whether non-binding exhortations for curriculum renewal are treated as more or less than this in local to national curriculum policy and development.

So in this spirit, what comes to constitute the contours for a non-technical introduction to the various themes, events and questions that have shaped curriculum for environmental education? Evaluation and reflection at World Environmental Education Congresses picks up on these considerations, as does the scholarly literature (see Reid & Dillon, 2017 for a compendium). Both readily illustrate a broad and contested range of accounts of the origins and traditions of environmental education around the world—including what is shared and distinctive in diverse locations, contexts and countries. But in both the 'public consciousness' and that of policy makers, so to speak, there is one particular watershed moment to recognise. As mentioned previously, in Dobson's (2016) short summary of key aspects of the history of this area, understanding why and how environmental and curriculum matters came to be more closely related than ever before has to take account of the work of the United Nations Conference on the Human Environment (UNCHE)—particularly if its roots are to be examined and any alternatives proposed for work in this area.

Held in Stockholm (1972), the Conference and its Declaration were emblematic of the mounting disquiet about environmental degradation across civil society and within government. In short, environmental reform was needed, not just awareness and acknowledgement of pollution. The UNCHE event positioned education as a key tool for policy-makers who were looking to raise public awareness about transnational pollution issues and their local manifestations (e.g. in Sweden, 'acid rain' as a result of emissions of air pollutants from the UK), and then, their possible management and resolution through legislation and other policy instruments.

In Principle 19 of 26 in the Declaration, and Recommendation 73 of 94, of its report, these were expressed as follows:

> Education in environmental matters, for the younger generation as well as adults, giving due consideration to the underprivileged, is essential in order to broaden the basis for an enlightened opinion and responsible conduct by individuals, enterprises and communities in protecting and improving the environment in its full human dimension. It is also essential that mass media of communications avoid contributing to the deterioration of the environment, but, on the contrary, disseminates information of an educational nature on the need to project and improve the environment in order to enable man [sic] to develop in every respect.
>
> Principle 19 (UNEP, 1972)

> It is recommended that Governments actively support, and contribute to, international programmes to acquire knowledge for the assessment of pollutant sources, pathways, exposures and risks and that those Governments in a position to do so provide educational, technical and other forms of assistance to facilitate broad participation by countries regardless of their economic or technical advancement.
>
> Recommendation 73 (UNEP, 1972)

While there is much to commend in such statements, early thinking about education's role in both fostering and restoring environmental quality didn't escape the charge of being limited in one way or another. Sometimes it drew on 'gestational' work on environmental education from a range of sources, including UNESCO (e.g. in Nevada, 1969, see IUCN, 1970), at other times it didn't (e.g. in countries not party to UNESCO's deliberations). Again, whether it was on how to inform or shape the consciousness of members of the public, government level policy-makers, school students or to curriculum experts, a key tension that emerges concerns the degree of constructive alignment of methods and ends-in-view. For example, would such goals be most likely achieved by educators through what we might now see as relying on relatively crude information gathering and dissemination activities on pollutants and environmental quality to foster environmental management or control, or by way of 'behaviour change' or 'social marketing' strategies? In other words, was education, and the work of schools in particular, being expected to offer a fix to yet another social problem (Jensen, 2004)?

It can be argued though that pursuing this line of critique too far (e.g. by ignoring work not associated with the UN) can lead to gross misunderstandings about the proto-DNA of the field, particularly its complex curriculum imaginary of citizen science-type, participatory and democratic approaches to education (see also Winn, 1970; Ginzbarg, 1971; UNESCO-UNEP, 1976, 1978; Weston, 1996; Jensen & Schnack, 1997; Heimlich & Falk, 2009; Wals et al., 2014). Nevertheless, as the collection shows, the span of reformist and conservative approaches to education were, and have largely remained, the main order of the day. Thus there are plenty of instances of those standard expectations of a renewed focus within school and university syllabus content to address environmental concerns, most notably though injecting a focus on understanding instances and patterns of environmental pollution and their immediate and long term impacts on quality of life for human and more-than-human communities (see Nash, 1976; Hungerford, 1980; O'Riordan, 1981). Coupled to this are often associated questions and priorities for using education to develop environmental quality and management strategies; but this is as if these were addressed primarily through 'instruction' rather 'education', and 'training' rather than 'education' (see Biesta, 1998).

Returning to the early days again, we note that in a report that laid the groundwork for UNCHE, Paragraph 3.11 of the 1971 Founex Report on Environment and Development (de Almeida, 1972) called for the integration of development and environment strategies in curriculum:

> It would also be helpful to compile all existing legislation regarding environmental control, including the regulations dealing with urban zoning, location of industries, protection of natural resources, and so on. This accumulation of information and knowledge should enable the developing countries to get a clearer perspective of their environmental problems and of the corrective action that they may require at different stages of development. Since public participation in any such efforts is vital, efforts should also be made to build the environmental concern into education curricula, and to disseminate it to the general public through media of mass information. We would like to stress once again the need for a good deal of careful research and study in this field, and the importance of avoiding hasty guidelines and action.

Thus a key question since the 1970s has been whether proponents of liberal to more critical forms of environmental education will ensure that any associated curriculum reform and education policy development does justice to what is at stake for 'people, profits and planet' around the world, rather than create another instance of a failed or faltering vision, be that on the part of government, local authorities, schools, teachers, or their communities of interest (see Walker, 1997, though, on the assumptions and critiques of socially critical theory in environmental education, and Biesta, 1998, on the 'impossible future' of critical pedagogy). Accordingly, a particular challenge to note in the collection's studies is the ways they address or question claims that the necessary action is sufficiently well conceptualised and happening, so as to redirect (environmental) education policy and practice—particularly in modes that are not simply undermined or contradicted by the aims, values and ends-in-view of the rest of the curriculum!

If we now consider the more recent climate for debate, we must also address *whether the discourse about sustainable development, or sustainability, has an effect on (environmental) education.* It has become an important question that runs throughout this collection and beyond, and often centres on another key expectation that came to prominence in the 1980s. Namely how does education internationally—and environmental education, in particular—find a secure place alongside other policy, change and advocacy imperatives on development and environment?

Education and sustainable development

Exemplars prior to the UN's Rio Earth Summit, 1992, or as it is more properly known, the United Nations Conference on Environment and Development (UNCED), include:

- the *World Conservation Strategy* (IUCN-WWF-UNEP, 1980), subtitled 'living resource conservation for sustainable development' (the first formal articulation of the term 'sustainable development'),
- the World Commission on Environment and Development (the Brundtland Commission, 1986), and its influential report defining sustainable development in *Our Common Future* (WCED, 1987) in terms of meeting the needs of current generations without compromising those of the future, as well as also signalled a shift from 'one earth to one world', and
- the update to the World Conservation Strategy, entitled, *Caring for the Earth: A Strategy for Sustainable Living* (IUCN-WWF-UNEP, 1991) which offered a prelude to the focus on strategy and action as offered by Rio's Agenda 21 (Hattingh, 2002).

In various ways, these reports and strategies teased out how education addressed legislation on environmental protection. For example, in section 13 of the World Conservation Strategy, on 'Building Support: Participation and Education' (IUCN-WWF-UNEP, 1980), its first, eighth and last paragraphs state:

> 1. Ultimately the behaviour of entire societies towards the biosphere must be transformed if the achievement of conservation objectives is to be assured. A new ethic, embracing plants and animals as well as people, is required for human societies to live in harmony with the natural world on which they depend for survival and wellbeing. The long term task of environmental education is to foster or reinforce attitudes and behaviour compatible with this new ethic.

> 8. Organizers of education programmes should determine the main target groups of such programmes, define precise programme objectives, and select the media and techniques that are most effective with the target groups. Results, together with the techniques and materials used, should be regularly evaluated against the stated objectives. The most important target groups are: legislators and administrators; development practitioners, industry and commerce, and trade unions; professional bodies and special interest groups; communities most affected by conservation projects; schoolchildren and students.

> 14. The need for environmental education is continuous because each new generation needs to learn for itself the importance of conservation. As such, individual campaigns and programmes should not be regarded as ends in themselves but as part of a long term, iterative process. It should also be recognized that any educational campaign is in competition for public attention with many others, including advertising campaigns. To win and retain as much of this attention as possible, it is essential for conservation to be seen as central to human interests and aspirations. At the same time, people—from heads of state to the members of rural communities—will most readily be brought to demand conservation if they themselves recognize the contribution of conservation to the achievement of their needs, as perceived by them, and the solution of their problems, as perceived by them.

Educating elites and not just the masses, then, was seen as a core strategy in the world of environmental education, to influence the hearts and minds of local to global decision-makers around the world, while they were in role and in the making.

This expectation presents a different challenge to curriculum design, making and evaluation than that usually targeted at schools. So another question to be tackled is, how directed and partisan might any associated curriculum, training and 'higher' education need to be in the face of pressing social and environmental problems, including associated educational inquiry (e.g. Jenkins, 1994), and how consistent should it be with what happens in schools, for example? Is it possible and desirable, for instance, to use educational interventions and programmes with elites to move away from defaulting to technocentric responses to 'wicked problems' such as climate change in education, by fostering a curriculum on 'transgressive learning' (see Wals et al., 2014)? Should that be prepared for in K-12 settings, and what should happen in relation to lifelong learning? More pointedly, when is it legitimate, for example, to blur the lines between education and advocacy in curriculum reform (Hopmann, 2003), as raised by questions of how all those who have been educated and continue to be educated in various ways can address pressing environmental and economic injustices, as in the UN's Sustainable Development Goals?

So is it an either/or, or both/and in relation to education and advocacy? Returning to an illustration of how this question has been addressed since the early days of the field, the United States National Environmental Education Act of October 1970 (PL 91-516), defined environmental education as 'intended to promote among citizens the awareness and understanding of the environment, our relationship to it, and the concern and responsible action necessary to assure our survival and to improve the quality of life'. It maintained this position in a renewed Act in 1990 too; but given the audience for this collection, we must also ask, what of the curriculum domain for such an environmental education?

Immediately prior to the 1970 statement of what should ideally take place—and using the heavily gender biased language of the time—we note the Act stated:

> Environmental education is an integrated process which deals with man's inter-relationship with his fellow man and with his natural and man-made surroundings, including the relationship of population growth, pollution, resource allocation and depletion, conservation, technology, and urban and rural planning to the total human environment. Environmental education is a study of the factors influencing ecosystems, mental and physical growth, living and working conditions, decaying communities and population pressures.

Edward Weidner, then Chancellor of the University of Wisconsin, reiterated the sense of urgency and crisis felt during that formative period not just in this field but in the world at large too (also noted at UNCHE that year).[2] On March 1, 1972, during the development of one of the US's first statewide environmental education plans, Weidner argued (Minnesota Environmental Education Council, 1972, p. 19):

> Environmental education is not just a new fad of education or a new fad of citizen concern. It is not just another element added to our curriculum. Environmental education is a fundamental element of education, of community action, and of life. The importance of environmental education can be deduced either through a consideration of the educational crisis or a

consideration of the environmental crisis. Whichever route is taken, a similar conclusion is reached. Education is essential to saving our environment, and an environmental focus makes for effective education.

Such hopes, justifications and expectations are typical of many others in and for an environmental education curriculum. In more detail, the Minnesota plan was one of the first to promote a 'non-disciplinary curriculum' (see also Covert, 1969; Moroni, 1978, and Cole, 2007), because amongst other things, there was a recognition that the 'unique nature of environmental education is that it encompasses all subjects usually found in the school curriculum' (Minnesota Environmental Education Council, 1972, p. 20). Also, it should 'encourage the teaching of skills through which values are formed rather than advocating any particular value or set of values that is, students should be taught how to think, not what to think'. In addition, it should 'deal with environmental problems in a rational and objective manner. Teachers and students should examine all sides of an issue.' Finally, it should 'begin with the students' local experiences but which lead ultimately to an expanded view of the total environment. The curriculum should deal with local, state, regional, national and international concerns' (ibid.). (This language is echoed in many statements of the North American Association for Environmental Education to this day, arguably the peak body that has carried forward the work of the Environmental Education Act in the USA—see www.naaee.org.)

We return to the logic, challenges and chequered fate of such recognitions and recommendations later on in the collection and in the accompanying essay and reflection. But before then, we note that in the 1980s, the goals of environmental protection and poverty alleviation were still expected to be those that were taken on together, and globally. They were visibly picked up within the aforementioned World Conservation Strategy (1980), and remain strongly associated with the key ideas associated with the World Commission on Environment and Development, including its report on sustainable development (*Our Common Future,* the Brundtland Report, 1987). For example, an often overlooked partner to a focus on environmental education that takes place in development education, is action on poverty alleviation (now often linked in the public mind to meeting a target of directing 0.7% of Gross National Income to international aid efforts, such as via the Millennium Development Goals).

Brundtland's oft-repeated definition of sustainable development is particularly significant in this regard, as for an associated curriculum, it places a strong focus on raising how education addresses *inter*generational and *intra*generational justice, i.e. is not just a futures-focused, or past-dominated education (including Romantic visions of lost Edens or restoring harmony between 'Man and Nature'—Russell, 2005). In this guise, and as adding to an already swollen array of imperatives for curriculum thinking and practice (Sterling, 1992), they also make a vivid appearance in further work on creating action in communities beyond the educational institution's gates, including through education geared towards sustainable development at local through to global levels, through for example, active and community engaged global (citizenship) education (see Hicks & Holden, 1995).

Nevertheless, for a (perhaps stereotypical) middling-aged highly educated Westerner, such ideas, imperatives and the associated work plans have tended to be most often remembered in relation to Rio de Janeiro and the UN Conference on Environment and Development (1992), and the emphasis betrayed by it being known as the 'Earth Summit'. However, contributors to the Summit and its key document, Agenda 21, put great effort into signalling that environmental action shouldn't be seen as the preserve of, say, 'the middle classes' or 'the privileged' in a 'minority' as opposed to 'majority world', and solely focused on questions of 'natural nature' as opposed to 'human nature' (e.g. NGO Forum, 1992). In fact, as early as the meeting at Founex (de Almeida, 1972), it was made clear that concern about the environment often sprang from that associated with the production and consumption patterns of the industrialised world, while many of the environmental problems in the world were a result of underdevelopment and poverty, and thus environmental justice was a key issue to be tackled by all, and for all.

However, rather than suggest a one-size-fits-all approach in terms of environmental protection and socio-economic development, it can be shown that attention was increasingly focused on pursuing a range of locally relevant 'development pathways' and 'sustainable livelihoods' through *Local* Agenda 21, be they in the global 'North' or 'South' (see Clover & Shaw, 2009). These concerns have been subsequently revisited in curriculum deliberations in relation to UNESCO's 'Education for All' agenda, and more recently, an 'Education for the World We Want', which give stronger attention to the local manifestation of key global issues, such as climate change, forced migration and other threats to social, cultural, economic and ecological resilience, or even attempts to map out how to provide learning objectives in 'Education for Sustainable Development Goals' (UNESCO, 2017).

While UNCED tried to cement education into a lengthy series of cross-cutting priorities for action in Agenda 21, it remains the case though that much of the discussion on education was in Chapter 36 of the action plan with its related concerns focused on training and public awareness, as well as establishing a taskforce on capacity building. As befits such a document, direct terminology wasn't avoided, as illustrated in the following extract:

> Both formal and non-formal education are indispensable to changing people's attitudes so that they have the capacity to assess and address their sustainable development concerns. It is also critical for achieving environmental and ethical awareness, values and attitudes, skills and behaviour consistent with sustainable development and for effective public participation in decision-making.
>
> (United Nations, 1992, n.p.)

So once more, we might consider how this is actually achieved, e.g. through 'instruction' rather 'education', and 'training' rather than 'education', as well as how education might be expected to function across the domains of *qualifying, socialising* and *subjectifying* on such matters (Biesta, 1998, 2012). To continue with our potted history, the associated work plan and strategy were subsequently revised and extended at the Johannesburg World Summit on Sustainable Development, in 2002. While, via various routes that lead us back to activities in Sweden again (e.g. the Gothenburg Recommendations on Education for Sustainable Development (GMV, 2009), as well as Swedish governmental support via funding for the Decade), the UN Decade of Education for Sustainable Development (2005–14) was positioned to draw on the key deliberations in Johannesburg, such that the summit's action plan initiated the various resolutions that led to the Decade's inauguration (Abe, 2014) and a major emphasis on both qualifying and socialising as the core functions of education on these topics, and education in general.

Less well-known local to global activities on education and the environment have taken place too, many of which took root and gained sustenance from the innovative work of the early to mid-1970s. These have usually promoted a range of alternative ideas and possible contours for environmental education curriculum and practice, and thus constellations of Biesta's functions (e.g. Ward & Fyson, 1973; Brennan, 1979; Hart, 1981; Huckle, 1983; Pepper, 1987; van Rossen, 1995; Agyeman, 1996; Taylor, 1996; Smith & Williams, 1999/2000; Russell et al. 2002; Nagel, 2005; Kahn, 2008). Prime examples are the emphases given to the development of 'urban environmental education', 'school eco-action groups', 'marine- and water-based environmental education', 'multicultural environmental education', 'place-based pedagogies' as well as the starkly termed, 'education for survival'.

However, it is clear that these don't feature as prominently in curriculum scholarship as those caught in the 'UN web'. UNESCO has spearheaded much of what amounts to the mainstream of work in establishing and promulgating environmental education and variations on this, including 'environmental education for sustainable development' (Tilbury, 1995), and now, Education for Sustainable Development. Given this, its statements, contributions and initiatives have proven to be a veritable goldmine for certain strands of environmental education curriculum critique (e.g. Sauvé et al., 2007). Most recently, attention has been given to the Aichi-Nagoya Declaration on Education for Sustainable Development (UNESCO, 2014) (noting too that the Japanese government had been the main promoter, and subsequently, an underwriter, to the UN Decade proposed in Johannesburg in 2002, see Abe, 2014). And here, we might also note the Declaration's preamble draws on the Brundtland definition and rearticulates the need for participants at the UNESCO World Conference on Education for Sustainable Development to:

> call for urgent action to further strengthen and scale up Education for Sustainable Development (ESD), in order to enable current generations to meet their needs while allowing future generations to meet their own, with a balanced and integrated approach regarding the economic, social and environmental dimensions of sustainable development.

As in the other examples of conferences, initiatives and documents mentioned above, UNESCO has sponsored, populated, scaffolded and encouraged various events, declarations, initiatives and projects to mainstream this work, as made visibly apparent in the prominence and content of its welcome messages and keynote addresses at, for example, the World Environmental Education Congress (see, for example, weec2015. org). More broadly, by securing financial support from the governments of

Japan, Sweden, Germany (and now, for education in the SDGs, Norway), amongst others, the work of UNESCO and UNEP has helped formalise and shape the various roles, goals, objectives, guiding principles, characteristics and audiences ascribed to environmental education in many corners of the world (e.g. since the late 1960s, in milestone events held in Carson City (Nevada), Belgrade, Tbilisi, Moscow, Thessaloniki, Ahmedabad, and Bonn—see for example, UNESCO, 2002; Lotz-Sisitka, 2008, 2009; and items collected in Reid & Dillon, 2017).

So, *why does this matter to curriculum scholarship?* Partly, there is a general point to be made: as a case study, it illustrates what can and can't become a 'mainstream' of curriculum (for an early take on this, see Goodson, 1983), while as the collection also shows, environmental education has ebbed and flowed around the world, and understanding why becomes crucial to reflecting on not just what people regard as 'mainstream' but also why 'eddies' and 'margins' in curriculum develop, and what to do about this, if anything—for a range of reasons.

Curriculum debates (and some considerable disquiet) are linked then to how environmental and sustainability education illustrate other vectors too, including through recognition of how the 'neoliberalisation' and 'globalisation' of environmental education curriculum might be linked (although not exclusively) to the work of UNESCO (see, for example, the special issue of *Environmental Education Research*, 2015, on 'Environmental Education in a Neoliberal Climate', which includes Huckle & Wals' (2015) critique of a lack of radicality and the inevitability of 'business as usual in the end', cf. Luke, 2001; Peters, 2001). There are also participants in these initiatives who query the terms of reference, evidence base and role of inquiry in advancing for this work, e.g. Stephen Sterling and colleagues on the priority given to 'scaling up' as a means to mainstreaming this work (Sterling et al., 2017). This is despite the fact that for those involved in Higher Education, the UN Decade of Education for Sustainable Development (DESD) Final Report recognised that 'more than scaling up of good practice' and 'greater attention to systemic approaches to curriculum change and capacity building for leaders will be needed' (UNESCO 2014, p. 31).

In this then, we must also recognise scholarly interests in understanding the particular features of the usual jostling for the centre of attention in any field of study or curriculum matrix (e.g. literacy discourses being both critically and uncritically used to promote various versions of 'ecoliteracy', or indeed accusations of neoliberalism being illiberally made!). Thus some of the more telling (and possibly faddish?) projects have become associated with a diverse set of interests in quite legitimate concerns. These include the prospects and priorities given to humane education, earth education, biophilia, ecofeminist and ecojustice pedagogies, but also whether these are distinct fields or hybridising agents for environmental education (see the citations above, and Grün, 1996, for an earlier commentary on this). Then there are those wrestling with yet other possible reorientations or turns in the curriculum field (e.g. to animals, post-materialisms, and speculative realisms, to illustrate some of the current options reported in the field's journals). But again, overarching these tends to be the broader education aspects of the UN's work on the Millennium Development Goals, and from 2015 onwards, the Sustainable Development Goals, #4 of which is focused on 'quality education', as raised in the general introduction.

So what does environmental education actually look like in practice, in what we might loosely call the everyday school? Looking for evidence of the realities of a wider public consciousness of environmental matters beyond those with enthusiasm for the topic can be frankly disheartening (Stevenson, 2007). For many of those in the world's Western or Westernised schools, environmental education may amount to little more than extra-curricular demonstration projects, eco-certification, or campaign days. Awareness may be stronger because of celebrating 'National Environmental Education Week' in April (a US initiative), 'World Environment Day' (a Stockholm initiative), or possibly 'World Oceans Day' in June or an 'Earth Hour' in April too. But while there is some value to this work (see Rickinson et al., 2016), it has to be said this is not what the likes of Weidner imagined as a deep level of engagement with environmental education that addresses the problems outlined above. Nor is the situation helped by the frequencies of ebbs rather than flows in environmental education provision between such events, leading to the persistence of absence rather than presence of environmental education in many curriculum experiences and priorities thereof, from early years to adulthood, as discussed in this collection.

So as another dimension to curriculum scholarship, we must consider whether these celebrations, outreach programmes and campaigns actually have much to do with the year-round and year-on-year curriculum, and their associated planning priorities, organisational challenges, assessment priorities and evaluation modes, as raised by the question of whether a whole-school approach is valuable or in evidence (e.g. Mogren & Gericke, 2017 a, b). Thus, further questions can be raised as to whether any environmental education associated with such events ends up being simply a matter of orchestrating largely *extra*-curricular activities that have little traction on the rest of a school, and the lives of those within it.

The question of margins and mainstream is also sharply realised when we consider what gets assessed and reported on in schools (Barratt Hacking, et al., 2010), especially in relation to environmental and sustainability matters. Are institutional leaders and stakeholders in education focusing their attention on data about assessments of learning; or the excursions and incursions on environmental education, such as fieldtrips to zoos, or visiting speakers; or holistic approaches to embedding sustainability priorities in not just the curriculum but the wider systems of the school (e.g. in procuring, and operations) (see Wals, 2003)? Or are these relegated to the sidelines with matters of standards and accountabilities focusing on other matters (Hopmann, 2008)?

While admittedly largely rhetorical in form, such questions can spur a further set of curriculum inquiries, such as how such events are approached in schools, and whether they disrupt, integrate with, or distract from the main(stream) curriculum. Is it, for example, essentially tokenistic as a form of environmental education in relation to curriculum given the expectations noted above, or perhaps little more than an example of 'greenwashing' (see

Greenall Gough, 1991), particularly for the various communities and stakeholders involved with education? Huckle (2013) is not the first to critique the role of corporate sponsorship of these events and programmes and the unlikely bedfellows that may appear: to illustrate, a National Environmental Education Week sponsored by Samsung Electronics America, Inc.—which tries to foster environmental education 'every day, in every way' by celebrating "the benefits of environmental education" and "its relevance for people of all ages and all walks of life" (https://www.neefusa.org/environmental-education-week)? And thus another raft of questions can also be asked of the degrees and forms of consensus and dissensus in the education community on environmental matters, including how to approach and secure that in, through and beyond the curriculum, through public and private sponsorship (including taxes and philanthropy) (Lai, 1998).

Another angle here relates to the still heated debates about how educators might respond to the politics and expressions of climate change denial in the face of overwhelming scientific consensus on the topic, including in curriculum texts (Román & Busch, 2016). Nevertheless, given ongoing debates about the 'curriculum wars' and how these are constructed and situated, a key challenge appears to be the scope and degree of consultation within school communities themselves on such matters. These include how their priorities and capacities are shaped as politicising and de-politicising the curriculum, particularly as they consider the significance of local to national sustainability issues and scales (Van Poeck & Östman, 2017). Put bluntly, when are these activities and initiatives actually more than a case of, say, eco-certification for certification's sake, or badging and grandstanding for marketing purposes? While to probe the deeper aspects, how authentic are such celebrations and commemorative activities in education and for education, politically and culturally?

So to return to the Sustainable Development Goals, we can ask how might or do these events and the prevailing cultures of educational institutions, for example, actually address such goals for 'quality education', expressed in 4.7, in terms of ensuring:

> . . . all learners acquire the knowledge and skills needed to promote sustainable development, including, among others, through education for sustainable development and sustainable lifestyles, human rights, gender equality, promotion of a culture of peace and non-violence, global citizenship and appreciation of cultural diversity and of culture's contribution to sustainable development.

In fact, given what has been sketched as a history of the field in this chapter, a related concern now five decades in is, the possible reinvention (if not elimination) of environmental education in the contexts for curriculum enaction, by certain shifts and twists in the conceptual and political terrains for education and environment (Jickling & Sterling, 2017). These include its blurring in many minds with 'education for sustainable development' (or 'education for sustainability') in policy and practice, as illustrated in target 4.7 (see Fien, 1991; Smyth, 1995, and Jickling & Wals, 2008, for a summary of key features and critical perspectives on these developments and shifts; and Reid & Dillon, 2017, for key papers that help document these changes since the 1960s, on environmental, and more recently, sustainability education, e.g. Volume 2, Part IV: Environmental Education and Sustainable Development). Moreover, given this is an environmentally focused set of concerns, how does a deeply anthropocentric ESD (Hattingh, 2002) square with advancing or departing from ecocentric notions of 'Nature' (Bonnett, 2007)? While as some might ask, can the conventional circles of community expected to take this on be reconfigured by considering their *ecological*, *intergenerational* and *interspecies* dimensions too (see footnote 2), including in educational institutions?

Thus as an example of a 'heavily dappled' discursive space, to continue with a painterly metaphor, smudging the boundaries has proven one way of distorting 'curriculum vision', while smearing them is another (see, for example, Bowers, 2000). More specifically, on the preceding points, when education in relation to sustainable development became both a visible concern and viable question for curriculum makers and commentators (particularly from the 1990s and Rio onwards), we might recall that a prominent environmental educationalist from that time, John Smyth (1998), drew on Churchillian terms to ask whether it signalled the 'end of the beginning', or 'the beginning of the end' for environmental education.

Papers in the *Journal of Curriculum Studies* as well as leading international specialist academic outlets (e.g. *Environmental Education Research* and *The Journal of Environmental Education*) continue to show how and why a variety of forms of environmental education still exist (as well as persist) in the curriculum, despite a certain degree of shifting ground for the field. Equally, regionally or nationally focused environmental education journals illustrate and debate whether there are distinctive approaches, qualities or characteristics given their contexts (e.g. in the *Australian Journal of Environmental Education*, what is uniquely Australian about it, asks Gough, 2011). But when it comes to the real world of contemporary environmental education, it can also be argued that what persists is now often framed by an emphasis on experiences from informal and non-formal activities, including those off-site or in the grounds of many schools, rather than in the classroom *per se* (see Dillon & Teamey, 2002). Put somewhat crudely, some may think such 'extra-curricular' experiences are ultimately a good thing (see Fen, 1961; Bonnett, 2007; Jardine, 1994), while others don't appear to mind the agglomeration or dis-locating of curriculum themes and experiences, be those within or outwith the walls of a classroom. This may be because they regard such boundaries as largely artificial anyway to understanding the lived experience, moments and digressions from 'official' curriculum and pedagogies, but also as these are a way of avoiding the curricular and pedagogical 'manipulations' and 'capitulations' that other subject areas seem to have undergone (see Huebner, 1967; Gruenewald, 2004; Cotton, 2006; Moroye, 2009).

To be clear, the broader value of, say, 'eco-', 'green', 'enviro-' or 'sustainable' schools initiatives, as a way of refuting or refusing too much emphasis on solely extra-curricular approaches isn't usually the issue here. This is despite the fact that these initiatives tend to address the environmental, social and economic sustainability dimensions of campus operations and school community, than the curriculum *per se*. Rather, the question is more that of how any such work relates to *curriculum making and reform* (if at all) (Rickinson et al., 2016). This is because we can recognise that curriculum energies and foci have variously shifted (or in some cases, been hijacked?) by other core 'priorities' or 'standards', including taking curriculum making out of the hands of many on-the-ground educators.

The reliance on external programs is a concern here, as is why environmental education no longer seems to have the profile that it entertained before the Decade of Education for Sustainable Development, and whether the two are related. For some, it depends on context: when environmental education is a cross-curricular theme or priority in curriculum, does it become, for example, largely a series of episodes of school-based incursions or excursions (see Ross, 2007). Others ask, particularly of the value(s) of the curriculum as a whole, does it not contribute something unique to notions of the 'learning virtues' across the lifespan (Foster, 2011)? On both, others raise, whether the Decade—its architects and emissaries (and dupes?)—are solely to blame (see Stevenson, 2006). Or are there other factors at play in how we understand curriculum continuities and change, particularly if it is to address the pressing need of mitigating anthropogenic climate change (Fahey, 2012)? What is, or should be, curriculum's role in this, or the role of curriculum studies? Or as some might put it, is environmental education part of the solution, or part of the problem (Orr, 1992)?

So Smyth's is a question worth revisiting after the Decade's closing and as a wider Post-2015 Agenda (GAP, the 'Global Action Plan') takes hold, particularly during a time of much more advocacy for climate change education (Aikens et al., 2016). In some countries, such as Australia, there has been clear resistance, if not a dismantling of both the scientific and educational community's capacity to engage both climate change and climate change education (Whitehouse, 2015). Such an approach makes it hard to see how certain citizenship-focused curriculum formations (Australian or in other state or national guises) can be made to address these or other set of targets associated with the UN, be that the 'World We Want' global consultation report (2013), or the Sustainable Development Goals (Henderson & Tudball, 2016).

Thus in moving from the scholarly perspectives of insiders to more those of outsiders, given the claims and arguments offered in the collection's studies and this revised and re-forming policy landscape, we might now ask, what will be a legitimate and most helpful focus of attention? Is it 'education', 'sustainability', 'environmental education', or say, 'education for sustainability' or 'sustainable education' (see UNESCO, 2002; Sterling, 2014; cf. Aikens et al., 2016)? What will (environmental) education be clustered with—or even integrated within, or positioned against—in terms of international guidance and policy on these matters, if it is to help broach matters of what is worthwhile or problematic to pursue here, e.g. targets on gender, health, civil society, consumption, governance . . .? SDG 12.8, for example, explicitly links its overarching goal of 'ensuring sustainable consumption and production patterns' to education: 'By 2030, ensure that people everywhere have the relevant information and awareness for sustainable development and lifestyles in harmony with nature.' And so, closely tied to all this is the wider question associated with the work of the Journal: how might this matter to curriculum and curriculum studies, be that in the context of a broad spectrum of educational provision and practice in the North and South?

Pulling this together, in light of these and related questions and challenges, preparing a collection of studies that illuminates these has required delving into the Journal's back catalogue of curriculum inquiries to locate possible responses, and understand what has framed and driven such work. This has led to the identification of a series of studies that illustrate the key curriculum challenges *from* and *for* the field of environmental education across a range of contexts and settings, and how their authors have sought to address them. These can be summarised as raising challenges that focus on how we might understand and debate:

(i) 'mainstream' versus 'alternative' curriculum orientations;
(ii) the place and critique of 'instrumentalism' in curriculum policy and practice; and
(iii) whether a particular form of 'extensionism' in reworking curriculum should be accepted or challenged, given that it presents a double-edged sword for any particular focus to curriculum.

We unpick these challenges further in the subsequent essay and reflection, but in brief, the papers show how the challenges often emerged amid what looks like various attempts to stretch curriculum to include yet another area of content or concern, e.g. treating environmental education primarily as an adjectival education, non-disciplinary in orientation, or possibly, as a whole-of-institution ethos. As UNESCO (2014, p. 51) puts it in commentating on the 'journeys' (UNESCO, 2011) nations have taken towards ESD:

> At the national level, many ESD policies in primary and secondary education gain legitimacy through their links with national educational priorities.
>
> . . . In such contexts, ESD policies inform and respond to national/local issues of educational quality, relevance, equity and inclusion. Many governments have also used ESD as an umbrella policy framework to integrate so-called 'adjectival' educations in primary and secondary schools: climate change education, health education, peace education, environmental education, human rights education, HIV and AIDS education, multicultural education, and so on. . . . As education policy-makers explore the relevance and purpose of education in society, they begin to adopt and integrate the broader lens of ESD and use that lens to reform educational policy, curricula, learning outcomes and skills attainment across all levels of education.

Crucially though, invoking the metaphor of an umbrella seems to be without necessarily addressing related questions of either the 'retiring' or 'compromising' of those features of curriculum and education already there. These include commitments to particular knowledge, skills and values that don't always sit together well, nor necessarily go with the grain of environmental or environmentalist concerns (as mentioned earlier, in relation to an unquestioning 'education for consumption'). Moreover, as much that is called environmental education curriculum work becomes essentially a meliorist rather than transformationist project, be it by a matter of default or design, reform or timing, any resulting whole-of-curriculum that takes account of environmental education means the curriculum is (somewhat paradoxically) *narrowed* as much as it is *broadened*, as these themes and issues are elevated for attention.

So an enduring risk in such flexing of curriculum content and boundaries is that of curricular *fracture*. It may occur in both the prevailing curriculum architecture and by way of a 'weeding' or 'thinning' of the 'textural experience' by creating what amounts to a curricular 'patchwork' (or perhaps at best, in an ensuing curriculum content 'mosaic'?) (Westbury, 2013). Whether it is for curriculum traditionalists, liberals or radicals, at stake is a question of emphasis: is it simply that of depth rather than breadth, and what is realistic as an 'ideal' (let alone tolerable) in these regards, particularly given related questions of the power and authority of one's own and others' viewpoints on the circumstances for curriculum development and possibly, continuity, renewal or innovation that meets environmental and sustainability challenges head on (see Postma, 2002)?

Another key point to recognise as we bring this non-technical introduction to a close is that the fracturing process will present in a variety of ways in the lives of students experiencing an environmental education, particularly if they are considered from the perspective of 'lifelong and lifewide learners' (Rickinson, 2006). This can also be understood as simply a reflection of how curriculum dynamics and shifts play out across the moments and trajectories of various lifecourses and generations. But it also raises an intergenerational educational question, given there are teachers both initially educated (and perhaps professionally developed) before and/or after the post-2015 Agenda, or Johannesburg, or Rio. . . . While at a more prosaic level, and echoing the mantra of the National Environmental Education Foundation, there's the niggling question of what it might mean to *engage* with one's environmental education-related studies day-by-day, week-by-week and year-on-year (as student or teacher). Put bluntly, is there a way of *embracing* rather than having to simply *endure* what curriculum gurus (or managers) have to offer as the latest (fad?) in curriculum thinking, innovation and reform in environmental education, however compelling or urgent it might be? Moreover, as raised in Reid and Dillon (2017), is there a clear line to be drawn between environmental and environmental*ist* education, and how important is it to determine or police that given the challenges raised above?

In sum, we trust that the contributions to this collection illustrate how such challenges offer a particular shape to—and set of possibilities for—responses in the face of various priorities ascribed to the curriculum, as these concern environmental and related areas at a particular moment and for a particular priority in education (Jensen, 2004). And together with these introductory observations and our additional commentaries, the diversity and depth of these challenges and their responses serve to underscore why we believe that taking a 'long view' is important to any critical engagement with contemporary international events such as the World Environmental Education Congress, the end and possible forms of continuance of a UN Decade of Education for Sustainable Development, and the elaboration, integration and progress of Sustainable Development Goals for nations around the globe that will have to feature a commitment to 'education'. All of these can—or *always should*—be engaged from a historical perspective on curriculum (Greenall Gough & Robottom, 1993), particularly if any 'Education for All' is to be sound, inclusive, rigorous and progressive across a range of perspectives on environmental education.

Notes

1. See Marsden (1971), reproduced in this collection, on the various intellectual aims, levels and expectations of subject integration, motivation, socialisation and vocation within the Environmental Studies syllabuses developed for Colleges in England in the late 1960s and into the early 1970s. It should be noted that this work took place before a formal commitment to environmental education at government level in the UK (e.g. tied to the work of HMI or the Schools Council), while a 'National Association' (previously Rural Studies Association, established in 1960), a 'Society' (shortlived) and a 'Council' for Environmental Education established in July 1968, had only just begun their work. These were informed by local and international perspectives (e.g. the UK's 'Countryside in 1970' conference held in 1965 at Keele University, the IUCN Nevada meeting in 1970, and work underway in NW Europe – see also Pritchard, 1968; Goodson, 1983, and Reid & Dillon, 2017).
2. University of Wisconsin is an institution that became well known for its subsequent support of environmental education, e.g. through its facilities at Stevens Point, and the professors, staff and students of the programs, outreach, leadership and research centred there. In some quarters, the sense of urgency and commitment noted continues to this day; see Sterling (2010). In the case of Wisconsin, it stretches back to such markers as Aldo Leopold's (1949) Sand County Almanac, describing the land around the author's home there in Sauk County. Articulating key themes for curriculum design and critique that we return to elsewhere in this collection, we note the Almanac crystallises a Land Ethic as: 'A thing is right when it tends to preserve the integrity, stability, and beauty of the biotic community. It is wrong when it tends otherwise' (p. 262), and (p. 242): 'The land ethic simply enlarges the boundaries of the community to include soils, waters, plants, and animals, or collectively: the land.'

References

Abe, Osamu (2014) ESD projects in Japanese schools and in non-formal education in Japan. In: Chi-Kin John Lee & Rob Efird (Eds.), *Schooling for Sustainable Development Across the Pacific* (pp. 125–139) (Dordrecht: Springer Netherlands).

Agyeman, Julian (1989) Into the 1990's: quality and equality. *Annual Review of Environmental Education*, 3, 23–25.

Aikens, Kathleen, McKenzie, Marcia, & Vaughter, Philip (2016) Environmental and sustainability education policy research: a systematic review of methodological and thematic trends, *Environmental Education Research*, 22:3, 333–359.

Ball, Stephen J., & Bowe, Richard (1992) Subject departments and the 'implementation' of National Curriculum policy: an overview of the issues, *Journal of Curriculum Studies*, 24:2, 97–115.

Barratt Hacking, Elisabeth, Scott, William, & Lee, Elsa (2010) *Evidence of Impact of Sustainable Schools* (London: Department for Children, Schools and Families (DCSF)).

Biesta, Gert J.J. (1998) Say you want a revolution... Suggestions for the impossible future of critical pedagogy, *Educational Theory*, 48:4, 499–510.

Biesta, Gert (2012) Becoming world-wise: an educational perspective on the rhetorical curriculum, *Journal of Curriculum Studies*, 44:6, 815–826.

Bonnett, Michael (2007) Environmental education and the issue of nature, *Journal of Curriculum Studies*, 39:6, 707–721.

Bowers, Chet (2000) Toward a cultural and ecological understanding of curriculum. In: William Doll, & N. Gough (Eds.), *Curriculum Visions* (pp. 75–85) (New York: Peter Lang).

Brennan, Matthew J. (1979) Where are we and what time is it? *The Journal of Environmental Education*, 11:1, 45–46.

Brennan, Andrew (1991) Environmental awareness and liberal education, *British Journal of Educational Studies*, 39:3, 279–296.

Clark, Edward (1975) Good education is environmental, *The Journal of Environmental Education*, 6:4, 1–5.

Clover, Darlene, & Shaw, Katie (2009) Re-imagining consumption: political and creative practices of arts-based environmental adult education. In: Jennifer A.Sandlin, and Peter McLaren (Eds.), *Critical Pedagogies of Consumption: Living and Learning in the Shadow of the "Shopocalypse"* (pp. 203–213) (New York: Routledge).

Cole, Anna Gahl (2007) Expanding the field: revisiting environmental education principles through multidisciplinary frameworks, *The Journal of Environmental Education*, 38:2, 35–45.

Cotton, Debby (2006) Implementing curriculum guidance on environmental education: the importance of teachers' beliefs, *Journal of Curriculum Studies*, 38:1, 67–83.

Covert, Douglas C. (1969) Toward a curriculum in environmental education. *Environmental Education*, 1:1, 11–12.

de Almeida, Miguel O. (1972) *Environment and Development: The Founex Report on Development and Environment* (New York: Carnegie Endowment for International Peace).

Dillon, Justin, & Teamey, Kelly (2002) Reconceptualizing environmental education: taking account of reality, *Canadian Journal of Math, Science & Technology Education*, 2:4, 467–483.

Dobson, Andrew (2016) *Environmental Politics: A Very Short Introduction* (Oxford: Oxford University Press).

Environmental Education Research (2015) Environmental education research in a neoliberal climate. [Special issue edited by David Hursh, Joseph Henderson & David Greenwood], 21:3, 299–505.

Fahey, Shireen J. (2012) Curriculum change and climate change: inside outside pressures in higher education, *Journal of Curriculum Studies*, 44:5, 703–722.

Fen, Sing-nan (1961) Education as growth of environmental consciousness, *Educational Theory*, 11:2, 85–92.

Fien, John (1991) Ideology, political education and teacher education: matching paradigms and models, *Journal of Curriculum Studies*, 23:3, 239–256.

Foster, John (2011) Sustainability and the learning virtues, *Journal of Curriculum Studies*, 43:3, 383–402.

Gayford, Chris (1986) Environmental education and the secondary school curriculum, *Journal of Curriculum Studies*, 18:2, 147–157.

Ginzbarg, Hana M. (1971) Environmental education for political action, *The Journal of Environmental Education*, 3:1, 26–27.

GMV [Centre for Environment and Sustainability] (2009) The Gothenburg Recommendations on Education for Sustainable Development, http://uf.gu.se/digitalAssets/1271/1271420_1270373_Gothenburg_recommendations_on_ESD_Adopted_Nov_12_2008.pdf

Goodson, Ivor (1983) Subjects for study: aspects of a social history of curriculum, *Journal of Curriculum Studies*, 15:4, 391–408.

Gough, Annette (2011) The Australian-*ness* of curriculum jigsaws: where does environmental education fit? *Australian Journal of Environmental Education*, 27, 1: 9–23.

Gough, Noel (1989) From epistemology to ecopolitics: renewing a paradigm for curriculum, *Journal of Curriculum Studies*, 21:3, 225–241.

Greenall Gough, Annette (1991) Greening the future for education: changing curriculum content and school organization, *Journal of Curriculum Studies*, 23:6, 559–571.

Greenall Gough, Annette, & Robottom, Ian (1993) Towards a socially critical environmental education: water quality studies in a coastal school, *Journal of Curriculum Studies*, 25:4, 301–316.

Gruenewald, David A. (2004) A Foucauldian analysis of environmental education: toward the socioecological challenge of the Earth Charter, *Curriculum Inquiry*, 34,1: 71–107.

Grün, Mauro (1996) An analysis of the discursive production of environmental education: terrorism, archaism and transcendentalism, *Curriculum Studies*, 4:3, 329–347.

Hart, E. Paul (1981) Identification of key characteristics of environmental education, *The Journal of Environmental Education*, 13:1, 12–16.

Hart, Paul (2003) *Teachers'Thinking in Environmental Education: Consciousness and Responsibility* (New York: Peter Lang).

Hattingh, Johan (2002) On the imperative of sustainable development: a philosophical and ethical appraisal. In: Eureta Janse van Rensburg, Johan Hattingh, Heila Lotz-Sisitka, Rob O'Donoghue (Eds.) *Environmental Education, Ethics & Action in Southern Africa* (pp. 5–12) (Pretoria: Human Sciences Research Council Publishers).

Hawkey, Kate (2014) A new look at big history, *Journal of Curriculum Studies*, 46:2, 163–179.

Heimlich, Joe E., & Falk, John H. (2009) Free-choice learning and the environment. In: John H. Falk, Joe E. Heimlich & Susan Foutz (Eds.), *Free-choice Learning and the Environment* (pp. 11–21) (Los Altos: AltaMira Press).

Henderson, Deborah J., & Tudball, Elizabeth J. (2016) Democratic and participatory citizenship: youth action for sustainability in Australia, *Asian Education and Development Studies*, 5:1, 5–19.

Hicks, David, & Holden, Cathy (1995) Exploring the future: a missing dimension in environmental education, *Environmental Education Research*, 1:2, 185–193.

Hillbur, Per, Ideland, Malin, & Malmberg, Claus (2016) Response and responsibility: fabrication of the eco-certified citizen in Swedish curricula 1962–2011, *Journal of Curriculum Studies*, 48:3, 409–426.

Hopmann, Stefan (2003) On the evaluation of curriculum reforms, *Journal of Curriculum Studies*, 35:4, 459–478.

Hopmann, Stefan (2008) No child, no school, no state left behind: schooling in the age of accountability, *Journal of Curriculum Studies*, 40:4, 417–456.

Huckle, John (1983) Environmental education. In: John Huckle (Ed.), *Geographical Education: Reflection and Action* (pp. 99–111) (Oxford: Oxford University Press).

Huckle, John (2013) Eco-schooling and sustainability citizenship: exploring issues raised by corporate sponsorship, *The Curriculum Journal*, 24:2, 206–223.

Huckle, John, & Wals, Arjen E. J. (2015) The UN Decade of Education for Sustainable Development: business as usual in the end, *Environmental Education Research*, 21:3, 491–505.

Huebner, Dwayne (1967) Curriculum as concern for man's temporality, *Theory Into Practice*, 6:4, 172–179.

Hungerford, Harold (1980) Goals for curriculum development in environmental education, *The Journal of Environmental Education*, 11:3, 42–47.

International Union for the Conservation of Nature (1970) International Working Meeting on Environmental Education in the School Curriculum. Final Report. IUCN/ UNESCO.

IUCN-WWF-UNEP (1980) World Conservation Strategy: Living Resource Conservation for Sustainable Development (Gland: IUCN). https://portals.iucn.org/library/efiles/documents/wcs-004.pdf

IUCN-WWF-UNEP (1991) Caring for the Earth: A Strategy for Sustainable Living (Gland: IUCN). https://portals.iucn.org/library/efiles/documents/cfe-003.pdf

Jardine, David W. (1994) 'Littered with literacy': an ecopedagogical reflection on whole language, pedocentrism and the necessity of refusal, *Journal of Curriculum Studies*, 26:5, 509–524.

Jenkins, Edgar W. (1994) Public understanding of science and science education for action, *Journal of Curriculum Studies*, 26:6, 601–611.

Jensen, Bjarne Bruun (2004) Environmental and health education viewed from an action-oriented perspective: a case from Denmark, *Journal of Curriculum Studies*, 36:4, 405–425.

Jensen, Bjarne Bruun, & Schnack, Karsten (1997) The action competence approach in environmental education, *Environmental Education Research*, 3:2, 163–178.

Jickling, Bob, & Wals, Arjen (2008) Globalization and environmental education: looking beyond sustainable development, *Journal of Curriculum Studies*, 40:1, 1–21.

Kahn, Richard (2008). Towards ecopedagogy: weaving a broad-based pedagogy of liberation for animals, nature and the oppressed peoples of the Earth. In: A. Darder, R. Torres and M. Baltodano (Eds.), *The Critical Pedagogy Reader* (2nd ed.). (pp. 522–540) (New York: Routledge).

Knapp, Doug (2000) The Thessaloniki Declaration: A Wake-Up Call for Environmental Education? *The Journal of Environmental Education*, 31:3, 32–39.

Kopnina, Helen (2014) Future acenarios and environmental education, *The Journal of Environmental Education*, 45:4, 217–231.

Krall, Florence R. (1979) Living metaphors: the real curriculum in environmental education, *Journal of Curriculum Theorizing*, 1:1, 180–185.

Lai, On-Kwok (1998) The perplexity of sponsored environment education: a critical view on Hong Kong and its future, *Environmental Education Research*, 4:3, 269–284.

Lee, John Chi-Kin (2000) Teacher receptivity to curriculum change in the implementation stage: the case of environmental education in Hong Kong, *Journal of Curriculum Studies*, 32:1, 95–115.

Leopold, Aldo (1949) The land ethic. In: *A Sand County Almanac: With Other Essays on Conservation from Round River* (pp. 237–264) (Oxford: Oxford University Press).

Lotz-Sisitka, Heila (2008) Utopianism and educational processes in the United Nations Decade of Education for Sustainable Development, *Canadian Journal of Environmental Education*, 13:1, 134–152.

Lotz-Sisitka, Heila (2009) How many declarations do we need? Inside the drafting of the Bonn declaration on education for sustainable development, *Journal of Education for Sustainable Development*, 3:2, 205–210.

Luke, Timothy W. (2001) Education, environment and sustainability: what are the issues, where to intervene, what must be done?, *Educational Philosophy and Theory*, 33:2, 187–202.

Marsden, William E. (1971) Environmental studies courses in colleges of education, *Journal of Curriculum Studies*, 3:2, 163–178.

Martin, Peter (1996) A WWF view of education and the role of NGOs. In: John Huckle & Stephen Sterling (Eds.), *Education for Sustainability* (pp. 40–51) (London: Earthscan).

McInnis, Noel (1975) What makes environment educational? In: Noel McInnis & Don Albrecht (Eds.), *What Makes Education Environmental?* (pp. 21–29) (Kentucky: Data Courier).

Minnesota Environmental Education Council (1972) *A State Plan for Environmental Education for the Citizens of Minnesota*. Office of Education (DHEW), Washington, D.C.

Mogren, Anna, & Gericke, Niklas (2017a). ESD implementation at the school organisation level, part 1 – investigating the quality criteria guiding school leaders' work at recognized ESD schools, *Environmental Education Research*, 23:7, 993–1014.

Mogren, Anna, & Gericke, Niklas (2017b) ESD implementation at the school organisation level, part 2 – investigating the transformative perspective in school leaders' quality strategies at ESD schools, *Environmental Education Research*, 23:7, 972–992.

Monroe, Martha C. (2012) The Co-Evolution of ESD and EE, *Journal of Education for Sustainable Development*, 6:1, 43–47.

Moroni, Antonio (1978) Interdisciplinarity and environmental education, *Prospects*, VIII:4, 480–494.

Moroye, Christy (2009) Complementary curriculum: the work of ecologically minded teachers, *Journal of Curriculum Studies*, 41:6, 789–811.

Morris, Marla (2002) Ecological consciousness and curriculum, *Journal of Curriculum Studies*, 34:5, 571–587.

Nagel, Michael (2005) Constructing apathy: how environmentalism and environmental education may be fostering 'learned hopelessness' in children, *Australian Journal of Environmental Education*, 21, 71–80.

Nash, Roderick (1976) Logs, universities, and the environmental education compromise, *The Journal of Environmental Education*, 8(2), 2–11.

NGO Forum (1992) *Environmental Education for Sustainable Societies and Global Responsibility* (Toronto: International Council for Adult Education).

O'Riordan, Timothy (1981) Environmentalism and education, *Journal of Geography in Higher Education*, 5:1, 3–17.

Pepper, David (1987) The basis of a radical curriculum in environmental education. In: Colin Lacey & Roy Williams (Eds.), *Education, Ecology and Development: The Case for an Education Network* (pp. 65–79) (London: The World Wildlife Fund & Kogan Page).

Peters, Michael A. (2001) Environmental education, neo-liberalism and globalisation: the 'New Zealand experiment', *Educational Philosophy and Theory*, 33:2, 203–216.

Postma, Dirk W. (2002) Taking the future seriously: on the inadequacies of the framework of liberalism for environmental education, *Journal of Philosophy of Education*, 36:1, 16–56.

Pritchard, Tom (1968) Environmental education – its social relevance in North-West Europe (Strasbourg: Council of Europe) (unpublished report CE/Nat (68) 67).

Reid, Alan, & Dillon, Justin (Eds.) (2017) *Environmental Education: Critical Concepts in the Environment* (London: Routledge).

Rickinson, Mark (2006) Researching and understanding environmental learning: hopes for the next ten years, *Environmental Education Research*, 12:3–4, 445–457.

Rickinson, Mark, Hall, Matt, & Reid, Alan (2016) Sustainable schools programmes: what influence on schools and how do we know? *Environmental Education Research*, 22:3, 360–389.

Román, Diego, & Busch, K. C. (2016) Textbooks of doubt: using systemic functional analysis to explore the framing of climate change in middle-school science textbooks, *Environmental Education Research*, 22:8, 1158–1180.

Ross, Hamish (2007) Environment in the curriculum: representation and development in the Scottish physical and social sciences, *Journal of Curriculum Studies*, 39:6, 659–677.

Russell, Connie (2005) Whoever does not write is written: The role of nature in post-post approaches to environmental education research, *Environmental Education Research*, 11:5, 433–443.

Russell, Constance L., Sarick, Tema, & Kennelly, Jacqueline (2002) Queering environmental education, *Canadian Journal of Environmental Education*, 7, 54–66.

Sandall, Klas, Ohman, Johan, & Ostman, Leif (2005) Selective traditions within environmental education. In: *Education for Sustainable Development* (pp. 155–168) (Lund: Studentlitteratur AB)¬.

Sauvé, Lucie, Berryman, Tom, & Brunelle, Renée (2007) Three decades of international guidelines for environment-related education: a critical hermeneutic of the United Nations discourse, *Canadian Journal of Environmental Education*, 12:1, 33–54.

Schinkel, Anders (2009) Justifying compulsory environmental education in liberal democracies, *Journal of Philosophy of Education*, 43:4, 507–526.

Smith, Gregory A., & Williams, Dilafruz R. (1999/2000) Ecological education: extending the definition of environmental education, *Australian Journal of Environmental Education*, 15/16, 139–146.

Smyth, John C. (1995) Environment and education: a view of a changing scene, *Environmental Education Research*, 1:1, 3–20.

Smyth, John (1998) Environmental education—the beginning of the end or the end of the beginning, *Environmental Communicator*, 28:4, 14–15.

Stapp, William B. et al. (1969) The concept of environmental education, *Environmental Education*, 1:1, 30–31.

Sterling, Stephen (1992) *Coming of Age: A Short History of Environmental Education (to 1989)* (Walsall: National Association for Environmental Education).

Sterling, Stephen (2010) Living *in* the Earth: towards an education for our times, *Journal of Education for Sustainable Development*, 4:2, 213–218.

Sterling, Stephen (2014) Separate tracks or real synergy? Achieving a closer relationship between education and SD, post-2015, *Journal of Education for Sustainable Development*, 8:2, 89–112.

Sterling, Stephen, Glasser, Harold, Rieckmann, Marco, & Warwick, Paul (2017) "More than scaling up": a critical and practical inquiry into operationalizing sustainability competencies. In: Peter Corcoran, Joseph Weakland & Arjen Wals (Eds.) *Envisioning Futures for Environmental and Sustainability Education* (pp. 153–168) (Wageningen: Wageningen Academic Publishers).

Stevenson, Robert B. (2006) Tensions and transitions in policy discourse: recontextualizing a decontextualized EE/ESD debate, *Environmental Education Research*, 12:3–4, 277–290.

Stevenson, Robert (2007) Schooling and environmental education: contradictions in purpose and practice, *Environmental Education Research*, 13,2: 139–153.

Swan, James A. (1969) The challenge of environmental education, *Phi Delta Kappan*, 51:1, 26–28.

Taylor, Dorceta (1996) Making multicultural environmental education a reality, *Race, Poverty and the Environment*, Winter/Spring, 3–6.

Tilbury, Daniella (1995) Environmental education for sustainability: defining the new focus of environmental education in the 1990s, *Environmental Education Research*, 1:2, 195–212.

UNEP (1972) Report of the United Nations Conference on the Human Environment. Available at: http://www.unep.org/Documents.Multilingual/Default.asp?DocumentID=97

UNESCO-UNEP (1976) The Belgrade Charter: a global framework for environmental education, *Connect*, I:1, 1–9, 23.

UNESCO-UNEP (1978) The Tbilisi Declaration, *Connect*, III:1, 1–8.

UNESCO (2002) Education for Sustainability – From Rio to Johannesburg: Lessons Learnt from a Decade of Commitment (Paris: UNESCO).

UNESCO (2011) National Journeys towards Education for Sustainable Development (Paris: UNESCO).

UNESCO (2014) Shaping the Future We Want: UN Decade of Education for Sustainable Development (2005–2014). Final Report (Paris: UNESCO). https://sustainabledevelopment.un.org/content/documents/1682Shaping%20the%20future%20we%20want.pdf

UNESCO (2017) Education for Sustainable Development Goals: Learning Objectives (Paris: UNESCO). http://unescodoc.unesco.org/images/0024/002474/247444e.pdf

United Nations (1992) United Nations Conference on Environment & Development Rio de Janerio, Brazil, 3 to 14 June 1992. https://sustainabledevelopment.un.org/content/documents/Agenda21.pdf

Van Poeck, Katrien, & Östman, Leif (2017) Creating space for 'the political' in environmental and sustainability education practice: a Political Move Analysis of educators' actions, *Environmental Education Research*, DOI: 10.1080/13504622.2017.1306835

Van Rossen, Judy (1995) Conceptual analysis in environmental education: why I want my children to be educated for sustainable development, *Australian Journal of Environmental Education*, 11, 73–81.

Vidart, Daniel (1978) Environmental education – theory and practice, *Prospects*, VIII:4, 466–479.

Walker, Kim (1997) Challenging critical theory in environmental education, *Environmental Education Research*, 3:2, 155–162.

Wals, Arjen (2003) Measuring environmental education – the meaning of standardisation and the language of instrumentalism. In: William Scott & Stephen Gough (Eds.) *Key Issues in Sustainable Development: A Critical Review* (pp. 179–181) (London: Routledge Falmer).

Wals, Arjen, Brody, Michael, Dillon, Justin, & Stevenson, Robert (2014) Convergence between science and environmental education, *Science*, 344: 6184, 583–584.

Ward, Colin, & Fyson, Anthony (1973) *Streetwork: The Exploding School* (London: Routledge & Kegan Paul).

Westbury, Ian (2013) Reading Schwab's the 'Practical' as an invitation to a curriculum enquiry, *Journal of Curriculum Studies*, 45:3, 640–651.

Weston, Anthony (1996) Deschooling environmental education, *Canadian Journal of Environmental Education*, 1:1, 35–46.

Whitehouse, Hilary (2015) Young people responding to the Anthropocene: re-considering active citizenship in a new epoch, *The Social Educator*, 33:2, 18–25.

Winn, Ira (1970) The education in environmental education, *Environmental Education*, 1:4, 140–141.

World Commission on Environment and Development (WCED) (The Brundtland Report) (1987) *Our Common Future* (Oxford: Oxford University Press).

Young, Michael (2013) Overcoming the crisis in curriculum theory: a knowledge-based approach, *Journal of Curriculum Studies*, 45:2, 101–118.

How to understand curriculum challenges *for* and *from* environmental education

ALAN REID

Introduction

In this brief essay, we continue the theme of 'Curriculum Challenges *for* and *from* Environmental Education' through a focus on how a curriculum framed in terms of the logic and language of adjectival educations and prepositional forms of education is variously understood, applied and critiqued in the field of environmental education. The essay shows how these frameworks return us to key questions in curriculum politics, planning and implementation, namely:

- which educative experiences should a curriculum foster and why?
- what should be the scope of a worthwhile curriculum and how should it be decided, organised and reworked?
- why is distinctive curricula provided to different groups of students? and
- how should (a) curriculum be best enacted and evaluated?

As mentioned in the general and non-technical introductions to this collection, the questions and associated challenges of *orientation, instrumentality* and *extensionism* sketched therein focus attention on:

(i) 'mainstream' versus 'alternative' curriculum orientations and movements;
(ii) the place and critique of 'instrumentalism' in curriculum policy and practice; and
(iii) how attending to a particular form of 'extensionism'—in essence, those attempts to stretch the curriculum to include another area of curriculum without removing or compromising attention to others—has come to shape key debates on the curriculum priorities and politics regarding environmental and related areas.

Of particular interest for this collection then, are statements of the kind such as those offered by UNESCO (2014, p. 51) during its commentary on the 'journeys' (UNESCO, 2011) nations may take from environmental education towards education for sustainable development (ESD), when it states that:

> At the national level, many ESD policies in primary and secondary education gain legitimacy through their links with national educational priorities.
>
> . . . In such contexts, ESD policies inform and respond to national/local issues of educational quality, relevance, equity and inclusion. Many governments have also used ESD as an umbrella policy framework to integrate so-called

'adjectival' educations in primary and secondary schools: climate change education, health education, peace education, environmental education, human rights education, HIV and AIDS education, multicultural education, and so on. . . . As education policy-makers explore the relevance and purpose of education in society, they begin to adopt and integrate the broader lens of ESD and use that lens to reform educational policy, curricula, learning outcomes and skills attainment across all levels of education.

As is often the case with environmental education and other 'adjectival educations', these complex challenges emerge because they typically require policy and curriculum makers to wrestle with the expectation that a growing plethora of social issues are represented in and by a curriculum (Jensen, 2004), even as any particular need for a place within a general curriculum will have to be addressed within an often already crowded and contested terrain (see Goodson, 1983; Greenall Gough, 1991; Gough & Gough, 2010; Lee, 2000; Cotton, 2006; Foster, 2011).

As Parts 4 and 5 for the collection show (available online and in the *Journal of Curriculum Studies*), high status knowledge in curriculum is (still) largely reflective of a distinct 'pecking order' of subject disciplines (Young, 2013). However, curriculum scholarship on environmental education might also suggest the field stands to offer much more than yet another focus for deliberations about status and value in curriculum circles (be those internationally, or nationally to locally). This is because they might raise a deeper set of challenges beyond those that have become 'stock questions' of curriculum construction, breadth, accommodation or integration (see Gough, 1989; Jardine, 1994; Bonnett, 2007; Moroye, 2009; Fahey, 2012).

Stephen Sterling (2010, pp. 215–217) crystallises these concerns in an essay called 'Living *in* the Earth: Towards an Education for Our Times' in the *Journal of Education for Sustainable Development*, as follows:

> During the 1970s and 1980s, the boundaries of what was meant by 'environmental education' (EE) expanded and became more inclusive. By the 1980s, the term 'EE' encompassed environmental studies and field studies, environmental science, environmental interpretation, urban studies, heritage education, conservation education and global environmental issues education, where different interests promoted different aspects, often through separate organisations and groups. At the same time, the boundaries of environmental education became less distinct and gradually more permeable. Various groups interested in aspects of 'education for change' and equity interpreted the transformative role of education differently. Hence the spectrum of what was sometimes referred to as 'adjectival educations' also included such emphases as development education, peace education, human rights education, antiracist education and futures education, whilst global education made a bid to represent them all.

> The proliferation in recent decades of 'adjectival educations'—each concerned with some aspect of social change—seems ironic (if understandable), given that many of those working in these areas seek a more holistic education than that offered by the compartmentalised and reductionist mainstream. This is why I have suggested 'sustainable education' (Sterling 2001) not as yet another adjectival education, but to suggest the need for, and bases

of, a changed educational paradigm. Any closed definition of education for change, whether EE, ESD or other manifestation, involves drawing conceptual boundaries. This carries the danger that all other educational policies, theories and practices appear to be outside or beyond these boundaries—and therefore actors outside the boundaries assume or perceive that this manifestation of sustainability education is not their concern.

Sterling's views on adjectival forms had been previously rehearsed in his Study Guide for the Masters in Environmental Education and Development Education at London South Bank University, which later became its Masters in Education for Sustainability (the longest running course with this particular title in the UK). Julian Agyeman's doctoral thesis awarded by the University of London's Institute of Education also traces these debates (see Agyeman, 1996, as well as Agyeman, 1994a, b) to argue for a shift in focus, as did Daniella Tilbury's (1995) doctorate at the University of Cambridge, leading to her argument that in the 1990s, the new focus of environmental education should be 'environmental education for sustainability'. However, as David Greenwood (2004, p. 73) argues:

> ... environmental education will be ineffective in advancing its own goal of creating an environmentally or ecologically literate (Orr, 1992) citizenry as long as it continues to discipline itself within the norms of general education. When it consorts with schools as an "adjectival" educational discourse (Martin, 1996), environmental education works to legitimize and reinforce problematic trends in general education; especially as environmental education is disciplined by science and conventional environmentalism, it tends to neglect the social, economic, political, and deeper cultural aspects of the ecological problem.

After applying a Foucauldian lens on power, knowledge and discourse to environmental education for analysing its educational practices, Gruenewald (2004, p. 72) ends up agreeing with many of Peter Martin's 'beliefs' in his contribution to Huckle and Sterling's (1996) course reader at South Bank, *Education for Sustainability*. Here, we include the rest of the quotation from Martin's (1996, (p. 51) conclusion to his chapter, using italics to replace the ellipses used by Gruenewald:

> *I believe that*, having become institutionalized, environmental education is a lost cause and should be phased out as soon as possible. *Its history and conventional wisdoms stand in the way of its morphosis into anything that can possibly achieve goals remotely akin to those set by the WCS [World Conservation Strategy]. Education for sustainability is a concept that can happily house all the necessary elements to be effective. However, I do believe that* the ultimate challenge is to remove all adjectival adjuncts *(for that is how they are perceived)* to education and develop the conventional notion that the education system must *as a key function*, prepare all people for their role as well-informed, skilled and experienced participators in determining the quality and structure of the world *we all inhabit.*

Gruenewald, however, continues (2004, p. 72):

> Perhaps Martin is too hasty in his call to abolish environmental education altogether, but he captures the issue succinctly: the institutionalization of environmental education has muted its potential as a transformative

educational discourse practice. The ultimate challenge for education, environmental or otherwise, is to prepare people with the skills and knowledge needed to identify and shape the quality of the world we share with others—human and nonhuman; in a multicultural and political world, this means education for cultural competence and political participation. Within the field of environmental education, both in and outside of schools, there is currently too much complacency toward problematizing the homogenizing standard practices of general education and too much caution around taking the political stands that will be needed to reform it.

How are we to understand this particular debate about environmental education and curriculum in general? Stepping back a little, we might revisit the terms used in the opening comments to this essay, in particular, how an adjectival terminology relays the ways in which conjunctional phrases might be used in Anglo-American accounts of curriculum, e.g. by arguing for a *development* education, *global* education, *citizenship* education, *human rights* education, and/or *sustainable* education. These conjoinings deliberately employ an adjective to signal a specific form and remit to education (which is primarily positioned in such phrases as a noun). Also, in this refocusing, they may serve the dual purpose of both *broadening* and *narrowing* curriculum provision and perspectives simultaneously too, away from some preexisting concerns, and more deeply into others.

Closely tied to such observations are that certain adjectival formats have come to be both prioritised and linked to particular models of education, as noted in the analysis and critique of *prepositional* forms of curriculum construction and commentary. Associated with, for example, the seminal work on educational concepts and slogans of Israel Scheffler, this requires an appreciation of the way an adjectival phraseology can be expected to work, in theory and practice, as an attempt to advance on a conjunctive logic, e.g. on what basis and in what form does the 'and' operate in linking education *and* environment? After Arthur Lucas (1972) and his empirically- and conceptually-based descriptors of a range of practices associated with the early days of the field, environmental education may be regarded as typically only ever an 'adjectival education' when it comes to the mainstays of much school-based curriculum practice (cf. Sterling, 2010, p. 215). In this, Lucas (1972) showed, it often requires: (i) a reversal of the adjectival form, and (ii) an adjective-to-noun shift when a prepositional form is employed. Then, (iii) the assorted prepositions available to the English-speaking world (such as *about/in/for*) can be used to signal and assess different modes and objectives in associated forms and modes of environmental education.

That this language for curriculum is often verbless should not be ignored, even as it is *educating for* 'global citizenship', 'intercultural peace', 'late capitalism' etc. that is expected, documented or suspected as taking place (see Huckle & Sterling, 1996, and Huckle & Wals, 2015). In fact, it is the forms of that *educating* (with perhaps due consideration of which ends-in-view too?) that has often become the crux of debates about *whose* and *which* curriculum and pedagogy is at stake in many of the curriculum challenges for and from *environmental* education, as illustrated in this collection and elsewhere (Foster, 2001; Gruenewald, 2004; see also Grün, 1996).

Revisiting the prepositional logic of environmental education

To illustrate the effects of adopting a prepositional format for this particular field, an 'education *about* the environment' can be shown through both common sense as much as by philosophical examination and empirical study to be the default position for how many schools, colleges and universities have addressed environmental topics since the 1970s. This is typically prosecuted by defining and delineating key concepts, content and topic areas primarily from the natural and social sciences, but increasingly from the 'environmental humanities' (see also Lucas, 1991, on priorities for understanding, skills, attitudes and values in environmental education). It is also familiar to those who regard civics education as centrally concerned with providing an education *about* citizenship, rather than, say, expecting education *through* citizenship.

For Lucas, 'education *about* the environment' typically conveys the intention to generate or provide information related to the biotic and abiotic aspects of environments in the curriculum. Students might be expected to develop, comprehend and interpret environmental data; analyse environmental systems, situations or predicaments into their component principles; synthesise explanations likely to account for an environmental phenomenon that is new to the student; and evaluate environmental data and phenomena, including, perhaps, the consequences of any proposed manipulation (e.g. 'environmental management strategy') in terms of likely environment responses (Lucas, 1972).

With this '*about*' construction as the predominant cornerstone to considerations of curriculum knowledge in environmental education, Lucas shows that it tends to be understood as offering something distinct from:

- an 'education *in* the environment'
  e.g. through a focus on ensuring students have access to fieldwork and other experiential learning opportunities largely outside the classroom,
- an 'education *with* the environment'
  e.g. by focusing on processes and infrastructures, as in pedagogical activities and decision-making designed to acknowledge and reduce ecological/carbon footprints, such as by asking what to consume during schooling; or how (if not, how far) to travel to learn; and
- an 'education *for* the environment'
  e.g. by focusing on a felt need and justifiable outcome to stimulate community action-taking within and/or beyond the educational institution, as when actions are undertaken to help resolve a pressing or important environmental problem affecting said community.

Equally, for some, 'education *in* the environment' becomes more a matter of pedagogical technique, rather than a set of educational goals, while for Robottom (1987), this construction betrays a distinctive purpose of 'education *in* the environment', that of promoting the ideal of a harmonious human-environmental relationship. In other words, instead of only identifying and transmitting knowledge about such typical environmental topic areas as ecosystems and biodiversity, environment-based activities are developed in places that emphasise the significance of a natural environmental context, and gaining experience there, such as through fieldwork (Robottom, 1987; cf. Fien, 1993) to develop pro-environmental values and behaviours, such as through taking part in place-conscious education (Gruenewald, 2003).

Put otherwise, the main assumptions of such an approach to environmental education are typically those of:

1) direct engagement with, or immersion in, a natural environment it-self is just as important in developing pro-environmental worldviews as the transmissions of environmental knowledge in formal education classroom settings;
2) interaction within environments is an effective way to develop environmental knowledge of the human and more-than-human world;
3) the development of an understanding, sensibility and feelings for the biotic and abiotic dimensions of an environment are important ways to prompt environmentally-conscious decisions to protect the environment; and
4) morally motivated actions are more effectively brought about engaging in behaviours that protect the environment than solely relying on the accumulation of generalised or abstract environmental knowledge.

Nevertheless, as one of the contributors to this collection, Bob Jickling (1997, p. 96), elaborates this isn't always enough, and with an 'education *for* the environment', we see an explicit aim:

> to develop critical thinking and enable problem-solving; to examine ideologies which underlie human-environment relationships; to criticise conventional wisdom; to explore material and ideological bases of conventional wisdom; to analyse power relationships within a particular society; to engage students in cultural criticism and reconstruction; to foster political literacy; to focus on real-world problems and participate in real issues; to open students' minds to alternative world views; to work and live cooperatively; and to realise that humans can act collectively to shape society.

Of course, each of these 'archetypal forms' for environmental education will be enacted and interpreted differently, as Lucas showed in his original study, and others have since then, including in the studies presented in this collection. They may also be 'co-enacted' during an educational experience or sequence (acknowledging too the likelihood of with some variation both within and across educational settings), and that this observation might be made more complex by having that deliberately experienced within or outwith an environmental education syllabus or event, e.g. through the lens and priorities of global citizenship education or outdoor education. As such, this reminds us of the possibility of *incidental* forms of environmental education, i.e. through other domains, such as through civics or citizenship education, environmental science, marine biology, home economics, arts-based approaches, or 'Studies of Society and Environment', depending on, for example, the prevailing curriculum framework (e.g. languages, vocabularies, grammars) available.

Towards curriculum critique

As the studies in this collection show in more detail, formulating the components of an adjectival education in such prepositional ways can also illustrate a range of contrasting and competing theoretical and practical interests and discourses about curriculum (e.g. Fien, 1991). For example, in some quarters, ideological orientations might be highlighted, as can (c)overt bias in curriculum (its prevailing and inherent anthropocentrism?), or the differentiated weighting of a range of learning outcomes and competencies expected through curriculum, e.g. to ask whether environmental education is really geared towards furthering rather than challenging or critiquing highly consumptive lifestyles (see Wals & van der Leij, 1997).

In other quarters, a critical appreciation of the status given to a prepositional form might help foster questions of the adequacies of particular and general curriculum and pedagogical coverage and depth, alongside matters of the adequacies and (in?)coherence in what is found within the margins through to the mainstream of curriculum. To illustrate from the collection, what is the role of experiential and participatory forms of engagement with contemporary and future oriented matters of 'ecological citizenship' (Goodson, 1983; Lee, 2000; Jensen, 2004; Cotton, 2006; Ross, 2007)? While in other studies, how does one argue the necessity of including or excluding a particular form or configuration of environmental education in curriculum for that particular moment, the immediate or broader political climate, or for a certain socio-ecological nexus (e.g. cross-curricular, infused, integrated, extra-curricular . . .) (see Agyeman, 1994b; Gruenewald, 2004; Martusewicz, 2009; Hart, 2010)?

So at a second level of complexity, in light of the above considerations and a closer reading of the associated studies in the collection, a key quandary about curriculum coverage and focus can be seen to emerge. It can be expressed in terms of a simple question about priorities: *which part of an adjectival or prepositional phrasing gets 'the upper hand'?* In the case of environmental education, will it inevitably boil down to whether it is a particular version or conjunction of the *environmental* or *educational*—and if so, *why*? As our introductions to the collection show, a corollary to this is the *how* aspect, if not to what end, e.g. in terms of education and/or advocacy: can both be balanced or integrated to some degree so that each aspect is properly 'heard', be that:

(i) within a specific course of study (e.g. is it largely in relation to one form of environmentalism, or . . .?), or

(ii) as a prepositional or adjectival construction (e.g. how much education 'about' or 'in' the environment is sufficient, and in what forms might it take if 'for' is the goal?), as well as

(iii) in relation to wider curriculum thinking, planning, provision and debate (e.g. as a cross-curricular theme or priority only, or largely?) that usually maintains rather than subverts the status quo (See Ball & Bowe, 1992)?

As the many volumes of the *Journal of Curriculum Studies* show, specialist and generalist curriculum scholars have routinely debated the broader aspects to such matters, including the sense of 'necessity and sufficiency' of prepositional distinctions and formulations for such 'adjectivized' areas of curriculum development and provision, as well as whether the whole is greater (or in fact, less) than the sum of its parts in achieving such ends-in-view. Indeed, as an ecological or Gestalt-like query, it is often asked *in* and *of* environmental education by insiders as much as outsiders, while in an aforementioned study published elsewhere by one of the collection's contributors, we note that Bob Jickling (1997) has probed the limits of both a strict and literalistic adherence to these forms in terms of his analysis of 'educational slogans' such as when education is understood to have education and no other as its legitimate end, when a more generous conceptualisation and interpretation may be in order.

Towards a bigger picture?

Even if it appears we are faced with largely conceptual exercises at this point [if not linguistic gymnastics!], a case can also be made that we would do well not to ignore the continuing shifts in the discursive framings and institutional frameworks for this area of curriculum and its debate. As Gough and Gough (2010, unpaginated) end their curriculum encyclopaedia entry on environmental education: 'While most would argue that we need "it", many still argue about what it is and where it can fit into an already overcrowded curriculum.' Probing why this is the case involves grappling with curriculum knowledge, experience, development, control and reform at school-based and government levels. Even more so, it is in recognising these interests that we might ask critical questions of vested interests and 'special interest groups', which might each look to imagine and shape environment-related curriculum in distinctly 'conservative' to 'progressive' ways (and not necessarily in that order!) (Gruenewald, 2004).

To regular or even occasional readers of the Journal, perhaps it is unsurprising that such questions and their answering require us to examine both explicit and veiled claims on the curriculum field, focusing in on its 'terrain' and 'territories' as much as its 'knowledges' and 'control(s)'. Examples can be readily illustrated through reference to the changes in environmental education curriculum thinking, priorities and practice, particularly in the Anglosphere, to which we now briefly turn.

In terms of the United Kingdom, we can start by briefly revisiting the chequered history of certain adjectival educations and their relation to both a subject-based culture and a legislative framework built around providing a 'national curriculum' for schools. A vivid example of this emerged in the mid-late 1980s to 1990s (see Gayford, 1986; Bonnett, 2007). As Nixon (1991), Ball and Bowe (1992) and Cotton (2006) show, this stemmed from largely Westminster-instigated work attempting to develop modern-day national curriculums for England, Wales, Scotland and Northern Ireland. The prospect and costs of this rallied the interests of many non-governmental organisations and (p)layers in the educational system in each of the countries, noting (albeit in passing for now), that each nation had distinctive cultural histories and curricular and pedagogical priorities too (on Scotland, for example, see Ross, 2007).

In revisiting the National Curriculum Orders in their original guise, we must recognise that their architecture was predicated on a 'grid' of core and foundation subjects in schools that was also intended to 'infuse', 'embed', and 'integrate'—but *not* 'compartmentalise'—cross-curricular themes 'related to the opportunities, responsibilities and experiences of adult life'. We should also recognise that the cross-curricular themes were expressed largely in adjectival and prepositional terms, under banner headings of immediate and anticipatory learning outcomes: *health education, careers education and guidance, economic and industrial understanding, environmental education* and *education for citizenship.* (For analysis at the time, including the failures of this policy and its underlying logic, see the previous cited studies, and Whitty et al., 1994 [1]).

Relocating the sphere of examples to the United States, questions about the profile and value of broadly equivalent environmental currents and their prospects in curriculum can be readily identified, particularly if we are familiar with the initial points of departure from the UN set out in its early statements on the field, and discussed at length in the introductions to this collection. As with Dobson (2016), these invite consideration of whether a range of environmentalist 'shades of green' exist (from a palette of light to darker hues), as well as whether their usual enactment is independently possible or supported, or largely occurs via (muddied) blends, given the predominantly 'red and blue' hues to the political and curricular spectrums in Anglo-American contexts (Huckle, 1983), including in teaching about 'controversial issues' (such as climate change?) (Cotton, 2006), as a case in point.

The brief examples that follow illustrate further matters of 'complexification', in that they are both a working *with* and *against* what those familiar with the sociology of education would recognise as different *classifications* and *framings* of educational knowledge, ideology and curriculum. They also suggest there may be other ways of accounting for this complexity, be that sociologically or politically. For example, in returning to the question of the 'upper hand', is an environmental education supposed to be compulsory or mandated in all education provided by the State, be that federally and/or in each state, province or territory (see Schinkel, 2009)? In more provocative terms, what of the institutions of education and the lead or examples they set: should schools and universities be 'carbon neutral', for example, and the curriculum too? While is there also a space for a 'free choice' of learners and teachers in any such setting, including evidence that they've experienced a range of traditions of environmental education (if only to ensure it is 'non-doctrinaire' on its environmental *and* educational aspects—Scott & Oulton, 1999; cf. Stevenson, 2006)?

On this latter point, language and its intended and implicit ideological loading becomes noticeable once more when we consider the slogan of 'creation care' (Hitzhusen, 2007) found in some quarters of the US environmental education movement. While it may also be popular amongst some who deny the realities of climate change (Jacques et al., 2008), is, for example, bald religious instruction better than none if this is the only possible framework for those being environmentally educated in particular contexts? Or to deconstruct this playfully, as when we consider further nouns in relation to 'dinner party' taboos circling on *religion, politics, sex* and *death*, is it actually 'religious instruction' or 'education *about* religion' that is politely or politically preferred, and why? (While perhaps somewhat cheekily as a thought experiment, does an equivalent logic apply when it comes to the aforementioned taboos, including in relation to their application with the other prepositions—namely, in/with/for—that are as equally possible as the 'lived curriculum', but not always desirable, or palatable; see Gough, 1994?)

Stepping back from some of the possible absurdities that can emerge here, to further illustrate some of the currents and contemporary patterns in environmental education curriculum design and their possibilities, we note there have been efforts in the US to:

(i) argue for both subtle to gross shifts of curriculum focus, such as to invert (and possibly subvert) 'No Child Left Behind' through its reworking towards a notion of 'No Child Left Inside'. This might be achieved by using, for example, 'Environment as an Integrating Context' for learning; or through meditative and therapeutic nature-based encounters, such as those typically used for 'youth at risk'. It might also be to better engage questions of what should and does count as 'national standards' for education in the face of wider environmental protection or poverty alleviation goals (Gruenewald, 2008; Hart, 2010). A prominent popular movement in this regard is associated with Richard Louv's work (e.g. his 2005 book, *Last Child in the Woods*) which advocates combatting 'Nature Deficit Disorder' among children (and adults) through the sustained provision of education out-of-doors[2], or simply, if not again, sloganistically, 'Vitamin N'.

Next, we might detect efforts to:

(ii) raise debate about when it is known—and with what qualification—that a particular adjectival education might turn out to be another 'incursion' or 'raid' into 'subjects', noting too their various 'tribes', pre-existing cultures and 'empires' and histories of prior conflicts (see Goodson, 1983). The terms used here are again most stark, as can be some of the motives: be they to challenge not just the sense of the 'purity' but also the 'plurality' of school subject discipline approaches and their traditions—as found in questions of the content, history and foci of the commitments in the curriculums of school geography, social studies, world history and civics (Agyeman, 1996; Jenkins, 2003). This is doubly so if curriculum is critiqued from place-based, ecological, intergenerational or ecojustice perspectives (be that in general and not just in terms of a 'speciality' or elective environmental education provision—see Gruenewald, 2004).

While in a final example on these topics and challenges that are also discussed in further detail in the collection, there might also be efforts to:

(iii) ask whether conceptual frames, such as metaphors and metonyms for thinking about school subjects and adjectival educations, can be 'rebooted' too? On this, we note that some environmental educationalists have tried to provoke the reconsideration of subject identities and boundaries with such statements as 'all environmental education is media education' (John Fien), if not 'all education is environmental education' (David Orr) [3]. Related questions highlight claims to aspects of the 'core' and 'periphery' of education and curriculum, with school science scholarship illustrating notions of *parasitism*, *mutualism* or *commensalism* in relation to environmental education. Is, for example, environmental education automatically a constituent of a broader STEM education and its various components *within* and *as* education as a part of a whole (see Gough, 2002; Wals et al., 2014), or is environmental education better off without science as a bedfellow (Ashley, 2000)?

Thus there are questions to ask of how such considerations actually untangle various lines of argument about the provenance and merit of established 'subject disciplines' and the priority of providing engagement with 'powerful' and 'worthwhile knowledge' in educational institutions. These are particularly important, it would appear, when scholars, curriculum designers and practitioners look to use metaphors or decipher why a certain lop-sidedness seems to dominate thinking and practice within and across various curriculum 'silos' and 'domains', when a more tensioned 'real worldly' situation horizon is a concern for environmental education. In other words, which particular orientations to the world would be regarded as more real and worthwhile by students and teachers alike? (For more on the examples above, see Le Grange, 2004; Sauvé et al., 2005; Hopmann, 2008; Marcinkowski, 2009; Biesta, 2012; Young, 2013; Penuel et al., 2014.)

What counts

As might be expected, the contributions to the collection show there is little by way of consensus on such thorny matters, including within or beyond the studies of curriculum reported in this Journal. So on this point, an additional 'lack' can be identified; namely, on what the various advocates and critics of curriculum and its scholarship have regarded as both intellectually compelling *and* the key aspects worth deliberating about in a public forum, such as in the pages of scholarly journals.[4]

A harder and more demanding task though, is to consider what is regarded as 'authoritative' rather than, say, 'authoritarian' in terms of what is argued to count as theoretical and/or practical concerns related to environmental education (cf. *Journal of Curriculum Studies*, 2013, special issue on 'The Practical, Curriculum, Theory And Practice: An International Dialogue On Schwab's The "Practical 1"', e.g. Westbury, 2013; cf. Gough, 1989). One particular response explored in this Journal—but not always in its articles on environmental education—has been through an attempt to expand the sphere of analysis to acknowledge and then include a deeper sense of educational polity. This approach has distinct advantages, particularly given the aims and scope of the Journal: it recognises not just the role of curriculum makers and policy makers, but also those of career academics, independent scholars, professional associations of educators and educationalists, single- and multi-issue groupings, school-based groups and networks, task forces and commissions, think tanks, foundations, and (not) for-profit organisations (amongst many others) in having a stake in what counts as curriculum.

Given this, it should be clear that in relation to environmental education or even in terms of 'subject disciplines', it is never simply a question of who it is that is (or isn't) in the 'big tent' erected for curriculum deliberation (see McEneaney, 2002). In fact, urgent and important matters are shown to be asking what they are all doing, have done and might do (separately and with/ to others), in light of wider socio-political, ecological, economic, ethical, moral—and pedagogic—challenges facing the world (ibid., but see also Kirk & Macdonald, 2001, on teachers' voice in curriculum change, in a study from Scotland; and elsewhere, Potter, 2009, on how US government spending might be (re)directed in the face of challenges associated with environmental pollution and a culture of 'malconsumption',[5] a commentary worth revisiting in the era of Trump and widespread cuts to the EPA's remit and budget).

Intensifying these concerns, we note again that not long after the World Environmental Education Congress in 2015, various international and national bodies were expected to debate the ways the replacements for the Millennium Development Goals—the 'Sustainable Development Goals'— might be understood, addressed, monitored and evaluated, including in relation to education, over the next 15 years (see SDSN, 2015). While in the context of a Post-2015 Agenda (the '2030 Agenda'), as we have shown in our introductions, all this will happen alongside questions of how these *do* or *don't* articulate productively with other matters of public—and pressing?— concern. These include most notably, on mitigation and adaption as key lines of response to the anthropogenic forcing of climate change and accelerating realities of global warming.

So, how (else) might curriculum studies and scholars respond?

One option that can be suggested from considering the scholarship on display in the collection is to critically engage with a range of thinking and scholarship about procedural and substantive matters in environmental education curriculum. Parts 4 and 5 of the collection bring together articles that have been selected to demonstrate a wide range of theoretical and practical *positions* (be they explicit or tacit, pre-dispositional or emergent), *strategies* (e.g. given various views and alliances on which knowledge is valuable and why for curriculum) and *tactics* (such as inflecting anthropological, social and ecological interests within curriculum considerations)—so as to then consider how these might be brought to bear on curriculum thinking and practice related to environmental education.

Yet in delving into such studies from this Journal (if not the wider literature, e.g. O'Donoghue & Russo, 2004), there is also an acute sense that any attempt at shaping both curriculum and curriculum debate about education and environment in one particular direction (including at the expense of others) imperils a sense of simplicity, clarity, and urgency found in much of the scholarly work on display in this collection. It can though emerge quite problematically in mismatches between ideals and praxis, including when a curriculum study of environmental education oscillates between yet another proverbial 'Scylla and Charybdis': that of *sincerity* and *hyperbole*. For example, arising from considerations of historic to recent claims and delineations of 'challenge' and 'import'—while also maintaining a long view of the field of curriculum studies about education and environment—if there is trouble in statements that pose 'the trouble is' variety of critique, it might actually be because there's something 'other' that appears to be on the horizon that continues to be ignored. In short, we must consider that in terms of the notions of critique and crisis sketched in our reflection on the collection, all these studies take place during the advent of the much vaunted onset of the 'Anthropocene'[6].

The challenge suggested by this argument at this particular 'planetary juncture' is compounded as much as destabilised by taking this on board in curriculum studies. This is because, amongst other things, its core claim suggests events like the World Environmental Education Congresses, as well as all modern-day curriculum scholarship, have taken place within this epoch, but that a certain 'geo-epistemological illiteracy' (if not 'fog') still seems to prevail. How to respond? First, it might be to examine questions of whether talk of the Anthropocene is to be regarded as more than (yet) a(nother) framing device to rally critical attention and activity in curriculum circles? Second, to ask is it sufficient to revitalise and redirect curriculum and curriculum studies related to environmental and other educations away from its historic interests (see Jickling & Wals, 2008; cf. Greenwood, 2014; Lotz-Sisitka, 2017)? Third, as has been argued before, while mindsets and praxis may properly shift on these and related watchwords and environmental(ist?) notions, might there also be some sense in maintaining that we are always already going to be enframed by an 'anthropo-' notion, given the decidedly anthropocentric heft of much curriculum, including curriculum thinking, practice and reform (see Bonnett, 2007). And finally, in reflecting on whether this is actually more about an *intensification* than a *redirection*, a counter view might be to raise whether it is more truthfully the case that we never were/are in 'the Anthropocene' anyway, given the twinned criticisms directed at the term's intrinsic hubris and anthropocentrism, no matter whether it is invoked portentously, or for that matter, somewhat ironically, in relation to an individual's lived praxis of their own curriculum (Scranton, 2015).

In other words, as we consider the trajectories of all life on the planet, by raising a range of questions regarding critical priorities, forms and substance, curriculum scholars must be given an open invitation to examine the ways and positions from which education, curriculum and curriculum studies can be understood, practiced and critiqued as anything other than more or less *anthropocen[tr]ic* (see Morris, 2002; Moroye, 2009; Foster, 2011).

To conclude, these are just some of the ways we might understand and engage curriculum challenges *for* and *from* environmental education. We trust readers' explorations of this collection will reveal or prompt further historic to contemporary curriculum questions and challenges that may also return us to some of the key interconnecting questions of the field of curriculum inquiry, debate and action; namely:

- which educative experiences should a curriculum foster and why?
- what should be the scope of a worthwhile curriculum and how should it be decided, organized and reworked?
- why is distinctive curricula provided to different groups of students? and
- how should (a) curriculum be best enacted and evaluated?

Notes

1. While see Fien (1991), for a critical account of similar initiatives in Australia to establish consensus on the scope and components of a 'national curriculum' during that period too; a debate that has re-emerged in updated form in the 2010s, in relation to what would be a distinctly 'Australian Curriculum' (e.g. Dyment, et al., 2015).
2. This too, might involve privileging appreciative, ecological, restorative, and place-based ways that show this is never really about arguing for increased budgets for sports, health or adventure provision, even if this wider canopy of curriculum provision is important. There is also a question of balance – is it on environmental awareness, appreciation, knowledge *or* stewardship, for example, as in the dimensions to Project WET (Water Education for Teachers)? We should note that the 'disorder' aspects have been subject to critique too, given much of Louv's account is based on reifying a rhetorical ploy most usually associated with Attention Deficit Disorder, that can also be often linked to a misdiagnosis (if not misuse) in terminologies.
3. As David Orr (2005, xi) has often put it, 'all education is environmental education . . . by what is included or excluded we teach the young that they are part of or apart from the natural world'. See Orr (2003, and 2004: 94–98) too, on the knotty challenge of distinguishing political conservatism from ecological conservationism as these relate to subject matter, curriculum and pedagogy – including commentary on the preconceptions versus evidence about their linkages.
4. On this, readers may want to consider the structure, organisation and depth of analysis of the studies in the collection, be that in their theoretical or empirical aspects, and even what is footnoted—see Jardine's contribution, for some curious illustrations.
5. Potter (2009) opens her piece with a section called 'Missed opportunities' after an epigram from the Science Advisory Board stating: 'When one generation's behavior necessitates environmental remediation in the future, a burden of environmental debt is bequeathed to its children just as surely as unbalanced government budgets bequeath a burden of future financial debt.' She continues by listing various misgivings about what the Environmental Protection Agency and various US agencies, reports and bills have (and haven't) achieved to address such a predicament.
6. In brief, the term foregrounds human agency as a cryospheric, hydrospheric, lithospheric, biospheric, atmospheric, geologic, and hence, planetary force. As such, it signals the wholesale transformation of the planet but also because in attributing this to the scale and scope of industrialisation processes unleashed and then intensified over the last 200 years, the burden of this notion is to show that 'something more' is occurring well beyond the immediate imaginary and scope of IPCC-style accounts of carbon and temperature shifts, the markers of which are primarily associated with anthropogenic *climate* [as opposed to planetary] modifications (see Crutzen & Stoermer, 2000; Crutzen, 2002; Crist, 2013; Garrard, Handwerk, & Wilke, 2014; Malm & Hornborg, 2014).

References

Agyeman, Julian (1994a) The converging strands of environmental education, *Annual Review of Environmental Education*, Special 25 Year Issue. Council for Environmental Education, p. 11.

Agyeman, Julian (1994b) The next step: education for participatory democracy? *Annual Review of Environmental Education*, Special 25 Year Issue. Council for Environmental Education, p. 52.

Agyeman, Julian (1996) An Alternative Approach to Urban Nature in Environmental Education at KS2. Doctoral Thesis, University of London Institute of Education.

Ashley, Martin (2000) Science: an unreliable friend to environmental education?, *Environmental Education Research*, 6:3, 269–280.

Ball, Stephen J., & Bowe, Richard (1992) Subject departments and the 'implementation' of National Curriculum policy: an overview of the issues, *Journal of Curriculum Studies*, 24:2, 97–115.

Biesta, Gert (2012) Becoming world-wise: an educational perspective on the rhetorical curriculum, *Journal of Curriculum Studies*, 44:6, 815–826.

Bonnett, Michael (2007) Environmental education and the issue of nature, *Journal of Curriculum Studies*, 39:6, 707–721.

Bowers, Chet (1990) Educational computing and the ecological crisis: some questions about our curriculum priorities, *Journal of Curriculum Studies*, 22:1, 72–76.

Cotton, Debby (2006) Implementing curriculum guidance on environmental education: the importance of teachers' beliefs, *Journal of Curriculum Studies*, 38:1, 67–83.

Crist, Eileen (2013) On the poverty of our nomenclature, *Environmental Humanities*, 3, 129–147.

Crutzen, Paul J. (2002) Geology of mankind, *Nature*, 415:6867, 23.

Crutzen, Paul J., & Stoermer, Eugene F. (2000) The 'Anthropocene,' *Global Change Newsletter*, 41, 17–18.

Dobson, Andrew (2016) *Environmental Politics: A Very Short Introduction* (Oxford: Oxford University Press).

Dyment, Janet, Hill, Allen, & Emery, Sherridan (2015) Sustainability as a cross-curricular priority in the Australian Curriculum: a Tasmanian investigation, *Environmental Education Research*, 21:8, 1105–1126.

Fahey, Shireen J. (2012) Curriculum change and climate change: inside outside pressures in higher education, *Journal of Curriculum Studies*, 44:5, 703–722.

Fien, John (1991) Ideology, political education and teacher education: matching paradigms and models, *Journal of Curriculum Studies*, 23:3, 239–256.

Fien, John (1993) *Education for the Environment: Critical Curriculum Theorizing and Environmental Education* (Geelong: Deakin University Press).

Foster, John (2001) Education as sustainability, *Environmental Education Research*, 7,2: 153–165.

Foster, John (2011) Sustainability and the learning virtues, *Journal of Curriculum Studies*, 43:3, 383–402.

Garrard, Greg, Handwerk, Gary, & Wilke, Sabine (2014) Introduction: Imagining anew: challenges of representing the Anthropocene, *Environmental Humanities*, 5, 149–153.

Goodson, Ivor (1983) Subjects for study: aspects of a social history of curriculum, *Journal of Curriculum Studies*, 15:4, 391–408.

Gough, Annette (2002) Mutualism: A different agenda for environmental and science education, *International Journal of Science Education*, 24:11, 1201–1215.

Gough, Noel (1989) From epistemology to ecopolitics: renewing a paradigm for curriculum, *Journal of Curriculum Studies*, 21:3, 225–241.

Gough, Noel (1994) Playing at catastrophe: ecopolitical education after poststructuralism, *Educational Theory*, 44:2, 189–210.

Gough, Noel, & Gough, Annette (2010) Environmental education. In: Craig Kridel (Ed.), *Encyclopedia of Curriculum Studies* (New York: Sage). http://sk.sagepub.com/reference/curriculumstudies/n188.xml

Greenall Gough, Annette (1991) Greening the future for education: changing curriculum content and school organization, *Journal of Curriculum Studies*, 23:6, 559–571.

Greenwood, David A. (formerly Gruenewald) (2008) A critical pedagogy of place: from gridlock to parallax, *Environmental Education Research*, 14:3, 336–348.

Greenwood, David A. (2013) A critical theory of place-conscious education. In: Robert B. Stevenson, Michael Brody, Justin. Dillon, and Arjen E. J. Wals (Eds.), *International Handbook of Research on Environmental Education*, (pp. 93–100) (New York: Routledge).

Greenwood, David A. (2014) Culture, environment, and Education in the Anthropocene. In: Michael P. Mueller, Deborah J. Tippins & Arthur J. Stewart (Eds.) *Assessing Schools for Generation R (Responsibility): A Guide for Legislation and School Policy in Science Education* (pp. 279–292) (Dordrecht: Springer).

Gruenewald, David A. (2003) The best of both worlds: a critical pedagogy of place, *Educational Researcher*, 32:4, 3–12.

Gruenewald, David A. (2004) A Foucauldian analysis of environmental education: toward the socioecological challenge of the Earth Charter, *Curriculum Inquiry*, 34,1: 71–107.

Grün, Mauro (1996) An analysis of the discursive production of environmental education: terrorism, archaism and transcendentalism, *Curriculum Studies*, 4:3, 329–347.

Hart, Paul (2010) No longer a "little added frill": the transformative potential of environmental education for educational change, *Teacher Education Quarterly*, 37(4), 155–177.

Hawkey, Kate (2014) A new look at big history, *Journal of Curriculum Studies*, 46:2, 163–179.

Hitzhusen, Gregory E. (2007) Judeo-Christian theology and the environment: moving beyond scepticism to new sources for environmental education in the United States, *Environmental Education Research*, 13:1, 55–74.

Hopmann, Stefan (2008) No child, no school, no state left behind: schooling in the age of accountability, *Journal of Curriculum Studies*, 40:4, 417–456.

Huckle, John (1983) Environmental education. In: John Huckle (Ed.), *Geographical Education: Reflection and Action* (pp. 99–111) (Oxford: Oxford University Press).

Huckle, John, & Sterling, Stephen (Eds.) (1996) *Education for Sustainability* (London: Earthscan).

Huckle, John, & Wals, Arjen E. J. (2015) The UN Decade of Education for Sustainable Development: business as usual in the end, *Environmental Education Research*, 21:3, 491–505.

Jacques, Peter J., Dunlap, Riley E., & Freeman, Mark (2008) The organisation of denial: Conservative think tanks and environmental scepticism, *Environmental Politics*, 17:3, 349–385.

Jardine, David W. (1994) 'Littered with literacy': an ecopedagogical reflection on whole language, pedocentrism and the necessity of refusal, *Journal of Curriculum Studies*, 26:5, 509–524.

Jenkins, Edgar (2003) Environmental education and the public understanding of science, *Frontiers in Ecology and the Environment*, 1:8, 437–443.

Jensen, Bjarne Bruun (2004) Environmental and health education viewed from an action-oriented perspective: a case from Denmark, *Journal of Curriculum Studies*, 36:4, 405–425.

Jickling, Bob (1997) If environmental education is to make sense for teachers, we had better rethink how we define it! *Canadian Journal of Environmental Education* 2, 86–103.

Jickling, Bob, & Wals, Arjen (2008) Globalization and environmental education: looking beyond sustainable development, *Journal of Curriculum Studies*, 40:1, 1–21.

Journal of Curriculum Studies (2013) The practical, curriculum, theory and practice: an international dialogue on Schwab's the 'Practical 1', *Journal of Curriculum Studies*, 45:3, 583–696.

Kirk, David, & Macdonald, Doune (2001) Teacher voice and ownership of curriculum change, *Journal of Curriculum Studies*, 33:5, 551–567.

Lee, John Chi-Kin (2000) Teacher receptivity to curriculum change in the implementation stage: the case of environmental education in Hong Kong, *Journal of Curriculum Studies*, 32:1, 95–115.

Le Grange, Lesley (2004) Embodiment, social praxis and environmental education: some thoughts, *Environmental Education Research*, 10:3, 387–399.

Lotz-Sisitka, Heila (2017) Decolonisation as future frame for environmental and sustainability education: embracing the commons with absence and emergence. In: Peter Corcoran, Joseph Weakland & Arjen Wals (Eds.) *Envisioning Futures for Environmental and Sustainability Education* (pp. 45–62) (Wageningen: Wageningen Academic Publishers).

Louv, Richard (2010) *Last Child in the Woods: Saving Our Children from Nature-deficit Disorder* (Chapel Hill, NC: Algonquin Books).

Lucas, Arthur M. (1972) Environment and environmental education: conceptual issues and curriculum implications. PhD Dissertation, Ohio State University. Available at: http://files.eric.ed.gov/fulltext/ED068371.pdf

Lucas, Arthur M. (1979) *Environment and Environmental education: Conceptual Issues and Curriculum Implications* (Melbourne: Australian International Press and Publications). [Reprint of Lucas, 1972]

Lucas, Arthur M. (1991) Environmental education. In: Arieh Lewy (Ed.) *The International Encyclopedia of Curriculum* (pp. 770–771) (Oxford: Pergamon Press).

Malm, Andreas, & Hornborg, Alf (2014) The geology of mankind? A critique of the Anthropocene narrative, *The Anthropocene Review*, 2053019613516291.

Marcinkowski, Thomas J. (2009) Contemporary challenges and opportunities in environmental education: where are we headed and what deserves our attention?, *The Journal of Environmental Education*, 41:1, 34–54.

Martin, Peter (1996) A WWF view of education and the role of NGOs. In: John Huckle & Stephen Sterling (Eds.), *Education for Sustainability* (pp. 40–51) (London: Earthscan).

Martusewicz, Rebecca A. (2009). Educating for "Collective Intelligence": Revitalizing the Cultural and Ecological Commons in Detroit. In: Marcia McKenzie, Paul Hart, Heesoon Bai & Bob Jickling (Eds.), *Fields of Green: Restorying Culture, Environment, and Education* (pp. 253–267) (Cresskill, NJ: Hampton Press).

McEneaney, Elizabeth H. (2002) Power and the knowledge produced by educationists, *Journal of Curriculum Studies*, 34:1, 103–115.

Moroye, Christy (2009) Complementary curriculum: the work of ecologically minded teachers, *Journal of Curriculum Studies*, 41:6, 789–811.

Morris, Marla (2002) Ecological consciousness and curriculum, *Journal of Curriculum Studies*, 34:5, 571–587.

Nixon, Jon (1991) Reclaiming coherence: cross-curriculum provision and the National Curriculum, *Journal of Curriculum Studies*, 23:2, 187–192.

O'Donoghue, Rob, & Russo, Vladimir (2004) Emerging patterns of abstraction in environmental education: a review of materials, methods and professional development perspectives, *Environmental Education Research*, 10,3: 331–351.

Orr, David W. (2003) Walking north on a southbound train, *Conservation Biology*, 17:2, 348–351.

Orr, David W. (2004) *Earth in Mind: On Education, Environment, and the Human Prospect*. London: Island Press.

Orr, David W. (2005) Foreword. In: Michael K. Stone & Zenobia Barlow (Eds.), *Ecological Literacy: Educating Our Children for a Sustainable World* (pp. ix–xiii) (San Francisco: Sierra Club Books).

Penuel, William R., Phillips, Rachel S., & Harris, Christopher J. (2014) Analysing teachers' curriculum implementation from integrity and actor-oriented perspectives, *Journal of Curriculum Studies*, 46:6, 751–777.

Potter, Ginger (2009) Environmental education for the 21st century: where do we go now?, *The Journal of Environmental Education*, 41:1, 22–33.

Reid, Alan, & Dillon, Justin (Eds.) (2017) *Environmental Education: Critical Concepts in the Environment*. London: Routledge.

Reid, Alan, & Dillon, Justin (2018) Beyond mutualism: exploring metaphors of commensalism between science and environmental education. In preparation.

Robottom, Ian (Ed.) (1987) *Environmental Education: Practice and Possibility* (Geelong: Deakin University Press).

Ross, Hamish (2007) Environment in the curriculum: representation and development in the Scottish physical and social sciences, *Journal of Curriculum Studies*, 39:6, 659–677.

Sauvé, Lucie, Brunelle, Renée, & Berryman, Tom (2005) Influence of the globalized and globalizing sustainable development framework on national policies related to environmental education, *Policy Futures in Education*, 3:3, 271–283.

Schinkel, Anders (2009) Justifying compulsory environmental education in liberal democracies. *Journal of Philosophy of Education*, 43:4, 507–526.

Scott, William, & Oulton, Christopher (1999) Environmental education: arguing the case for multiple approaches. *Educational Studies*, 25:1, 89–97.

Scranton, Roy (2015) *Learning to Die in the Anthropocene: Reflections on the End of a Civilisation* (San Francisco: City Lights).

Smyth, John C. (1995) Environment and education: a view of a changing scene, *Environmental Education Research*, 1:1, 3–20.

Sterling, Stephen (2001) *Sustainable Education—Re-visioning Learning and Change*, Schumacher Briefing no. 6 (Dartington: Schumacher Society/Green Books).

Sterling, Stephen (2010) Living *in* the Earth: towards an education for our times, *Journal of Education for Sustainable Development*, 4:2, 213–218.

Stevenson, R. (2006) Tensions and transitions in policy discourse: re-contextualizing a de-contextualized EE/ESD debate. *Environmental Education Research*, 12:3/4, 277–290.

Sustainable Development Solutions Network (2015) Indicators and a Monitoring Framework for Sustainable Development Goals: Launching a data revolution for the SDGs. UN. http://unsdsn.org/resources/publications/indicators/ Accessed 15 May 2015.

Tilbury, Daniella (1995) Environmental education for sustainability: defining the new focus of environmental education in the 1990s, *Environmental Education Research*, 1:2, 195–212.

UNESCO (2011) National Journeys towards Education for Sustainable Development (Paris: UNESCO).

UNESCO (2014) Shaping the Future We Want: UN Decade of Education for Sustainable Development (2005–2014). Final Report (Paris: UNESCO). https://sustainabledevelopment.un.org/content/documents/1682Shaping%20the%20future%20we%20want.pdf

Wals, Arjen E. J., & van der Leij, Tore (1997) Alternatives to national standards for environmental education: process-based quality assessment, *Canadian Journal of Environmental Education*, 2:1, 7–27.

Wals, Arjen, Brody, Michael, Dillon, Justin, & Stevenson, Robert (2014) Convergence between science and environmental education, *Science*, 344: 6184, 583–584.

Westbury, Ian (2013) Reading Schwab's the 'Practical' as an invitation to a curriculum enquiry, *Journal of Curriculum Studies*, 45:3, 640–651.

Whitty, Geoff, Rowe, Gabrielle, & Aggleton, Peter (1994) Discourse in cross-curricular contexts: limits to empowerment, *International Studies in Sociology of Education*, 4:1, 25–42.

Young, Michael (2013) Overcoming the crisis in curriculum theory: a knowledge-based approach, *Journal of Curriculum Studies*, 45:2, 101–118.

Environmental education and the issue of nature

MICHAEL BONNETT

Much official environmental education policy in the UK and elsewhere makes scant reference to nature as such, and the issue of our underlying attitude towards it is rarely addressed. For the most part such policy is pre-occupied with the issue of meeting 'sustainably' what are taken to be present and future human needs. This paper considers several issues posed by this anthropocentric approach and explores the view that environmental education—indeed *any* education—worthy of the name needs to bring a range of searching questions concerning nature to the attention of learners, and to encourage them to develop their own on-going responses to those questions. It is argued that our present environmental predicament not only provides an exciting opportunity to re-focus education on the issue of human relationship to nature, but also requires the exploration of this issue for its long-term resolution. Extensive implications for the curriculum and the culture of the school are raised.

It is difficult to think of a set of issues more important now to the welfare of us as human beings than those concerning the environment. Problems of climate change, pollution, and the depletion of natural resources are now only too familiar—as is the putative remedy of 'sustainable development'. And the curricula of many national education systems, at least in their rhetoric, are attempting to address this area of concern, particularly, but not exclusively, those nations that were signatory to Agenda 21 of the 1992 Rio Earth Summit, United Nations Conference on Environment and Development (UNCED 1992). Although there is considerable variation in detail of the curricula approaches taken to environmental education, it remains the case that 'sustainable development' is a key orientating idea. In this paper I raise some critical questions about this orthodoxy on the grounds that it can incline educators to pay too little attention to the key issue for environmental education: our understanding of nature and our relationship to it. In particular, I examine two interwoven issues:

- the general issue of the meaning of 'nature' and its importance to education, and
- what taking nature seriously suggests for the aims and curriculum of environmental education.

I am keenly aware that these issues have been extracted from an extensive web of inter-related concerns and perspectives. Although there is a larger story to tell, I hope to show that the strands that I have selected are central to understanding the curriculum implications of our current environmental predicament.[1]

Education and nature

Much official educational policy, including that which relates to the environment, makes scant reference to nature. This is reflected, for example, in English National Curriculum documents where few references are made to nature as such as a focus of learning.[2] The relative invisibility of nature is perhaps nowhere better illustrated than in the case of science education, which is clearly taken to be the place where any understanding of nature is to be undertaken. For example, the learning objectives of the English National Curriculum for 5–11-year-olds (Department for Education and Employment 1999) are pre-occupied with learning 'investigative skills' and testing scientific ideas. This is very much in line with the priorities recently specified by a group drawn from 'the expert community' of science educators that centre on students learning such things as the need to make predictions, collect relevant evidence, and for experiments to be replicated, the difference between causality and correlation, and how to present and write up findings (Collins *et al.* 2001). While not denying a certain importance to such goals at one level, seen from the standpoint of elucidating the underlying spirit of science education they reflect a worrying banality, lacking any cognisance of science education as seeking to develop an *appreciation of nature*. A largely analytic/instrumental/invasive rationality dominates.

Although it would be stretching the point to suggest that little has changed since Francis Bacon's (1561–1626) advocacy of modern experimental science at its inception as needing to 'hound' nature 'in her wanderings' and for the scientist to 'torture nature's secrets from her' (Capra 1983: 40–41)—in more contemporary parlance, we might speak of science as seeking to explain, predict, and control aspects of nature—modern science is sometimes still willing to intervene quite aggressively to achieve this. For example, it is prepared to subject organisms to laboratory conditions, kill them in order to dissect them, inject foreign (often toxic) substances to study their effects, etc. On the whole, this is all regarded as quite acceptable by the scientific community and many outside.

It is not impossible, therefore, to recognize echoes of Bacon's recommendations in much (but by no means *all*) science as currently practised, particularly when we acknowledge the volume of science funded by bodies whose underlying objectives are primarily commercial or military. And we can draw a vivid contrast with the scientist of yore as 'natural historian' and science education as 'nature study', a view that some might belittle as mere 'stamp-collecting' because of its passivity. However, given our current environmental situation, which in part has been fuelled by some two centuries of ascendancy of science of a more aggressive kind, the general demotion of the ancient motive of appreciation in studying nature must be open to question.

This apart, with the increased consciousness of environmental problems, the widespread ignoring of nature as a topic in itself is, at the very least, disappointing. Indeed in my view it represents a certain dereliction of duty. I will argue that environmental education in particular, and education in general, should have at their heart the ambition to bring a range of searching questions concerning nature to the attention of learners, and to encourage them to develop their own on-going responses to those questions. I have in mind here questions such as:

- What is nature and what is our place in it?
- How can we know nature and what should be our attitude towards it?
- Against what criteria should humankind judge its progress/success/ flourishing in relation to the natural world?

Such questions are educationally relevant because they represent important ways of articulating our understanding of the human situation, and the present environmental predicament provides an exciting opportunity to re-focus education on them (I say *re*-focus, because, of course there was a time, e.g. in the 18th century, when they would have been a natural part of education). Also, I will argue, such questions are central to addressing the environmental predicament itself.

However, any such re-focusing will not come easily. The contemporary highly instrumental stance within Western society at large necessarily diverts attention from issues concerning the meaning of nature by simply ignoring them or by making them sound 'purely academic', esoteric, even frothy. And this stance is not without allies in academia, where raising such questions can be inhibited by denying the language in which they can be meaningfully articulated; for example, by dismissing 'nature' as simply a social construct that has no objective reality and is redolent with ideological bias (e.g. Haraway 1991, Giddens 1994).[3] Viewed from another standpoint, the effective proscribing of talk of 'nature' appears more as a form of what, following Lyotard (1984: xxiv), one could describe as 'linguistic terrorism'. And, sadly, there are reasons for believing that the situation is not helped by the way that our current understanding of environmental issues has become orientated around the concept of sustainable development.

What is wrong with the idea of sustainable development?

I will not rehearse here its history, but for brevity will simply refer to the well-known and highly influential definition of sustainable development from the report of the World Commission on Environment and Development (Brundtland Commission 1987: 43): 'a development that meets the needs of the present without compromising the ability of future generations to meet their own needs'.

This is a very seductive notion as, on a generous interpretation, it seems to marry two highly desired goals: first, the idea of conserving those aspects of nature that are valued (i.e. in some sense 'needed') but that are currently endangered by human agency; secondly, the idea of accommodating on-going human aspirations to 'develop', that is, in some sense to have more or better. However, on such an interpretation criticisms concerning the extreme ambiguity of the term immediately arise. For example: precisely *what* is to be sustained, at what level, and over what time span? Precisely *whose* needs are to be met, how are they to be prioritized, and according to what criteria? And so forth. The problem is not that answers cannot be given to such questions, but that the definition does not do so, and that when they are provided they are highly contestable.[4]

As things stand, its ambiguity allows sustainable development, so defined, to be something that almost everyone can subscribe to without too much inconvenience—from enlightened captains of industry to eco-warriors. And trading on such ambiguities has enabled the rhetoric of some policy-makers and commercial enterprises to give the impression that they are concerned to do one thing, such as sustain natural ecosystems, while in fact attempting something quite different, such as sustain conditions for the continuance of their own, often narrowly defined, economic growth. Similarly, such vague-ness also enables sustainable development to be interpreted into a set of prac-tices that are reasonably congenial in almost *any* social/political context: in the UK, predominantly in terms of energy efficiency and recycling that are compatible with consumerism. Here there is relatively little talk of having *less*, and generally a continuing expectation of having *more*. Under current arrangements the health of our economy simply demands this. (Consider the stock market despondency that occurs when Main Street sales falter.) One cannot help thinking that a certain sleight-of-hand is in play when the ulti-mately unsustainable assumption of continuous (material) economic growth is apparently brought into harmony with a much vaunted eco-friendliness. It seems to me that these are just some examples of very significant problems for the idea of sustainable development as a policy.

However, most serious of all from the perspective that I wish to develop in this paper, Brundtland-type definitions of sustainable development reflect highly anthropocentric and economic motives that lead to nature being seen essentially as a *resource*, an object to be intellectually possessed and physically manipulated and exploited in whatever ways are perceived to suit (someone's version of) human needs and wants. That is to say, they are redolent with the general metaphysics of mastery that informs modernity and is precisely the root cause of our current environmental predicament. With humanistic hubris nature is constantly to be challenged, set in order, re-engineered, etc., to meet human needs—and often, not even this, but merely human convenience. The underlying attitude is implicit in the meta-phors sometimes employed to describe our achievements and aspirations: man *conquered* Everest, *tamed* the jungle, needs to *manage* the oceans, etc. Nature is frequently conceived as the terrain where frontiers are to be extended, boundaries overcome, its integrity penetrated, which, of course, once invited the romantic reaction from the poet William Wordsworth that 'we murder to dissect'. And we start early in encouraging this hubris. For example, in children's literature the highly regarded UK author Philippa Pearce (1920–2006) sometimes has portrayed animals as objects to satisfy human emotions.

However, seriously, why bother with the issue of nature as such? Why not just get on pragmatically, exercising a degree of enlightened self-interest? It seems to me that there are two main reasons for reviewing such an approach. First, precisely this attitude has led to many serious threats to our physical well-being. While not for one moment wishing to deny the many benefits that some scientific activity has provided, such as potable water and anaesthetics, the remorseless assaults on nature that an underlying mastery motive energizes are a heavy contributor to current environmental problems. And the implicit assumption that we can somehow 'manage' nature on an increasingly grand scale is false; we simply do not—*and never could*—know enough. Little surprise, then, that the history of attempting to do so is largely a history of unintended consequences. The example of the use of DDT and certain subsequent attempts at the biological control of pests provide vivid illustration of this.[5]

The second reason is equally important. Acting extensively out of pragmatic self-interest embodies a stymieing ignorance that brings a spiritual impoverishment, that in turn diminishes our sense of ourselves and what it is to live well. It prevents us from seeking to know the world as it is, itself—in its intrinsic value (celebrated in so much of our art and music)—and therefore truly understanding our place within it. Humankind prides itself on its ability to think, to rise above instinctive reaction and to see beyond what is immediate and to evaluate situations. In the light of its track record, it needs to think about *itself* and its actions, and perhaps to take seriously the previously mentioned questions concerning prevailing assumptions about what counts as flourishing and progress. With our now massive impact on the planet, these are questions that need addressing, but hubris and preoccupation with practical gain lead to them too rarely being pursued in ways that have outcomes commensurate with their importance. If, however, we allow them to stand, and in the context of environmental concern, such questions raise a key issue: *What should be our underlying attitude towards nature?*

The answer to this is far from straightforward. For example, one recurring suggestion is that we should learn to *love* nature. However, in what sense does one love something that is completely indifferent to us—that currently on the whole supports us, but may at some point destroy us either locally, as with hurricane Katrina, or eventually globally, as when the Sun desiccates planet Earth or some chance asteroid strikes? What exactly is the appropriate attitude towards something that for the most part does not—that is *incapable*—of caring for us in the slightest? However, perhaps framing the question in this holistic way is simply misguided, making unhelpful assumptions about the time-span and level of generality with which we conceive nature. Surely, there are important distinctions to be drawn between individual events and organisms at one end of the spectrum and nature as a whole—the great scheme of things—at the other. It might be said that trying to lump such a variety of phenomena together under one overarching idea just brings obfuscation when thinking about what our relationship to nature should be. For example, the love of one's dog is different from the love of one's garden, a tree, or a landscape. It might be said that nature is not *one* thing and that, rightly, our attitude will depend upon the particular aspect we have in mind.

So maybe we love our pets and hate the malaria bacillus? However, arguably the rejection of holism here is itself a facet of an underlying instrumentalism—loving only what we believe meets our needs and desires—which in turn is an expression of the metaphysical mastery that has issued in an atomistic domination of nature. Furthermore, and just as importantly, it has also cabined our conception of what the world has to offer and who we are.

What is nature?

While not wishing to deny our ability to discriminate within the natural world, I suggest that there is some virtue in holding onto a holistic sense of nature—but *not* now only in the prevalent ecological sense of a vast inter-connecting system in which human beings are nested, rather in the sense of a key *quality* implicit in our experience of all things we perceive as natural. In our experience of them, we come across natural things as standing there independently of us—i.e. as pre-eminently having their *own* being that we can affect but of which we are not the author. That is to say, we experience nature as '*self-arising*'. This essentially non-artefactual quality of the standing forth from out of itself is a definitive feature of our experience of nature, whether it be a star, a mountain, or an amoeba. And this self-arising quality can stimulate us to find even the ability of harmful organisms to function autonomously and self-replicate a source of wonder. At this point it is perhaps important to note a distinction that can be drawn between the quality of our *experience* of nature and nature as a *concept*. Elsewhere, I (Bonnett 2004a) have argued that concepts of nature (of which, there are of course a number) arise in the context of human practices and are part of a web that to some degree mediates all experience. Thus, as with all other concepts, we can say that they are socially produced. However, within this socially mediated experience we encounter nature as precisely *not* socially produced. Furthermore, I argue that this experience of nature as the self-arising is deeply constitutive of our form of sensibility. Our experience as human beings is absolutely founded on the assumption of an independently existing world, i.e. a world of which we are not sole author and that at some level and at any particular time we have simply to accept as given—and of which, therefore, some things are true and some things are not. I argue that this intuition cannot be expelled from our form of sensibility—nor can it seriously be treated as merely some optional description or (dead) meta-phor, as Rorty (1980) and some other post-modernists claim. *Nature as the self-arising* thus provides certain salient features of our idea of an underlying reality. I will focus on just two of these features and some of the issues they raise.

The first of these features I have already alluded to: nature's *otherness*, in the sense that we are not its author. Nature lies ever before and beyond our intentions, and although we can *affect* it in all manner of ways, we do not ulti-mately *determine* it. For example, in the case of our own bodies, which clearly can be affected by our choices and actions, we maintain (or destroy) our health by interacting with powers of which ultimately we are not the author and that are beyond our ability to transform. There is a nature, an order, recognized as external to our will with which we have to find a harmony, or at least an accommodation.

This relates very closely to a second feature, namely nature's *epistemological mystery*. Although in one sense something with which we may on occasion feel ourselves to be intimately involved, and therefore may be intimately known, as self-arising, nature is that which can never be fully known, intellectually possessed. It is composed of (expressed in) open, many-faceted *things*—each with its own unique history and drawing towards its own open future, and capable of exhibiting an infinity of profiles and countenances, only *some* of which we will ever witness. This contrasts with defined *objects* of thought, whose being is exhausted by the characteristics of category membership that we devise and impose—as when, say, we sum up a beech tree in terms of static objective properties such as leaf shape and colour, flower type, growth habit, etc. (and which thus reckoned up, could readily be entered into some data-base), *rather than* see it as this unique thing whose limbs possess an unfathomable massiveness, yet that dance in the breeze, and whose colour and shape constantly change with every nuance of light and shade as night falls. Generalized causal explanations and scientific 'laws' say nothing about the *sheer existence* of natural things—give no insight into the experience of their individual standing forth in their suchness and their ability to affect us in unique and never wholly predictable ways. As one commentator put it: 'Surprise is the general reaction of the attentive walker in natural space' (Grange 1997: 96). And the appropriate response to self-arising nature then is not to seek summative knowledge, but to allow a sense of the ineffable.

The Italian post-Enlightenment philosopher Giambattista Vico (1668–1744) argued that man can only have full knowledge of things of which he is the author, for those who create something can understand it as mere observers of it cannot. Thus, we are capable of having a more definitive understanding of ourselves and the human world than of nature, because we have an 'insider's' view. We know what it is like to *be* a human being in a way in which we cannot know what it is like to be a lion or a tree. With the latter, we are merely passive observers looking on from outside, capable only of 'dark speculation' about the inner lives or goals of what we see (and if indeed there are any such goals); capable only of seeing the 'surfaces' of things and events that are essentially alien and mysterious. As Berlin (2000: 13) puts it, for Vico the laws of the natural world are *knowable* but not *intelligible*. Thus, for example, one can know of the beliefs of another person or the practices of another culture, yet they may remain unintelligible until some further explanation is given, communicated—and even with this one may fail to achieve the kind of 'internal' understanding to which Vico refers. And, of course, in the case of non-human nature no such further explanation can be forthcoming. On this kind of account, the anthropomorphism that is so ubiquitous in our perceptions of nature destroys its sheer otherness, and a proper attitude to nature cannot simply be some kind of extension of a human ethic. Yet we can, and often do, have a sense of what would count as the well-being, even 'interests', of things in nature. Stone (1974), a lawyer, has argued that it is at least as plausible to allow that 'natural objects' have interests that can be legally represented as it is make this claim on behalf of a comatose person or a corporation.

How can we know nature?

Amongst other things, the motive of mastery that I have argued to be implicit in 'sustainable development' tends to reinforce the conventional curriculum attitude of regarding nature as most properly revealed through the prism of science, where the subjectivity of the knower is minimized in the process of coming to know. In contrast to this approach, on the account of nature that I am offering, what is needed is a kind of knowing in which personal, moral, and aesthetic dimensions are embedded, i.e. a knowledge of things in which 'fact' and 'value' are not separated out because things are perceived in their *life*, wholeness, and inherent mystery. This suggests that we perhaps need to rehabilitate the notion of '*knowledge by acquaintance*' into the curriculum, where the character of the acquaintanceship is akin to (but not identical with) the sense in which we may become acquainted with a person—a direct, intimate, tacit knowledge that *affects* and is capable of engaging all the senses. In other words, we seek an enriched, vitalized, sense of knowledge, something of the essential poetic character of which is suggested by Thoreau (1949: 394):

> Live in each season as it passes; breathe the air, drink the drink, taste the fruit, and resign yourself to the influences of each. ... Open all your pores and bathe in all the tides of Nature, in all her streams and oceans, at all seasons. ... Grow green with spring, yellow and ripe with autumn ...

Or again, as Merleau-Ponty (1962: 318) put it, in a less metaphorical vein: 'what I call experience of the thing or of reality—not merely of a reality-for-sight or for-touch, but of an absolute reality—is my full co-existence with the phenomenon, at the moment when it is in every way at its maximum articulation ...'.

Undoubtedly science provides an important access to nature, but of a reductionist kind, and therefore there are reasons for not *privileging* the access it provides. For example, why should the knowledge of nature that it enables—a knowledge ever increasingly articulated mathematically—be regarded as 'truer' or more authentic than, say, the knowledge achieved through the experience of helming a sailing boat in which one is acutely alert to, and in harmony with, the subtle nuances of wave and breeze?—where one *feels* directly the sheer power and sublime delicacy of nature. The degree of mutuality with its object of such knowledge contrasts strongly with the disengagement both from its object and the subjectivity of the knower that is valorized by Western rationality. It asks us to reassert the value of a knowledge that is existentially embedded, to exhibit patience and humility in the face of that which is both 'other' and intimately felt, to employ a genuine attentiveness.

There are some resonances in this view with the position of McDowell (1996) on the re-enchantment of nature, both in terms of the kind of thing nature has to be in order to constitute a source of validation of empirical beliefs and by implication what would be involved in knowing nature. The key point is a rejection of equating nature with the logical space in which science locates it, a space of 'blind' universal laws. If experience were to be construed as made up of impressions, 'impingements by the world on a possessor of sensory capacities' (p. xv), and this world is the world described by the natural sciences—i.e. it is the logical space in which they function—it is *different in kind* from the normative relations that constitute the logical space of reasons (where, for example, there is talk of one thing's being warranted, or correct, in the light of another). And if the logical space of reasons is *sui generis* as compared with the logical space of nature, it is impossible for experience to act as tribunal for empirical thinking, and the idea of empirical thinking itself becomes incoherent. Hence, McDowell argues that reason in the sense of the operation of concepts goes all the way down to the level of our most primordial experience of things. There is no prior apprehension of a pre-conceptual 'given' upon which concepts then operate. How could they obtain a purchase or participate in an alien logical space? Concepts are involved in our apprehension of nature from the bottom up, and nature is therefore no longer divested of everything normative.

It may seem that this account runs counter to my emphasis on the otherness of nature and its foundational role, nature as a 'given'. It is true that McDowell wishes to repudiate what he terms the 'myth of the given'. However, here it is important to be clear about what is meant by 'the given'. If we mean by the given some sort of absolutely pristine sensory experience (such as was, for example, postulated by 'sense data theory' of Moore (1953), Price (1932), and Ayer (1940)) there is no more room for it in my account than in McDowell's. We are always in the world understandingly and even that favourite example of sense-data theorists of isolated after-images of coloured patches are always understood as occurring in a certain context and possessing a certain significance. To borrow a term taken from a different philosophical tradition, they—and all experience—are always had within some meaning-giving horizon. However, this does not make the being of nature transparent, deny its inherent mystery in its self-arising. Quite the reverse. Mystery is only possible in the logical space where significances are in play. Mystery, too, is just such and so. Of course, this still leaves room for debate as to how developed, abstract, reifying, and systematized concepts need to be, but the central point that nature occurs within, as I would put it, a form of sensibility that is shot through and through with human significances and is in that limited sense rational, and a cultural product has to be granted. Only on this premise can nature be apprehended at all, and so construed it is capable of possessing a rich founding and normative dimension. It can be rightly construed now as a 'given' in the sense that it is both not simply a product of our decision-making and choice—as McDowell (1996: 10–13) concedes, there is an essential element of human passivity in our perception of it—and that as an orientating idea, it is so deeply embedded in our form of sensibility that it is constitutive of our way of seeing and understanding the world both cognitively and affectively. It is not, for us, a disposable idea. Furthermore, by extricating it from the logical space of science it is made clear that the kind of unity and transcendence that nature possesses is precisely *not*, fundamentally, that of highly abstract laws and conceptual schemes in terms of which the natural sciences characterize it. The important curriculum implications of this argument will be explored presently. First, a further epistemological point needs to be made.

For a range of reasons, traditionally the canons of intellectual understanding have set the tone for formal education, but previous argument suggests that there is a need to re-evaluate the knowledge that we possess through bodily contact with the world. In feeling the resilience of this piece of grass underfoot, this piece of earth to the spade, this piece of wood to the chisel, in feeling the growing chill in the air and apprehending the brooding presence of storm clouds, we engage with the world less through a cognitive ordering and more through a receptive *sensing* that is less susceptible to abstract generalization and objectification. Here we have a form of acquaintanceship in which there is a knowing of the embodied by the embodied that, at its deepest level, apprehends the other in its alterity and not primarily as a vehicle or obstacle to satisfying our desires. Taken thus, it can be seen to constitute a kind of love. As Murdoch (1959: 51) once put it:

> Art and morals are, with certain provisos … one. Their essence is the same. The essence of both of them is love. Love is the perception of individuals. Love is the extremely difficult realization that something other than oneself is real. Love, and so art and morals, is the discovery of reality.

It strikes me that this idea of love/caring as itself a way of knowing is seminal. Both art and morals involve a certain 'letting be' of the other that at the same time is a creative knowing. For reasons previously indicated, for us moderns this requires a certain preparation of heart and mind to receive what is offered—a suspension of the mastery motive and the adoption of an attitude that is neither an indifference nor a possessive desiring, but rather a *dialogical openness* that incorporates a sense of the well-being of things themselves. The self-arising cannot reveal itself to the eye that primarily seeks to organize, to manage, and to manipulate. Let me cite a negative illustration from the anthropologist Sharp (1988). In contrast to the other-oriented etiquette of the Chipeweyan Indian, he described White Canada's interaction with the non-human world in the following terms:

> White Canada does not come silently and openly into the bush in search of understanding or communion, it sojourns briefly in the full glory of its colonial power to exploit and regulate all animate being. … It comes asserting a clashing alien causality certain in the fundamentalist exercise of the power of its belief. It talks too loudly, its posture is wrong, its movement harsh and graceless; it does not know what to see and it hears nothing. Its presence brings a stunning confusion heard deafeningly in a growing circle of silence created by a confused and disordered animate universe (p. 145).

Implications for the curriculum and the culture of the school

The general upshot of this line of argument is that we, as educators, need to be concerned with a gradual change in how we apprehend the world at a fundamental level—that is, we need to explore the possibility of a different *metaphysical* basis to education. And this immediately prompts a key question: *what kinds of knowing and learning should education encourage?* What should be their underlying spirit or attitude towards the world? The relevance of this issue is emphasized when we recall that the primary agendas of many traditional school subjects reflect a social history that was not only largely innocent of environmental problems, but whose underlying motives included the subordination and exploitation of nature. To hark back to the example of science, it was not only Bacon, but other luminaries of the new science of the 17th century such as René Descartes (1596–1650), Joseph Glanvill (1636–1680), and Robert Boyle (1627–1691) who advocated the mastery of nature as its prime goal. This both fed into and reflected the general anthropocentrism of Enlightenment humanism which came to modulate all areas of understanding. Thus, it becomes important to ask what projects towards the world different kinds of knowledge and learning express. Clearly, the defining qualities of nature as the self-arising elucidated above—a fluid world of open, many-faceted things in constant, and often mysterious, interplay—simply becomes invisible to an encounter preoccupied with intellectual (and material) possession achieved through the deployment of increasingly highly systematized and ossifying conceptual schemes. Self-arising facets of the world are simply occluded by teaching that has this orientation.

The implications of this extend not simply to the propositional content of a subject area, but more essentially to the manner in which such content is held and explicated. The value of a more intimate, intuitive, non-logical style of encounter with the world needs to be acknowledged, one whose rigour derives less from adherence to superimposed rules upon experience and more from an open attentiveness to the things experienced. This will involve an attitude of participation, celebration, and a willingness to be affected as contrasted with an overweening drive to disengage from the immediately present so as to set it to order, to control it, to be 'effective'. Here then, the issue of environmentally adequate knowledge and its curriculum organization is raised: Should it be subject-based and, if so, what kinds of subjects? Should it be a holistic, integrated curriculum, and, if so, according to what structural principles?

With regard to the latter, it has been argued that it would be a mistake to attempt to conceive of environmental education as some holistic cross-disciplinary element (Stables and Scott 2002). Erroneously, this would imply that there is, as it were, some single environmental grand narrative to be conveyed. Instead, it should be developed from *within* the differing perspectives that existing school disciplines have to offer—which has the further practical advantage that as we are not in a position to regenerate the education system (particularly teachers' expertise and attitudes) from scratch, we could build upon existing strengths. Such a view provokes a number of important responses.

To begin with, it must be granted that environmental education should not take the form of some totalizing cross-curricula alternative to the disciplines. Not only would this carry with it the danger of a certain eco-fascism, it would overlook the fact that a concern with the self-arising, as essentially mysterious, can hardly be hived off into some discrete theme. Nor, especially, can it be given some detailed, pre-specifiable content. As previously argued, genuine openness to situations is not enhanced by seeking to impose all-embracing systematic conceptualizations. On the assumption that they express enduring strands of our form of sensibility—and hence constitute part of the intellectual and cultural capital upon which we must ultimately draw in addressing *any* deeply rooted set of problems—traditional subjects as conceived, say, in the Oakeshottian (Oakeshott 1962) sense of on-going conversations, have, potentially, an important role to play. However, there are certain caveats.

First, given the point made earlier concerning the historical social/ intellectual milieu of traditional subjects, proper account will need to be taken of the danger of motives inherent in a discipline that may be covertly hostile to self-arising nature and, therefore, that have a propensity to construe environmental problems in a manner that veils their own contribution to them. Thus, an important area for investigation is opened up: *What motives and attitudes towards nature are implicit in different areas of the school curriculum?* I have referred to some versions of science in this regard, but equally one could ask this question of art, literature, design and technology, geography, history, religious education—even, and especially, I have argued (Bonnett 2004a: 161–166), aspects of information and communications technology—*as they are taught in schools*. Secondly, there is raised the desirability of inviting pupils to locate the claims to knowledge made by the disciplines against a broader backcloth of metaphysical understanding. This might enable them to challenge, say, approaches to understanding the world that require them to conceive a local pond or hedgerow as essentially part of an abstract deterministic causal network, or energy flow, or information system, rather than, say, as sensuous things experienced as having an intimate place and value in their own life world. Disengaging them from their own intuitions of the real, encouraging them to feel that such knowledge of the world is inferior to an abstract disengaged approach, not only further separates them from what has been argued to be essential to our experience of nature, but also raises the danger of leaving them disaffected. Hence, authentic discussion that celebrates differing languages for articulating environmental issues needs to be an important aspect of the curriculum. Such discussion can be refined and extended by drawing upon the arts and humanities as much as the sciences. Indeed, as far as our relationship with nature goes, it is far from clear that Newton has as much to offer as the English poet Gerard Manley Hopkins (1844–1889). In curriculum terms, this particular example raises important questions about the general character of English as a subject and the spirit in which literature is presented to pupils. Current emphases in the UK on a 'literacy' that for many takes the form of a routine of pre-specified exercises hardly encourages the kind of creative engagement with poetic language that would enable lived encounters with nature (or anything else).

Closely related to this, environmental education would need to include a (again, ultimately metaphysical) critical investigation into current social/ economic practices—to identify and evaluate the motives that energize them and the ways in which we are all, to some degree, implicated in them. As previous argument would suggest, I do not have in mind here some formal course, but rather an encouragement to students to reflect in this way on familiar experiences and practices, and issues that concern them— supported, perhaps, by documented case studies where appropriate and an understanding of larger enveloping life contexts provided by literature and art. Taken altogether, the picture of the curriculum that arises is one of *emergent engagements*, whose unity is not the result of pre-formed inter-disciplinary connections determined by academics distant from particular sites of learning, but the result of an evolving inter-play of consciously felt demands arising from a receptive participation in the issues and listening to the call of the as yet unknown.

Finally, if education is to help pupils to live in such a way that they can flourish in an authentic—that is, poetic—relationship with the self-arising, it will need to help them to learn properly *to love the self-arising in themselves*, so as not to be prepared to sell themselves cheaply to a global economism that requires them to be ever on call to produce and ready to consume. Thus, in many ways the issue is not primarily one of formal curriculum content as of the general culture of the school (and, of course, society). It is a matter of the underlying versions of human flourishing and the good life that are implicit in the ethos and practices of the school as a community and how they connect with life 'outside'. This ethos both invites direct participation in certain ways of going about the world and conditions the spirit in which the curriculum is taught and received. Only as it begins to reverberate to a different metaphysics can a space arise for those kinds of intimate experience of the presence of nature in which the power and subtlety of otherness and the elemental are felt and allowed to matter.

To conclude

The essence of my argument is that our environmental predicament is a crisis not simply of our physical survival, but of our spiritual survival—that is, our understanding of what we are and how we should relate to the world around us. This is a crisis that is as much of human feeling as it is of the intellect; it is a crisis of our whole mode of sensibility. From this standpoint it is possible to draw two broad educational implications:

(1) Environmental education must have two agendas:
    (a) a *short-term* pragmatic agenda of damage limitation that would focus on the cautious but imaginative use of science and technology to monitor and help ameliorate undesirable outcomes of the impact of human behaviour on nature. This agenda is now widely being addressed.
    (b) a *long-term* agenda of developing a sense of a right relationship with nature as the self-arising—this gradually, but increasingly, informing and orientating the more immediate agenda above. This, the most important agenda, is constantly peripheralized and subverted by the dominant metaphysics of our time that can only permit it as a façade, a public relations exercise.[6]
(2) Environmental *education* is much richer and more profound in its aspirations than the idea of sustainable development encourages. It is essentially concerned with an understanding and appreciation of the environment and the significance of the natural order, including

our place in it. At the heart of this will be an attempt to characterize, and develop in life, what should count as a right relationship with nature and thus a fuller understanding of what truly should count as human flourishing. Human well-being remains a central concern, but its interpretation is not restricted to the economic, and its achievement is understood as involving an understanding of our own nature and an appreciation of nature's value that truly transcends the instrumental.

Notes

1. I have attempted to explore further philosophical aspects of this story in Bonnett (2004a, b). Also, lack of space has precluded reference to a range of other issues in the field. For an extensive general review of the research literature on environmental education, see Rickinson (2001).
2. This strikes an interesting contrast with countries such as Norway where the core curriculum is described in terms of reflecting 'the joy of nature' and that 'Education should enkindle a sense of joy in physical activity and nature's grandeur, of living in a beautiful country, the lines of the landscape, and in the changing seasons' (Royal Ministry of Education, Research and Church Affairs 1999).
3. I have attempted to demonstrate the deficiencies and ultimate incoherence of this stance in *Retrieving Nature* (Bonnett 2004a: 42–69). See also Soper (1995) and Soulé and Lease (1995).
4. I am grateful to John Foster at the University of Lancaster for suggesting to me that the problem here with sustainable development is not so much that it is ambiguous, as that it is indeterminate. It offers to posit a determinate baseline (e.g. for levels or values of critical natural capital), but because we construct the science on which the relevant metrics depend, and the social readings or interpretations of those metrics, the model never brings us up against any genuinely constraining limits. There is no fundamental distinction between the sustainability obligations we are under and the ones we are comfortable under. I think that this notion of the indeterminance of 'sustainable development' presents an interesting and very pertinent perspective. It is also one that chimes well with my general concern to (re-)establish a genuine encounter with the essential otherness of nature as the self-arising. To acknowledge this is not, though, in my view to dismiss important elements of ambiguity in the notion—indeed, now, for example, in terms of how 'critical natural capital' is to be construed, 'critical' being relative to sets of needs that can be defined in a variety of competing ways.
5. See Carson (1962) for a classic account of the former.
6. See Blühdorn (2000).

References

Ayer, A. J. (1940) *The Foundations of Empirical Knowledge* (London: Macmillan).

Berlin, I. (2000) *Three Critics of the Enlightenment: Vico, Hamann, Herder,* ed. H. Hardy (London: Pimlico).

Blühdorn, I. (2000) *Post-Ecologist Politics: Social Theory and the Abdication of the Ecologist Paradigm* (London: Routledge).

Bonnett, M. (2004a) *Retrieving Nature: Education for a Post-Humanist Age* (Oxford: Blackwell).

Bonnett, M. (2004b) Lost in space? Education and the concept of nature. *Studies in Philosophy and Education*, 23(2–3), 117–130.

Brundtland Commission (1987) *Our Common Future* [World Commission on Environment and Development] (Oxford: Oxford University Press).

Capra, F. (1983) *The Turning Point: Science, Society, and the Rising Culture* (London: Fontana).

Carson, R. (1962) *Silent Spring* (Boston: Houghton Mifflin).

Collins, S., Osborne, J., Ratcliffe, M., Millar, R. and Duschl, R (2001) What ideas about science should be taught in school science? A Delphi Study of the 'expert' community. Paper presented at the annual conference of the National Association for Research In Science Teaching, 26–29 March, St Louis, USA. Available online at: http://www.kcl.ac.uk/content/1/c6/01/52/90/ideasaboutscience.pdf, accessed 13 April 2007.

Department for Education and Employment (1999) *The National Curriculum Handbook for Primary Teachers in England: Key Stages 1 and 2* (London: Her Majesty's Stationery Office).

Giddens, A. (1994) *Beyond Left and Right: The Future of Radical Politics* (Cambridge: Polity Press).

Grange, J. (1997) *Nature: An Environmental Cosmology* (New York: State University of New York Press).

Haraway, D. J. (1991) *Simians, Cyborgs and Women: The Reinvention of Nature* (London: Free Association Books).

Lyotard, J.-F. (1984) *The Postmodern Condition: A Report on Knowledge,* trans. G. Bennington and B. Massumi (Manchester, UK: Manchester University Press).

McDowell, J. (1996) *Mind and World* (Cambridge, MA: Harvard University Press).

Merleau-Ponty, M. (1962) *Phenomenology of Perception,* trans. C. Smith (London: Routledge & Kegan Paul).

Moore, G. E. (1953) *Some Main Problems of Philosophy* (London: Allen & Unwin).

Murdoch, I (1959) The sublime and the good. *Chicago Review,* 13(3), 42–55.

Oakeshott, M (1962) The voice of poetry in the conversation of mankind. In M. Oakeshott *Rationalism in Politics and Other Essays* (London: Methuen), 197–247.

Price, H. H. (1932) *Perception* (London: Methuen)

Rickinson, M. (2001) Learners and learning in environmental education: a critical review of the evidence [Special issue]. *Environmental Education Research,* 7(3).

Rorty, R. (1980) *Philosophy and the Mirror of Nature* (Oxford: Blackwell).

Royal Ministry of Education, Research and Church Affairs (1999) *The Curriculum for the 10-year Compulsory School in Norway* (Oslo: Royal Ministry of Education, Research and Church Affairs). Available online at: http://www2.udir.no/L97/L97_eng/index.html, accessed 10 May 2007.

Sharp, H. S. (1988) *The Transformation of Bigfoot: Maleness, Power, and Belief among the Chipewyan* (Washington, DC: Smithsonian Institution Press).

Soper, K. (1995) *What is Nature? Culture, Politics, and the Non-Human* (Oxford: Blackwell).

Soulé, M. and Lease, G. (eds) (1995) *Reinventing Nature: Responses to Postmodern Deconstruction* (Washington, DC: Island Press).

Stables, A. and Scott, W. (2002) The quest for holism in education for sustainable development. *Environmental Education Research,* 8(1), 53–60.

Stone, C. D. (1974) *Should Trees Have Standing? Towards Legal Rights for Natural Objects* (Los Altos, CA: William Kaufman).

Thoreau, H. (1949) *The Journal of Henry David Thoreau, Volume 5,* ed. B. Torrey and F. H. Allen (Cambridge, MA: Houghton Mifflin).

United Nations Conference on Environment and Development (UNCED) (1992) *Agenda 21: A Programme for Action for Sustainable Development* (New York: United Nations).

'Littered with literacy': an ecopedagogical reflection on whole language, pedocentrism and the necessity of refusal

DAVID W. JARDINE

> No matter the distinctions we draw, the connections the dependencies, remain. To damage the Earth is to damage your children. (Wendell Berry, *The Unsettling of America* [1977])

Prelude: on 'ecopedagogy'

The term 'ecopedagogy' is meant to re-awaken a sense of the intimate interconnection between ecological awareness and pedagogy. This interconnection is not the outcome of a concerted application of the principles and practices of one domain (ecology) to another domain (pedagogy). Such questions of 'domains' and 'application' between ecology and pedagogy inadvertently assume the separation of these two disciplines, and assume, as well, that the connections between the sustainable generativities of the Earth and the generativity represented by children and embraced by pedagogy are somehow ours to be made, not made or unmade. As the above-cited passage from Berry (1977: 13) suggests, such matters are not at our disposal or discretion to connect or disconnect. More pointedly, these matters are not at our disposal to ignore under the guise of 'working in another area'. Even, say, in the realm of the mathematics curriculum, ecopedagogical images and profound ecological consequences and choices have already been made and are always already at work (Jardine 1990).

Ecopedagogy assumes that there is always and already a deep, ambiguous kinship at work between the real, Earthly life of children, the tasks of pedagogy (including our understandings of various curriculum specializations, how we envisage and practise a relation between the old and the young, our conceptions and embodiments of knowledge and our images of ourselves and our tasks as teachers) and the Earth's 'limits of necessity and mystery' (Berry 1983: 13). This ambiguous kinship can be simply framed, although its implications pose enormous questions to our lives as pedagogues and to the telling tales that underlie our practices:

- to the extent that ecology considers the conditions under which life can go on (Smith 1988a), it is always already intimately pedagogic at its heart;
- to the extent that the task of pedagogy is to usher children into those understandings of the Earth's ways required for life to go on in a full and healthy and wholesome and sustainable way, it is already intimately ecological at its heart.

Finally, at the heart of the notion of ecopedagogy hides a traumatic twist. It is possible to work with full confidence and fully gracious intent in the area of pedagogy and yet betray an unintended ecological insanity. We can speak with grand aspirations of our hopes for our children while remaining unaware of how those very aspirations, in their ecological assumptions and consequence, might work against the actual breath required to utter them. In the face of this potential ecological insanity, the task we face is to consider our pedagogical theories and practices (including various curriculum areas) in light of some sense of their ecological possibility and sustainability. This task is becoming more and more urgent because, as the desperate state of our Earth attests, we can do the impossible: we can unintentionally work against the real, Earthly conditions under which pedagogy is actually possible.

Ecopedagogy and the unsustainable extremities of language-arts curriculum

These broad formulations of the nature of ecopedagogy can be particularized to a particular curricular area in the following way: what constitutes, for example, a sustainable and generative understanding of the ways of language and texts and reading and writing? Such an understanding of language curriculum must consider both how new growth can be nurtured, encouraged and sustained and what must be conserved/preserved of the old growth and the old rooting soils for this nurturance, encouragement and sustenance to remain possible. Language curriculum theories and practices must not become too enamoured of the new and the young or we may lose sight of the conditions under which the new can thrive. Such theories and practices must not become too enamoured of the old or we might lose sight of the ways in which the old requires the fecund regenerativity and transformation that the young provide.[1]

In the area of language-arts curriculum, we have tended to 'ride the pendulum' (Stahl 1990) between these extremities in search of a clear and solid and fixed foundation for our theories and practices. Ecology suggests, however, that there is no such securable ground to the living practices of human life, only ongoing, shifting, ambiguous nests or communities of interrelations which are constantly in need of renewal, regeneration, rethinking. And this suggests that in the area of pedagogy (Smith 1988a: 176), 'The old unilateral options of *gericentrism* (appealing to the authority of age, convention, tradition, nostalgia) and *pedocentrism* (child-centred pedagogy) only produce monstrous states of seige which are irresponsible to the matters at hand, i.e., to the question of how life is mediated through relations between the old and young'. Human life thrives only in this 'belly of a paradox' (Smith 1988a: 175) – the ongoing nests of interrelations between the old and the young, the established and the new. Both gericentrism and pedocentrism can thus be understood as breakdowns in this living nest or community of relations – attempts to anchor educational theory or practice to a fixed point (e.g., 'the child is the centre of the curriculum' or 'back to basics') instead of in the mediated set of relations themselves. The child at the centre will always have to confront what is basic in curriculum if it is truly basic to the course of human life; what is basic in curriculum will always have to be understandable to the child, invigorating of the child's life and invigorated by their taking it up anew. Out of relation, both of these extremities become monstrous.

Healthy language-arts theories and practices must consider how to embrace this continuously mediated and remediated state of interdependency that inheres in the living nature of language itself and how to become tolerant of the difficulty and ambiguity that inheres in it (and this in spite of the fact that a certain clarity and cleanness would result if we simply severed these interdependencies and fell prey to either gericentrism or pedocentrism). The life of language occurs in the tensive, mediated interplays between the forms and disciplines and established wisdoms of/in language and the newly erupting voices of the young. These extremities *depend upon one another*, and each finds both its (re)source and its limit in its opposite. Without the renewal that the young provide (Arendt 1969: 187), the senatorial (e.g., a piece of literature considered by the wisdoms of age and convention and tradition to be worthwhile, a piece of literature that is thus ripe for reconsideration, open and ready to support and nurture and teach the young) can become simply the senile (e.g., already fixed and established answers in the back of the book which foreclose on the necessity of children ever needing to consider this work deeply, since the 'answers' seem to be already on hand, not requiring their consideration). Similarly, without the deep-rooting soils that age and time and wisdom and discipline provide, the beautiful voice of the individual child (e.g., the child as author/originator [Harste *et al.* 1988: 5]) can end up merely puerile. Put more colloquially, the young and the old are dependants, kin, relations/related, and only *in* these relations does each avoid its weakest aspect (senility, puerility).[2] Each finds comfort (common fortitude, strength) in the other.

Rather than attempting to *solve* this precarious and tensive interplay between the old/established and the young/new (as if it were a mistake we might wish to fix), healthy and sustainable language-arts theories and practices attempt to bear this ambivalence in the understanding that 'ambivalence is the adequate reaction to the whole truth' (Hillman 1967: 15) about language as a living system. The task then is not to solve this ambivalence, but to resolve ourselves to 'make meaningful and beautiful th[is] primary paradox that human beings *have* to live with' (Snyder 1980: 29–30).

Understood this way, what at first appears to be a failure to figure out, finally and once and for all, the language-arts curriculum for young children in a solid and fixed and clear fashion, complete with a consistent and comprehensible theory and a package of fail-safe, teacher- and child-proof practices just might turn out to be our greatest success. It may be a sign of the deeply (inter)*dependent* character of language that lives beneath our desires for fixity and clarity and centration (the central urges of literalism). Accountable and warrantable action in the language-arts classroom always *depends*. It proceeds as a living response to the living, deeply dependent interactions that pertain in that classroom. This child's work may call for a more honed and careful attention to some of the disciplines of language; this child may need to be encouraged to be more free and ebullient in what he or she writes; it may be the appropriate time to introduce these children to that author. And all of these actions are never necessarily accountable in all cases, as if what is needed in order to write well never *depends* on the living relations in which each child finds him or herself immersed.[3]

This may *appear* to be child-centred but it is neither this nor its opposite. Warrantable action does not depend only on the child, but, for example, on the child's ability to communicate to this audience in that fashion about these topics. Thus, we are always confronted with the child *in relation* to the disciplines and conventions of communication, *in relation* to the varieties of audience (to which their writing must be obedient-*ab audire*: audience and obedience are rooted together etymologically), *in relation* to this type of writing (e.g., a letter) and its forms and features, and *in relation* to these topics which have probably already been written about in a myriad of ways (ways which themselves bespeak a myriad of relations) – and all of this *in relation* to what the child has already come to understand about these matters, *in relation* to the classroom setting and the tasks currently under way. Although the child surely figures strongly in all these relations, he or she is not the 'centre' of these relations except in the most porous of senses. Sometimes, for example, the demands of convention will decentre the child, requiring that he or she learn more or differently. Sometimes, for example, convention will say 'no' to the child – gently, one hopes, with care and tact, full of pedagogic intent, but 'no' nevertheless. This 'no' is an attempt to keep the child *in relation*, to keep them rooted in the dependencies needed to sustain their ongoing, generative efforts. This 'no', this refusal, is thus not precisely *against* the child but is rather against allowing, in the name of the child and of the generativity they bring, the despoilment of the conditions under which their ongoing, generative efforts can sustainably continue.

In this way, language arts becomes a delicate, ambiguous, ecological matter.

'Littered with literacy': the ecologies of whole language

Whole language can be understood as an effort aimed at 'restoring [the] life [of language] to its original difficulty' (Caputo 1987) – the original difficulty that pertains in its deep and generative state of (inter)dependency. This attempt to restore generative relations, dependencies, connections, is also an ecological effort in its premise and consequence.

It is thus no coincidence that the continuing growth of interest in whole language is occurring now, in the midst of a growing interest in ecological issues of wholeness and health. Ecology reminds us that the Earth is a living system constituted by a vast interweaving and interconnected web of dependencies. To live well on the Earth is to learn to live in and with these dependencies. After all, as Berry (1977: 14) suggests, 'the care of the earth is our most ancient and most worthy and, after all, our most pleasing responsibility. To cherish what remains of it, and to foster its renewal, is our only legitimate hope.' Likewise, whole language tells us that the vibrancy and life of language is found in its living interdependencies, not in its lifeless fragments (writing cut off from reading, phonics, or graphics cut off from meaning, authorship and originality cut off from discipline, ebullience cut off from craft, and the like). A pedagogy that is becoming of such interdependencies is likewise 'our only legitimate hope', and the 'continuity of attention and devotion' (Berry 1977: 14) that such a pedagogy requires cannot help but be 'our most pleasing responsibility'. This perhaps bespeaks the deep pleasure and relief expressed by many teachers upon their introduction to whole language – a 'release' of sorts into the living beauty of language as lived, and a freedom from its deadening calcifications. This is not the pleasure of triviality and ease and simplicity and mindless enthusiasm: it is the pleasure of coming upon 'the real work' (Snyder 1980) of language and its real 'original difficulty'.

Framed in this way, it is clear how 'whole language' is not offering yet another theory about language or another 'method' (Newman 1985). Whole language wishes to turn our attention to how it is that we *already live* in language. As with ecological awareness, what is at stake is not a *theory*. It is our living that is at stake: how we actually live in the world (of language) and what such living requires of us if it is to be sustainable and healthy and whole. What is first and foremost is not some clear and consistent theory, but the real, living practices of living readers and writers in the world (Edelsky *et al.* 1991: 22, Harste *et al.*. 1988: 3) with all the contradiction, paradox and living difficulty that such a world requires and invites. It is no coincidence, then, that Weaver (1990: 106) uses ecological metaphors to describe the character of a whole-language classroom: 'The classrooms are "littered with literacy": that is, they offer a rich variety of environmental print as well as books, magazines, and newspapers. They are print-rich environments.'

We have here a wonderful image of a classroom bursting full of language in all its living forms. This image works against a certain old image of cleanness (an image often linked to clarity [see Jardine 1992b: 31–51, Turner 1987: 7]) and evokes an 'untidy', ambiguous, Earthy sense in which 'litter' is a necessary precondition of healthy growth. Ecologically speaking, such a littered language classroom is not organized to suit some need *other than* this living world of language itself (e.g., ease of institutional/logico-mathematical accountability – one could consider the clarity and cleanness of a phonics worksheet and the 'mess' created by linking phonics back to *this* child's desire to communicate well about *this* topic to *this* friend or enemy). Such a littered classroom is, however, very orderly, orientated to the indigenous orderliness of the living world of language itself: in the *life* of language, phonemic and graphic conventions exist on behalf of making communication with others more 'convenient'. The mastery of such conventions does not disappear in the whole-language classroom. On the contrary, such conventions are regenerated and re-enlivened by being linked back to the living reason(s) for such mastery. Such mastery makes a difference in how we can live (in language).[4] Whole language names this living order of a living system and the whole-language classroom is thus constituted by *living in* this order.

This is a deeply ecological sense of 'littered with literacy': the sense in which, for example, trees spontaneously litter every year. The material which trees regularly scatter around is precisely that which is required to sustain the tree itself and sustain the possibility of producing new life.[5] In and through littering, it is building its own humus and even in the passing of this particular tree, its own 'death becomes potentiality' (Berry 1983: 73) as it adds its own life to the conditions for the possibility of new life.

Here is an image for understanding the community established in the whole-language classroom (Calkins 1986, Graves 1983, Harste *et al.* 1988) as an ecosystem in which: 'the community is an order of memories preserved consciously in instructions, songs and stories, and both consciously and unconsciously in *ways*. A healthy culture holds preserving knowledge *in place* for a *long* time. That is, the essential wisdom accumulates in the community much as fertility builds in the soil' (Berry 1983: 73). The litter of literacy in such a classroom allows children and teachers alike to learn the ways of language by generatively participating in and contributing to these ways. Such a classroom is constituted by the interplays between the generativity and liveliness and authority/originality[6] of the young and the ways and stories and order of memories needed to sustain such generativity. Whole language requires envisaging language not as an inert set of instruments that are at our beck and call, but as something that houses us, something we are in, something which responds to us and something to which we are responsible. We are 'in' the community of language. And once we envisage language as a place in which we dwell, the emphasis of whole language on 'communication' becomes immediately obvious, communication being the generative case of community. Also, understood as a 'community', language is understood to have a life and a history and a wisdom that both goes beyond the individual and that needs the individual ('the new blood', to use an archaic term common to the initiation of the new ones into the community) for its own renewal. As a sustainable community or place it includes both the old/established and the young/new and provides a place for each to find its source and limit, its comfort, in the other.

The hidden ecological irony in the use of the notion 'littered with literacy' should be ignored, however, because it signals a potentially dangerous misunderstanding of whole language. 'Littered with literacy' is *intended* to signal richness, fullness and diversity. But 'litter' can also evoke images of abandonment, excess and irresponsibility. If 'littered with literacy' takes on these more colloquial connotations, whole language can easily become confused with a sort of linguistic free-for-all that at best ignores and at worst despises age, tradition, convention, discipline and the like and simply stuffs the classroom full of 'any old garbage'. It suggest an image of whole language as a form of *pedocentrism* which 'leaves children to their own devices'. Even more strongly put, such a classroom community is constituted by 'abandonment and betrayal' (Arendt 1969: 188, 196).

Let us look at the roots of this misconstrual of whole language as a form of pedocentrism. If we consider the whole-language classroom as a place where children might be at home in language, consider the following passage in relation to Weaver's (in Berry 1977: 51) description of her classroom:

> It is impossible to divorce the question of what we do from the question of where we are – or, rather, where we think we are. That no sane creature befouls its own nest is accepted as generally true. What we conceive to be our nest, and where we think it is, are therefore questions of the greatest importance.

A classroom 'littered with literacy' must be understood as a living place, a nest. The question then becomes one of how to resist the (more colloquial) sense of 'litter' that suggests befouling this nest. There is a sense in which a classroom can become all too easily 'littered' with garbage that does not lend itself to the life of language in its wholeness and health and that therefore (intentionally or not) works against an environment which will sustain and nurture the young. A nest works against itself if it becomes too centred on the young; it is properly and sustainably 'centred' on the bearing and nurturing and protecting and teaching and raising *of* the young. It is thus not precisely 'centred' but 'decentred' outwards into ecopedagogical relations between the young and the ways of raising, the young and the limits of protecting, the young and the teachings and so on. And these relations are always such that, for example, the teachings will tell us something about the young, but the young, too, will tell us something about the teachings. Each will invigorate and strengthen the other. This is the sense in which ecology is understandably conservative in one sense (wishing to conserve the conditions under which life can go on) and not conservative in another (realizing that those conditions require the regeneration, renewal, revision, transformation that the young provide). These 'relations', if they are healthy, are always right in the midst of being rethought, renewed, re-established. The ongoing 'conversation' between the old/established and the young/new is thus not an error. What *is* an error is not allowing this conversation to go on in a healthy way. 'Garbage' is what works against this nest of relations by existing 'out of relation': garbage is always 'out of place'.

'No sane creature befouls its own nest'. If we look again at this example of the littering tree, it becomes evident how we are unlike it. For the most part, the tree cannot work against its own sustenance and renewal. It is ecologically impossible for it to 'litter' in the sense of producing garbage. Unlike the tree, we are possessed of what at first glance appears to be a certain freedom. We, unlike the tree, can do the impossible; as Heidegger (1973:109) writes:

> The unnoticeable law of the Earth preserves the Earth in the sufficiency of the emerging and perishing of all things in the allotted sphere of the possible which everything follows and yet nothing knows. The birch tree never oversteps its possibility. It is [human] will which drives the Earth beyond the sphere of its possibility into such things that are no longer a possibility and are thus the impossible. It is one thing to just use the Earth, another to receive the blessing of the Earth and to become at home in the law of this reception in order to shepherd the mystery and watch over the inviolability of the possible.

Human action, human will, can, so to speak, spiral out of order, out of proportion, breaking the Earthly 'limits of necessity and mystery' that might delimit our prerogative to what is ecologically possible (i.e., to what is, in the long run, generative and sustainable). Our truly sane, human prerogative is thus not commensurate with what we *can* do (a very abstract notion of what it is 'possible' for us to do), assuming 'that the human prerogative is unlimited, that we *must* do whatever we have the power to do. What is lacking [in such an assumption] is the idea that humans have a place and that this place is limited by responsibility on the one hand and by humility on the other' (Berry 1983: 54–55).

Simply filling a classroom with all possible forms of print and allowing any and every form of language-arts activity assumes that everything we *can* fill the classroom with is 'litter' and not 'garbage'. However, as a practised teacher knows, littering the classroom with literacy names a profound nest of *decisions* on the part of the teacher, decisions taken in full awareness that we can do the impossible. We can work against the health and wholeness of language. Because we can do the impossible, we must understand the deep interrelations that inhere in the wholeness and health of language and we must actively and consciously refuse that which might violate or overstep those limits – that which would turn litter into garbage. 'Knowledge of these limits and how to live within them is the most comely and graceful knowledge that we have, the most healing and the most whole' (Berry 1977: 94). And such comely and graceful knowledge is had only by the teacher him- or herself living in the midst of language as a real writer, a real reader, someone attuned both to how language responds to us and how we are responsible to it. Under the surface gloss of occasional 'relentless enthusiasm' (Smith 1988b: 236), the whole-language teacher develops a deep and considered sense of restraint that is gained from living well in the interior of language itself and nurturing in oneself such comely and graceful knowledge, holding it in place for a long time and *living with the consequences* (Berry 1977, 1983). This is the living, generative sense of the restraint that inheres in dependents/dependence/(inter)relations. 'By restraint they make themselves whole' (Berry 1977: 95).

Whole language and the necessity of refusal

North American culture is not one especially enamoured of restraint or dependence, born as it is out of ecologically naïve images of freedom and independence. And it is important to consider how whole language was itself born in part as a response to earlier language-arts practices which suppressed the individual's desire to write under the weight of the disciplines of language. In its first flush, it is too easy to imagine that whole language involves the overturning of such weight and the unabashed acceptance and confirmation of everything and anything without restraint, without limit or measure, without disciplined and considered refusal. It does not. And to believe so is to confuse whole language with pedocentrism. There is, however, a sense in which whole language easily lends itself to such confusion. Again, the lessons of ecology are telling in this regard.

There has been a disturbing loss in the area of ecological awareness. Currently, in Canada at least, there are three 'Rs' to environmentalism: reduce, re-use and recycle (some have introduced a fourth 'r': recover). Several years ago there was another, different fourth 'R' which has since gone missing: refuse.

It is vital to not misread this missing fourth 'R'. It is not simply 'refuse' in the sense of 'garbage'. It also suggests refusal. The most potent form of ecological action is simply saying 'no' to those elements of our lives and our ways that are unsustainable, that befoul our nest. Saying 'no' to the garbage. Refusing. The loss of this fourth 'R' – the loss of the power and potency and responsibility involved in the act of refusal – is, unfortunately, not very mysterious. It leaves us with a vision of ecology which does not demand that we take responsibility for our own consumptive desires except *after* they are fully satiated. We can consume anything we want as long as we deal with the garbage *afterwards*. We are not required to consider how it may be that much of what we consume is *itself* garbage and how our relentless consumptiveness – our inability to say 'no' – might itself spell ecological disaster. We live in an economy geared to saying 'yes' without hesitation, geared to growth without restraint, geared to the giddy sense of consumptive vitality that such a headlong rush provides.

In a horrible twist of logic, the relinquishing of the power of refusal leads to precisely that sense of rootlessness and powerlessness and futility that makes one susceptible to becoming a relentless consumer who is unable to refuse, as Berry (1977: 24) writes:

> People whose governing habit is the relinquishment of power, competence and responsibility, and whose characteristic suffering is the anxiety of futility, make excellent spenders. They are the ideal consumers. By inducing in them little panics of boredom, powerlessness, sexual failure, mortality, paranoia, they can be made to buy virtually anything that is 'attractively packaged'.

I cannot help but read this passage in relation to the lurid advertisements that haunt educational magazines, offering purchasable whole-language kits and packages for instant classroom use. In an almost inevitable turn, whole language has become 'the latest thing', purchasable and consumable with, it seems, little cost, little agony and little real work:

> Dr Terry Johnson will show you (quickly and easily) how to turn your classroom into a whole language showplace. You'll learn everything you need to know to profoundly increase your whole language teaching skills (And. . .we'll even buy you lunch!). (Johnson 1990: 32)[7]

As with ecology, however, whole language at its heart works against this unsustainable 'modernism' and its giddy, puerile enamourment with the easily and quickly consumable; as Berry (1983: 13) writes:

> What we call the modern world is not necessarily, and not often, the real world, and there is no virtue in being up-to-date in it. It is a false world, based upon economies and values and desires that are fantastical – a world in which millions of people have lost any idea of the materials, the disciplines, the restraints, and the work necessary to support human life, and have thus become dangerous to their own lives and to the possibility of life. The job now

is to get back to that perennial and substantial world in which we really do live, in which the foundations of our life will be visible to us, and in which we can accept our responsibilities again within the conditions of necessity and mystery.

Whole language evokes this archaic, ecological sense of the perennial and substantial world of language as lived – a world which is not quick, not easy, not constituted by the surface fashionabilities (and rapid outdatability) of a showplace and in which one *never* learns everything they need to know. Against a comely and graceful knowledge of the life and limits of language, we are able to refuse the giddy fantasy versions of 'whole language' that in fact despoil it. Whole language, as a way of living in language (and therefore precisely *not* as an easily purchasable and consumable thing), provides a way to say no, to refuse in a strong and considered way.

So here is the rub. In its first flush, whole language found itself out from under the gericentric anonymity of work-sheets, lovingly ushering the ebullient child into the heart of language. But in this adrenalin rush of vigour into language-arts practices, we can too easily confuse this influx of vigour with unabashed pedocentrism. Consider the case of a 10-year-old boy who, in response to the question 'are you a good writer?' says 'I don't know. My teacher says everything is good.' As Calkins (1986: 165) warns, 'out of fear of "taking ownership" [we can] desperately avoid teaching'. This timidity is what Arendt (1969:181) calls the feeling of 'standing helpless before the child', the sense of being 'out of relation', having 'relinquished our identities as teachers in order to give students ownership of their craft' (Calkins 1986: 165). The teacher who says that 'everything is good', even if they do so in the name of confirming the child, displays, however unwittingly, a lack of knowledge of language and its traditions that might make considered resistance possible and generative for the child. By the teacher being unable (or unwilling) to refuse – considerately and tactfully and delicately, one hopes – this child is abandoned to his own devices. Worse still, this child is abandoned in the full knowledge that the teacher's relentless confirmations are simply garbage. And, finally, with no good example of considered refusal, the child is not taught how to say 'no' for himself or herself to the unsustainable garbage that surrounds him or her. This is not *whole* language, and its supporters must avoid the allure of such grinning, pedocentric (one might even say 'puerile') enthusiasm. Whole language does not require such timidity in the face of the child.

Not surprisingly, if we relinquish the power of refusal, we also relinquish any strong and considered sense of acceptance. I recall reading the first few pages of *A Wizard of Earthsea* (LeGuin 1968) to a group of elementary school teachers and the turmoil caused when the suggestion was made that it was a good book, worthy of the attention of children and adults, strong and well drawn in its characters and so on. Again, this turmoil suggests a sort of timidity and rootlessness: in the name of affirming everyone's right to their own opinion, all opinions about a text are equally 'uprooted' ('That's just what *you* think of the book') and therefore none can be strongly affirmed as a good opinion of the text, an opinion that might have the strength to 'stand' public debate and scrutiny. Once the power of refusal is relinquished, any notion of a good or strong opinion becomes understandable only as the *disconfirmation* of the individual and their right to *their own* opinion. We become unable to voice any strong preference for LeGuin's book because preference becomes scattershot into always and only *personal* preference. In such a context, whole language easily blurs out into a type of expressionist rhetoric of 'private visions' and 'personal opinions' (Berlin 1988) which are always only directed towards affirming themselves ('this is my opinion': whole language confused with *self-affirmation*). This results in the puerile and ecologically disastrous belief that nothing *pertains* beyond the individual and his or her 'personal preferences'. Just as we can easily abandon and betray the child by leaving them to their own devices, this despoiling of LeGuin's work in the name of the child and their right to their own opinion abandons and betrays the world by saying, in this case, that any possible opinion (and hence nothing in particular) pertains to it. Here, LeGuin's work is not allowed to have the power of refusal, the power of saying 'no' to certain opinions. Again, this is not *whole* language.

In this way, then, the loss of the ability to refuse is twinned with the loss of the ability to accept. Against these twin timidities, whole language wishes to recover a strong sense of refusal and acceptance. It does not wish to do so according to some geriphilic notion of law and order, nor to some pedophilic fantasy of the preciousness of the young, but through a strong and vital and vibrant sense of community and a sense in which that community needs the new blood of the young. This necessitates a strong sense of our dual responsibility as teachers:

> Education is the point at which we decide whether we love the world enough to assume responsibility for it and by the same token, save it from the ruin which, except for renewal, except for the coming of the new and the young, would be inevitable. And education, too, is where we decide whether we love our children enough not to expel them from our world and leave them to their own devices. (Arendt 1969: 196)

Only in this ambivalent twinning of responsibilities is the life of language possible in its wholeness.

Concluding reflections

Gericentrism wars against the wildness of youth and pedocentrism rails against the calcification of age. In such states of seige, each extreme reads its opposite in its weakest aspect. Whole language can be conceived as an effort directed towards the mending and healing of these besieged relations.[8] It wishes to read these opposites in relation to each other and therefore to read each in its strength. As such, whole language is a deeply ecological phenomenon. But, as with ecology, there is a sting to this issue.

It is *my own* wholeness and health that is at issue in whole language. Ecopedagogical reflection places the issue of language-arts theories and practices back into the sphere in which we actually live – the sphere of my understanding and care and attention to language, my love of good books and good writing, my joy at the ebullience of children's work and the hard work they can pursue in the name of communication/community, my patience and frustration and their (and my own) occasional puerility and lack of discipline and care, the wisdom and originality they can display, and the pleasures that can come from delving into the wisdoms of the world and measuring one's life against them, learning from them. Ecopedagogical reflection places the issue of language back into the writing of this very paper and the gericentrisms and puerilities to which it may have fallen prey.

Ecopedagogy in the area of language-arts practices inevitably centres the 'continuity of attention and devotion' (Berry 1977: 14) that a generative and deep understanding of writing and reading and texts requires *of me* and *of the life I actually live* (*in language*). Such attention and devotion is not a trick or method or technique: it is a form of *life* dedicated to developing a deep sense of the wholeness and health of language and therewith developing my own wholeness and health *in* language. Because I can do the impossible, such wholeness and health requires that I develop a strong (and generative) and considered (and considerate) sense of acceptance and refusal.

Acknowledgements

I would like to acknowledge the contributions of Annette Gough regarding the ecological notion of 'litter'. I would also like to acknowledge the support of the Social Sciences and Humanities Research Council of Canada, grant no. 410-91-586 for the project 'Critical Considerations of Innovative Language Arts Practices at the Elementary School Level' (James C. Field, co-investigator).

Notes

1. This relation between the fecund instance and the already established is at the heart of Gadamer's (1989) conception of hermeneutics. In such a conception, tradition and age are essential to understanding but the fecund instance is essential to tradition. This instance is not simply additive to but transformative of tradition. That is to say, the relation between the instance and tradition is interpretive, not a matter of simple application (of established wisdoms to the new instance) or addition. Therefore, just as this paper suggests that there is an inner affinity between whole language and ecology, so too, there is an inner affinity between these and the interpretive disciplines (see Jardine 1992a, b).
2. Hillman (1967: 20–21) provides a wonderful (and occasionally frightening) description of the images of the old cut off from the young, a description full of pedagogic consequence:

   The negative senex is the senex cut off from its own puer aspect. He has lost his 'child'. The complex now split loses its inherent tension, its ambivalence, and is just dead. Without the enthusiasm and eros of the son, authority loses its idealism. It aspires to nothing but its own perpetuation, leading but to tyranny and cynicism; for meaning cannot be sustained by structure and order alone. Such spirit is one-sided, and one-sidedness is crippling. Time – called euphemistically 'experience', but more often just the crusted accretions of profane history – becomes a moral virtue. The old is always preferred to the new. Sexuality without young eros becomes goaty; weakness becomes complaint; creative isolation only paranoid loneliness. Cut off from its own child, and fool the complex no longer has anything to tell us. Without folly it has no wisdom, only knowledge – serious, depressing, hoarded in an academic vault or used as power. The integration of personality becomes the subjugation of personality, a unification through dominance, and integration only a selfsame repetition of firm principle. Because the complex is unable to catch on and sow seed, it feeds on the growth of [others]', as for instance the growth of one's own children.

Hillman goes on to artfully counterbalance this picture of the senex gone senile with an equally disturbing picture of the puer gone puerile – flightiness, relentlessness, thoughtlessness, naïvety, over-emphasis on the personal and private and the like. If whole language is to remain whole, it must somehow avoid both of these crippling and one-sided monsters.

3. It is clear from this that the need for more discipline or the need for more ebullience is not identical to the chronological categories of child and adult. Adults are capable of profound puerility as much as children are of wisdom. Children can be as caught in the senilities of 'having to spell every word correctly or I won't write' as adults can be caught in relentless and giddy self-annunciation.

The task of the teacher in the whole-language classroom is therefore not simply representing the wisdoms of age or representing and making room for the new voice, but *living in the middle of this living maelstrom of relations between the two* – what Hillman (1983) tellingly and accurately portrays as an 'agony' one *lives through*, not a set of polarities one can simply peruse.

But another consequence is essential to bring out as well. Whole language suggests that there is nothing wrong with puerile, undisciplined adult writing, *providing it is understood in context*. Diary- or journal-writing, first drafts and brainstorming, notes to myself, joyous or angry exuberances or outbursts, personal stories and conversations can all lend themselves to writing unleashed from the restraints of age, wisdom and tradition, unleashed, therefore, from any considered notion of 'good writing'. (As to the ways in which whole-language theory and practice are implicated in the North American enamourment with individuality and 'talk show' self-announcement wherein having *anything* to say is grounds for public display – these issues remain to be explored in full.)

But this is not to say that whole language contains no deep sense of 'good writing' and that it is always on behalf of nothing but exuberance (this again confuses whole language with pedocentrism). Rather, it suggests that we read and write with an eye to the living whole, the living context(s) which make *this* or *that* writing what it deeply is.

This point is vital since some adult writing in the area of whole language is, from the point of view of age, wisdom and tradition, not very good writing. For example, the poem 'What is a Whole Language Teacher?' by Mickelson (1990) ought not to be read as an example of 'good adult writing' (it is not). It ought to be read in the context of the joyous exuberances that whole language can bring in teaching children and as an example of what Mickelson preaches: everyone has the desire and the right to write what they deeply feel. It is thus 'good' in *this* sense. However, the detractors of whole language must not be allowed to suggest that this is what comes of disciplined, careful, considered adult writing if one believes in whole language. (Perhaps Goodman *et al.* (1990), the editors of the text in which Mickelson's work appears, should have been more careful to contextualize this writing in their editing of their text. It would be tactless of me to assume that they did not recognize that this poem is too easily readable as 'garbage' by those who do not understand the notion of contextuality in whole language.)

4. Smith (1988b) suggests that if we take Wittgenstein's (1968) notion seriously that forms of language bespeak forms of life, a shift in our image of and dwelling in language bespeaks a shift in the life we live. Thus, the shift to whole language has potentially profound political and ecological consequences.

Again, this suggests the interlacing of ecology, whole language and the interpretive disciplines. Newman (1985) claims that whole language involves a shift in our philosophy in its deepest sense – a profound shift in how we live in the world. And Heidegger (1971: 57) suggest that:

> To undergo an experience with language means to let ourselves be properly concerned with the claim of language by entering into it. If it is true that we find the proper abode

of our existence in language – whether we are aware of it or not – then an experience we undergo with language will touch the innermost nexus of our existence.

And, equally, the shift in our experience of language foretold by whole language will effect or require a shift in our lives.

5. I might add here that the litter produced by a tree can also act to *prevent* new life from crowding it out. The old growth therefore also protects itself from the 'onslaught of the new' (Arendt 1969). Ecologically speaking, if this protection of the old is done too well, the forest will not renew itself. But if the old pass too easily and too quickly, rooting soils will be lost and, again, the forest will not renew itself. It is the delicate, ongoing 'balance' between the old and the young that makes renewal possible. This is a fascinating metaphor for considering how traditions in education resist and prevent the eruption of the new and how such resistance is not necessarily a bad thing in and of itself. Such resistance becomes unregenerative only when it passes over a certain ambiguous limit, and the problem of where this limit lies cannot be decided upon in general or in principle.

6. Regarding the identification of originality and authorship, see Jardine (1992c). In some early elementary school classrooms (in Canada at least), the name of 'author' is often given away too easily and too cheaply. 'Everyone in my classroom is an author' is a common teacher's announcement that is *intended* to say that every child has the ability and right to participate in writing in the classroom and that each child possesses an 'original' voice (this point is itself debatable and debases the difficulties that authors can encounter in the struggle to 'find' such an original voice: originality is perhaps confused here with sincerity).

    As Harste *et al.* (1988) suggest, 'author' can be defined conventionally and in an exclusionary way as someone who has produced a recognized literary work for public consumption. On the other hand, however, we have the ebullience of the individual voice: 'authorship' in the sense of 'one that originates or gives existence' (Harste *et al.* 1988: 5). Each child, in speaking and writing with 'their own voice', can be considered an author, an 'originator/original'. In its extreme, any expression of such uniqueness/originality is automatically the achievement of authorship. This sense of 'author-as-original' is predominant in whole language texts and practices.

    However, giving away the name of 'author' in this way tends to suggest that *it makes no difference to be named an author* since everyone is (apparently indiscriminately) so named. Jardine (1992c) explores how authorship might be understood as making a deep difference and deep decision in one's life: the decision to be *known for what you write*. This is not a decision that all people (adults and children alike) are willing or able to make. Giving away the name 'author' too easily and too cheaply helps precipitate the slide of whole language into a form of pedocentrism.

7. It is *essential* to curb this by saying that this is not precisely *this author's* problem. What this passage displays is *our* problem, one we *all* face. We are all enamoured and allured by the tendencies that this text bespeaks. My own rather manic desire to get this paper published is itself caught up in this allure – publish more and more in better and better journals. This desire is part and parcel of our profession's relentless proliferation of information and its relentless consumptiveness of such information. I expect that we all know something of the exhaustion and bewilderment that this can produce.

    Johnson himself can be most graciously understood as bearing (and baring) some of these tendencies on our behalf and as giving us a hint of their ecological insanity and unsustainability. His text thus provides the opportunity for self-understanding and a reconsideration of the tendencies that *my own living* may unwittingly support. Consider:

    > Often, in education, a programme becomes the latest fad, and is widely implemented before being evaluated. Then, when there is disillusionment with the results, the programme is discarded, good and bad aspects both, to be replaced by another package. This disillusion appears to be inevitable, given the overselling necessary to get the widespread implementation in the first place. (Stahl 1990: 141)

    Disillusion with whole language is inevitable if it is based on the illusions expressed in Johnson's advertisement. This is not to say that there is nothing to be learned in his seminars and workshops. It is to say-as I am sure Johnson, as a real reader and real writer, understands – that living well in language requires a dedication and discipline that is not easy, quick or especially fashionable. And this lesson needs to be interlaced some-how with academe's 'publish-or-perish' fetish which, in the case of this paper, is itself an instance of precisely the unsustainable consumerism that it then goes on to critique.

8. There is a trail of interrelations here that bespeak the times. Wittgenstein (1968: 32, 36) characterizes the ambivalent, generative intertwining of language as being constituted by 'family resemblances' or 'kinships' which are not reducible either to the univocity of a single meaning (one image of gericentrism) or to the multivocity of 'everyone has their own meaning' (one image of pedocentrism). Both gericentrism and pedocentrism bespeak breakdowns in the fabric of language. We thus get a telling sequence of implications. Forms of language are understandable as forms of life and as nests of family resemblance, such that breakdowns in language (both gericentric and pedocentric) are 'family breakdowns'/community breakdowns. And it is precisely these breakdowns that whole language addresses and wishes to 'heal' in its emphases on communities of readers (Edelsky *et al.* 1991), communities of writers (Graves 1983) and communication.

References

ARENDT, H. (1969) *Crisis in Education* (New York: Penguin).

BERLIN, J. (1988) Rhetoric and ideology in the writing class. *College English*, 50 (5): 477–494.

BERRY, W. (1977) *The Unsettling of America: Culture and Agriculture* (San Francisco: Sierra Club Books).

BERRY, W. (1983) *Standing by Words: Essays* (San Francisco: North Point Press).

CALKINS, L. M. (1986) *The Art of Teaching Writing* (Portsmouth, NH: Heinemann).

CAPUTO, J. D. (1987) *Radical Hermeneutics: Repetition, Deconstruction and the Hermeneutic Project* (Bloomington: Indiana University Press).

EDELSKY, C., ALTWERGER, B. and FLORES, B. (1991) *Whole Language: What's the Difference?* (Portsmouth, NH: Heinemann).

GADAMER, H. G. (1989) *Truth and Method*, 2nd revised edition, trans, revised by J. Weinshe-imer and D. G. Marshall (New York: Crossroad).

GOODMAN, K. S., GOODMAN, Y. M. and BIRD, L. S. (1990) (eds) *The Whole Language Cata-logue* (Santa Rosa, CA: American School).

GRAVES, D. H. (1983) *Writing: Teachers and Children at Work* (Portsmouth, NH: Heinemann).

HARSTE, J. C., SHORT, K. G. and BURKE, C. (1988) *Creating Classrooms for Authors: The Read-ing-Writing Connection* (Portsmouth, NH: Heinemann).

HEIDEGGER, M. (1971) *On the Way to Language*, trans. P. D. Hertz (New York: Harper and Row).

HEIDEGGER, M. (1973) *The End of Philosophy*, trans. J. Stambaugh (New York: Harper and Row).

HILLMAN, J. (1967) Senex and puer. In J. Hillman (ed.) *Puer Papers* (Dallas: Spring Publica-tions), 3–53.

HILLMAN, J. (1983) *Healing Fiction* (Barrytown, NH: Station Hill).

JARDINE, D. W. (1990) On the humility of mathematical language. *Educational Theory*, 40(2): 181–191.

JARDINE, D.W. (1992a) 'The fecundity of the individual case': considerations of the peda-gogic heart of interpretive work. *British Journal for Philosophy and Education*, 23 (1): 51–61.

JARDINE, D.W. (1992b) *Speaking with a Boneless Tongue* (Bragg Creek, Alberta: Makyo Press).

JARDINE, D. W. (1992c) Naming children 'authors'. *Readings in Canadian Literacy*, 10 (4): 168–173.

JOHNSON, T. (1990) Announcing . . . these dynamite workshops to help you on the road to becoming a whole language pro (Advertisement). *Reading Today*, 8 (1): 32.

LEGUIN, U. K. (1968) *A Wizard of Earthsea* (New York: Penguin).

MICKELSON, N. (1990) What is a whole language teacher? In L. B. Bird, K. S. Goodman and

Y. M. Goodman (eds) *The Whole Language Catalogue* (Santa Rosa, CA: American School), 383.

NEWMAN, J. M. (1985) *Whole Language: Theory in Use* (Portsmouth, NH: Heinemann).

SMITH, D. G. (1988a) Children and the gods of war. *Journal of Educational Thought*, 22 (2A): 173–177.

SMITH, D. G. (1988b) On being critical about language: the critical theory tradition and implications for language education. *Reading–Canada–Lecture*, 6 (4): 243–248.

SNYDER, G. (1980) *The Real Work: Interviews and Talks 1966–1979* (New York: New Directions).

STAHL, S. (1990) Riding the pendulum: a rejoinder to Schickendanz, McGee and Lomax. *Review of Educational Research*, 16 (1): 141–151.

TURNER, V. (1987) Betwixt and between: the liminal period in rites of passage. In S. Foster, M. Little and L. C. Mahdi (eds) *Betwixt and Between: Patterns of Masculine and Feminine Initiation* (LaSalle, IL: Open Court), 3–22.

WEAVER, C. (1990) *Understanding Whole Language: From Principles to Practice* (Portsmouth, NH: Heinemann).

WITTGENSTEIN, L. (1968) *Philosophical Investigations*, 3rd edition, trans. G. E. M. Anscombe (Oxford: Basil Blackwell).

From epistemology to ecopolitics: renewing a paradigm for curriculum

NOEL GOUGH

In recent years much has been said and written about 'paradigms' and especially about the emergence of (or the need for) a 'new' and 'higher' paradigm that unites the knowledge claims of 'science, philosophy–psychology, and religion–mysticism . . . a type of truly unified world view' (Wilber 1983: 1). Much that has been written in this context is deeply critical of the materialistic and atomistic world view that has long been dominant in Western industrial society. Thus some writers assert that the Green political movement in Europe and elsewhere is a significant manifestation of a new paradigm (e.g. Ash 1980, Capra 1983, Spretnakand Capra 1984). This version of the new paradigm is characterized by such emphases as whole-system perspectives, ecological consciousness, feminism, transmaterialist spirituality, cultural pluralism, non-violent change, decentralization of decision-making, human-scale technology and solidarity with developing countries (Harman 1985: 319–321). Other advocates of a new paradigm are less specific about its characteristics and suggest instead that new ways of being and of understanding are immanent in scientific theories emerging from recent work in cosmology, quantum mechanics and the thermodynamics of nonequilibrium systems (Briggs and Peat 1984, Prigogine 1980, Prigogine and Stengers 1984).

These different versions of a new paradigm are compatible in many respects (Fox 1986) but they are by no means congruent; for example, those who base their vision of a new paradigm on scientific theorizing tend to be more enthusiastic about high technology than are those who embrace Green politics (Michael and Anderson 1986). Nevertheless, the cross-cultural pervasiveness of many of the common elements of new-paradigm thinking has led to the suggestion that some kind of 'global mind change' is occurring (Harman 1988), affecting developing countries as well as Western industralized nations.

Such suggestions invite speculation about the implications of a paradigm shift for education in general and for curriculum work in particular. Some of these implications appear to be obvious if one holds to the common-sense view that education is, or should be, shaped by conceptions of what one takes to be 'real' in a philosophical sense (i.c. on that which is presumed to exist independently of human imagination). According to this view, a paradigm is a metaphorical equivalent of architectural foundations: buildings are designed with certain assumptions about the underlying clay, sand or rock in mind and, in much the same way, social institutions – like education – are built upon understandings of reality, nature and human nature that are taken for granted in our culture. If the soundness of these understandings is challenged, then all that has been founded upon them may be called into question. Thus, for example, the feminist critique of patriarchy has quite properly provoked an educational response that has gone beyond superficial issues of curriculum content to address deeper and more complex structural problems of access and equity. Less attention has been given so far to the misrepresentations of physical reality which underlie many educational orthodoxies, though both Emery (1981) and Doll (1986, 1988) have described some of the deleterious consequences of founding education on assumptions about reality that have been derived from Newton's physics.

But the implications of a paradigm shift for education go beyond the expectation that educational systems and programmes should reflect changing conceptions of reality, nature and human nature. Social institutions like education also arise from humans imagining that which is non-real, supernatural and transcendental. Thus paradigms are also metaphorically equivalent to myths – stories that embed individual experiences in a larger framework of shared values, meanings and purposes and that persist in a culture over relatively long periods of time. The urge to tell our children new stories, such as 'environmental education' and 'peace studies', instead of (or in addition to) some of the older stories, such as 'science' or 'history', may be a superficial manifestation of deeper and more subtle changes in underlying cultural myths. This mythic sense of paradigms is more inclusive than the sense of paradigms as foundations. Conceptions of 'reality', for example, can more readily be subsumed within myths than vice versa, especially if one accepts the non-realistic image of the universe provided by quantum mechanics (wherein it is claimed that consciousness is a necessary – and perhaps sufficient – condition for the existence of the universe; see McCusker and McCusker 1988: 78). More importantly, the mythic sense of paradigms emphasizes their transtemporal qualities. That is to say, the mere demonstration of deficiencies in the now dominant world view should not impel us to embark – as some of my colleagues would have it – on 'the quest for a new educational paradigm' (Dufty and Dufty 1988). Before considering such a quest, it might be prudent to explore the possibility that any paradigm shift that is occurring or may be desired involves not a 'new' paradigm, but rather the renewal of existing (albeit suppressed) myths.

My personal position is this: I have little sympathy with the materialistic and atomistic world view which still dominates formal education in Western society; I see great virtue in Green politics and many challenging ideas emerging from the new frontiers of science; but I have little enthusiasm for any 'quest' for a 'new' educational paradigm. In part, this lack of enthusiasm reflects my antipathy towards the quasi-religious zealotry of such a quest: 'The new paradigm story is a postmodern version of ancient millennarian cults that predicted the imminent coming of a new order, a paradise on Earth' (Michael and Anderson 1986: 119). I simply cannot reconcile much of the rhetoric of new-paradigm thinking with the kind of curriculum work that I want to do now and in future (for example, I am suspicious of the quality of life after quests: what does one actually *do* after one has *found* the Holy Grail?). Certainly, this sort of rhetoric is too pretentious for the kind of work I am doing here, now. I am not attempting to write a chapter in one of the Great Books, I am writing a work-in-progress report – a short story – with the intention of engaging you, the reader, in a further exploration of the world it signifies: 'What we need is not great works but playful ones. . . . A story is a game someone has played so you can play it too' (Sukenick 1969, quoted in Waugh 1984: 34).

What follows, then, is, in essence, a story about a story: ecopolitics could very well be the name of the game I play in my everyday work as a curriculum scholar and teacher educator. This is a story about why I have come to play it that way.[1]

Ecopolitics and education 1: ecological theories of perception

Recent developments in the study of human perception have provided the basis for questioning many of the theoretic foundations of Western industrial society's educational orthodoxies. Emery (1981) provides a very useful synthesis of these developments and their implications.

Emery suggests that many of the most entrenched educational practices in our society can be traced to the empiricist theories of perception and knowledge advanced by several eighteenth-century philosophers, notably Locke, Berkeley and Hume. However, their arguments remain reasonable only if we agree that the world is as Newton depicted it and that the transfer of information from an object to a viewer obeys Euclid's geometry. According to this view, light reflected from an object to the retina yields only a 'chaotic two-dimensional representation of reality . . . any useful knowledge of a three-dimensional world (such as stops one falling off cliffs) would have to come from some sort of intellectual inference' (Emery 1981: 2). Locke, Berkeley and Hume 'proved' that in a Newtonian world, based on Euclidean space, individuals could have no sure knowledge of a world outside them — that stimuli could yield no direct and immediate information about a three-dimensional world of solid, persistent objects and causal relations. To cut Emery's long story short, Herbart spelt out what this implied for educational practice, and the further refinements made by behavioural psychologists like Pavlov, Thorndike, Hull and Skinner allowed Lockean theories 'to be preserved in the face of Darwinian challenges as to how such incompetent perceptual systems could have had survival value' (Emery 1981: 3). As a consequence, educational practice since the onrush of positivist science has not valued an individual's perceptions as a source of knowledge. The meaning of perceptions is held to emerge from intellectual processes of analytic abstraction and logical inference (hence the now taken-for-granted separation of perception from cognition) and the prime task of education is to distribute the socially validated knowledge that has been so gained. Learning has thus come to be seen as a process of guided induction into bodies of organized propositional knowledge, in the workings of formal logic and in the skills of textual expression and comprehension (through which organized propositional knowledge is accumulated and accessed).

It should be clear that any educational theories that derive from eighteenth-century conceptions of the physical universe should be treated with considerable caution. In the early part of this century Newton's physics and Euclid's geometry were displaced by quantum mechanics and its attendant probabilistic mathematics. It is not an idle question to ask: what if human perceptual organs are geared to the kind of time–space continuum envisaged by Einstein rather than to Euclidean space? It seems reasonable to assume that our perceptual systems have evolved so as to be adapted to the universe *as it is* rather than to an approximate (and perhaps distorted) social construction of it, yet learning in formal settings is geared to the materialistic, deterministic, atomistic, reductionist and objective vision of the universe that has been rendered unsupportable by quantum mechanics. Indeed, it would seem that formal education teaches us to distrust our own perceptual systems rather than to exercise the perceptual skills bequeathed to us by natural selection. For example, Piaget and Inhelder (1956) observed that the pre-school child's concept of space is topological but that by the age of 12 it is Euclidean.

A direct challenge to neo-Lockean theories of perception comes from the work of Fritz Heider and others whose research questions the assumption that the meaning of perceptions (such as the perception of order) can only arise from intellectual cogitation. Heider's papers were written in Berlin during the late 1920s (which, perhaps coincidentally, was also the time at which Schrödinger, Heisenberg, Bohr, Pauli and others were formulating quantum mechanics) but were not translated into English until 1959, by which time his research had been paralleled by J. J. Gibson. Heider (1959) and Gibson (1979) demonstrate the plausibility of what the latter calls an ecological approach to perception. Their research suggests that the environment has an informational structure at the level of objects and their causal interactions and that human perceptual systems have evolved to detect and extract that information. Other researchers in human perception have drawn similar conclusions:

> there is ample evidence that the senses are not only genetically preattuned but become more sensitively calibrated to pick up those exigencies of the environment that bear directly on the survival, success and well-being of the perceiver – what has sometimes been called the 'education of attention' (Shaw and Pittenger 1977: 107)

Neo-Lockean theories of perception have led us to believe that 'real knowledge is locked up in the storehouses of knowledge that are so jealously guarded by a priesthood of scholars and scientists' (Emery 1981: 7) and that the best way to gain access to that knowledge is through years of schooling in the disciplines that have been our means of organizing the contents of these 'storehouses'. Ecological theories of perception suggest that limitless information is present in our personal, social and physical environments and that with an 'education of attention' we can access as much of it as we need: 'It is an education in *searching* with our own perceptual systems not an education in how to someday *research* in the accumulated pile of so-called social knowledge' (Emery 1981: 7). Learning to attend to the informational structure of environments seems to be very similar to what Edward de Bono calls 'generative thinking'. From his work with young children and adults de Bono concludes that generative thinking about our environment and our place in it is a matter of perception, of seeing things in context, rather than a matter of puzzling over abstractions in our minds: 'The teaching of thinking is not the teaching of logic but the teaching of perception' (de Bono 1979: 77).

Ecological theories of perception liberate teachers from being, as it were, tourguides in 'the accumulated pile of so-called social knowledge'. The 'education of attention' means guiding learners in the many and various ways of enhancing their capabilities for extracting information from their environments. Emery (1981: 15) suggests that this involves 'recentering' teaching by shifting the focus of a teacher's activity from the teacher–learner relationship to the interrelations between learners and environments. That is, learners' own perceptions of their environments are often disregarded by teachers, who see such perceptions as distractions from the transmission of socially validated knowledge – a process within which the teacher's authority is central. Ecological theories of perception suggest that teaching which is centred on the teacher–learner relationship may inhibit learning because learners will be distracted, by teachers, from attending to what is before them in their environments.

Emery argues that the educational ramifications of the breakdown of eighteenth-century empiricist theories of perception are such as to warrant thinking in terms of a paradigm shift and he uses the term 'ecological paradigm' to distinguish a 'new' and 'emerging' paradigm from the one he refers to as 'old' and 'traditional'. Some of these terms are a little misleading because the educational paradigm rejected by Emery has dominated merely the last two centuries of Western industrial civilization, which, in relative terms, seems too recent a period (and too limited a location) of human history and culture to be called 'old' or 'traditional'. Terms like 'old', 'new', 'traditional' and 'emerging' may be useful in identifying changing fashions in education but they are unlikely to reflect deeper continuities or more enduring positions.

It may be more meaningful to refer to a shift away from an 'epistemological' paradigm. I prefer this term for two reasons. First, as commonly used in education, 'epistemology' means the origins and method of knowledge, and the core assumptions of Western industrial society's systems of education rest on an epistemology – on a particular set of theories about how humans gain knowledge of themselves and their world. Secondly, I take 'epistemological' to refer to the *kinds* of knowledge that are valued most highly in the paradigm. The ancient Greek word '*episteme*' referred to 'theoretic' knowledge and the kinds of knowledge which have dominated Western education for two centuries are those which have been structured by Western society's pervasive form of theorizing, namely positivist empirical science.

Ecological theories of perception provide some compelling reasons for rethinking educational paradigms. However, the political reality is that Heider's and Gibson's conclusions are unlikely to convince most educators, parents and employers that an epistemological paradigm is a baseless myth or a deeply flawed foundation upon which to build educational systems and programmes. Existing systems of mass education may have been founded on spurious theories of perception and knowledge, but they are supported now by entrenched social interests and powerful elites. Furthermore, those of us who have already been inducted into the 'priesthood of scholars and scientists' are unlikely to turn our backs on the storehouses of theoretic knowledge with which we are so familiar and it would thus lack conviction for us to encourage our students to do so. But ecological theories of perception provide a context for these storehouses which should enable us to see them as a *part* (but by no means the whole) of the personal, social and physical environments that the 'education of attention' allows us to search.

Ecopolitics and education 2: transtemporal connections

Empirical evidence in support of the view that a 'new' paradigm is emerging includes survey data (cited by Harman 1985) which confirms that there has been a recent strengthening of 'inner-directed' values (ecological, humane, spiritual) in Western industralized countries, together with a deeper and more subtle shift in beliefs 'away from the confident scientific materialism of the earlier part of this century' (Harman 1985: 325). As part of his examination of the global significance of Green politics, Harman notes that a parallel shift in developing countries is also away from Western materialism and towards a reassertion of native cultural values and beliefs: 'The change in both cases is fundamentally a shift in our attitude toward our inner, subjective experience, affirming its importance and its validity' (*ibid.:* 325).

The cross-cultural attitude change to which Harman refers may represent a departure from the norms of the recent past but the strengthened beliefs are no novelty, especially when one considers the longer-term history of Western culture and education prior to the scientific revolution. For example, in the Aristotelian scholastic curriculum which predominated in Europe until the eighteenth century, no strong distinction was made between matters of fact and matters of value (Reid 1981). The ideal of scientific detachment, or of any attempt to eliminate human values from supposedly 'objective' world views, was foreign to this scholarly tradition, regardless of whether one was studying nature, human nature or the supernatural. Thus the recent strengthening of beliefs in the value of inner, subjective experiences may not so much be evidence of a 'new' paradigm, but rather can be seen as the re-emergence of a deeper continuity in our culture.

Prior to the scientific and industrial revolutions, the disciplines of the medieval scholastic curriculum were conceived as practical arts rather than as theoretic 'sciences'. That is, the purposes of studying literature, religion, natural history or social history were essentially similar: to help resolve the practical problems faced by humans when their desires fail to match their circumstances. These disciplines focused on the interrelationships between human moral purposes and the personal, social and physical environments in which they were seen to be situated. The goal of scholarship in these disciplines was therefore practical, that is 'to perform good works', rather than theoretic, that is to discover or demonstrate some final good or universal truth (McKeon 1977: 208). This goal changed under the influence of 'scientific method' and many of the humane disciplines were reconceived as social 'sciences'. These sorts of distinction can be traced back to Aristotle's conceptions of *episteme* (theoretic knowledge or 'knowing that. . .'), *phronesis* (practical judgement or 'knowing I/we should. . .') and *techne* (technical knowledge or 'know-how'). But, while these concepts can readily be distinguished from one another for scholarly purposes, it does not necessarily follow that scholarship itself should be organized around the principle of their separability. It may be significant that Western culture has expanded specialist studies of 'objective truth' and 'know-how' to such a degree that terms like 'epistemology' and 'technology' have become generic while 'praxiology' has not; practical judgement is often assumed to be no more than an 'application' of science or technology. The increasing popularity of the term 'praxis' (which connotes practical *action* rather than a *logos* – a subject of study) improves this situation to some degree, in spite of its appropriation by neo-Marxist scholars (see Gough 1988b).

It is worth noting that Aristotle used the same words to describe what we would now call 'practical' as he did to refer to what we would now call 'political' (see Lobkowicz 1967). The common ancestry of these terms points to some complementarities between contemporary ecopolitics and the relatively recent revival of interest in a neo-Aristotelian conception of 'practical' curriculum study, initially explicated in Joseph Schwab's seminal series of papers on 'The practical' (1969, 1971, 1973), and refined further by such scholars as William Reid (1981) and Maurice Holt (1987) under the name of the 'deliberative' approach to the study of curriculum. These scholars assert that curriculum problems are in essence practical rather than theoretic and I would add that such problems can also be characterized as being 'ecopolitical'. That is, curriculum problems can only be resolved in the light of complex human–environment interrelationships which must be treated holistically rather than analytically and which necessarily involve subjectivity rather than just the 'objective' methods of the sciences and technologies.

Thus I think it can be demonstrated that much of what is being referred to as a 'new' paradigm for education is a confluence of concerns to make education more practical, more ecological and more ecopolitical (and note that in this context these terms can be taken to have both their everyday meanings and the kinds of specialist meaning that are alluded to elsewhere in this paper: 'practical' has both its connotations of 'hands-on learning' and its neo-Aristotelian meanings; 'ecological' refers to the study of the interrelationships among organisms and to the theories of perception discussed by Emery, above, and so on). This confluence puts any paradigm shift which may be occurring into a longer-term historical perspective.[2] If an emerging ecopolitical paradigm for education can legitimately be conceived (at least to some extent) as a renewal of a scholarly tradition which spans more than two thousand years, then the domination of education by an epistemological paradigm for a mere two centuries may best be seen as a relatively recent aberration.

There is, however, at least one difficulty with such a view. Some of the most valuable contributions to contemporary Green politics have been made by feminist scholars, yet they may be among the least likely to agree that the kind of scholarship, or the kind of education, suggested by ecopolitics has (or needs to have) some of its historical roots in ancient Greece. For example, in a review of Maurice Holt's *Judgment, Planning and Educational Change*, Pagano (1988) commends and defends many of Holt's arguments and conclusions (which, as she observes, are situated firmly in an Aristotelian moral universe), with the following reservation:

Because of the depth of my agreement with Holt. . . I wish he had left Aristotle to the Greeks, and I must challenge his conclusion that educational practice and curriculum planning built on the foundations of a liberal arts education (at least as such is presently conceived) will achieve our goals.

Aristotle's notion of virtue belongs to the noble *man*. Virtue is the realization of the *telos* of *man* – not women, not children, not slaves, all of whom are imperfect. . . 'Man' can never include others. Nor can a liberal education dedicated to persistent questions of *mankind* tell a story which will empower others. . .

As I interpret narratives of education, I read texts in which only man is present, in which it is against his needs and desires that all is to be judged. That was Aristotle's project. . . It is the foundation of liberal education. (Pagano 1988: 288)

Pagano is absolutely right: the story of liberal education *as it is presently conceived* is a story of exclusion. But it is also a form of exclusion to leave Aristotle to the Greeks. Transtemporal discourse may be difficult but so too is cross-cultural conversation and both are worth the effort. In any case, as Pagano's review illustrates, it is a bit difficult to reach agreement with Holt without including Aristotle (as he appears in Holt's story) in the conversation. With two thousand years of hindsight we can see that Aristotle's concepts of 'man' and the *polis* are too narrow, but his assertion of their interrelationships retains its wisdom, namely that man is the political animal who becomes what he is capable of becoming in the context of the *polis*. We now know that we must expand 'man' to 'human' and the *polis* to the *ecopolis* – to the larger context of the evolving biosphere. Thus Aristotelian assumptions can be transcended and, similarly, a liberal education needs to be reconceived in the light of contemporary contributions to *episteme* (such as ecological theories of perception) and contemporary forms of *praxis* (which is what I understand feminism to be).

Ecopolitics and education 3: the re-emergence of more practical and more holistic subjects for study

A shift from an epistemological world view in education towards one which is more ecopolitical is manifested, albeit superficially, in the plethora of new subject matters which have competed for inclusion in school curricula in recent years, such as computer studies, consumer education, development education, environmental education, health education, legal studies, media studies, outdoor education, peace studies, personal development, STS (science, technology and society), traffic safety education and women's studies, to name but a few.[3] While these subject matters have developed in idiosyncratic ways some commonalities among them can be discerned. The impetus for each subject's development usually has included perceptions of deficiencies in curricula based on what are currently regarded as the traditional academic disciplines, deficiencies which lead to the subject matter in question being ignored, undervalued or distorted. The most common criticisms are of curricula which are 'too theoretic' or 'too compartmentalized' and, implicitly, these are also criticisms of what I have called an epistemological paradigm – an atomistic world view which has abstracted reality, nature and human nature into 'a world composed of separate entities such as atoms, individuals, academic departments, corporations, cities and nations' (Michael and Anderson 1986: 115).

Thus these new subject matters are generally considered by their developers to be more 'practical' and more holistic than the studies they are intended to complement, supplement or replace in school curricula. For the most part, these subjects are 'practical' only in somewhat limited senses of the term: some are so deemed because their subject matters are thought to be 'useful' or 'relevant' (e.g. health education, legal studies), others because their characteristic learning experiences are 'experiential' or 'activity-based' (e.g. computer studies, outdoor education). Many of these subjects teach the skills for achieving goals which are taken for granted in terms of their desirability and, therefore, are technical rather than practical in the neo-Aristotelian sense of being oriented to 'political life' or 'doing good works'. The significance of this difference is alluded to in the following comment on computer studies, which was made, however, in relation to peace studies:

> there is enormous propaganda about the positive effects of computer technology, and computer companies have spent millions of dollars involving schools in their activities. So computing courses proliferate, with 'hands on' experience with your 'friendly' and 'personalised' computers. . . But you will find little analysis of the more sinister effects [of] large political–industrial computerised human systems. . . making decisions that affect not only the everyday personal lives of individuals, but the survival of communities, and ultimately of the whole planet. (Wilson 1987: 13)

Some of these 'sinister effects' include economic and political oppression in Third World countries, the contribution that 'over-developed' countries like Australia make to world poverty, and the economic and political reasons for the arms race and the arms trade (Wilson 1987: 13). Wilson's comments illustrate that at least some of the new subject matters (e.g. peace studies and development education) are conceived in terms which reflect a shift from a world view based on a theoretic and atomistic epistemology towards the more practical and holistic perspective of ecopolitics. I suspect that some educators would stereotype the leap in Wilson's reasoning (from 'hands on' computer courses to global politics) as an alarmist – and probably Marxist – exaggeration of the extent to which sinister ideological ghosts lurk in innocent machines. I prefer to think of such leaps as examples of the holistic emphases that are characteristic of the changing world view: 'The most striking feature of the postmodern world is its systemic character, its astounding proliferation of linkages among once-separate cultures, governments, economies and ecosystems. . . everything is connected to everything' (Michael and Anderson 1986: 115).

Holistic emphases are conspicuous by their absence in conventional schools, which reinforce an atomistic world view through virtually every aspect of their design, construction and modes of operation. Age-graded children study separate subjects with specialist teachers in specialized classrooms, with their activities synchronized to a timetable which both symbolizes and brings about a fragmented world view. One of the very few countervailing tendencies is the increasing recognition of the value of 'interdisciplinary' or 'integrated' studies, which is often part of the rhetoric supporting the introduction of the new subject matters discussed above. Such studies are consistent with a shift towards a more holistic world view because, by focusing on interconnections between what were once studied as separate entities, they can be seen to have taken some form of ecological understanding as their subject matter. The STS movement in science education is a case in point: part of the rationale for STS is that students should be able to study the ways in which, say, the properties of matter and the workings of machines and human societies are interrelated, rather than studying each of them separately in subjects like chemistry, physics and history. Many of these new subject matters can fairly be described as 'ecological' even though they may not have the attributes which are stereotypically associated with this term. For example, environmental education and media studies can equally well be regarded as studies in human ecology, which the *Oxford English Dictionary* defines as a 'study of interaction of persons with their environment'. Environmental education and media studies are similar in that each attempts to increase the learner's understanding of human interactions with *some* of our environments. They are different only in so far as environmental education focuses chiefly on human interactions with the 'natural' environment whereas media studies is more concerned with the environment created by some of humankind's most pervasive inventions – the texts and technologies of the mass media.

I began this section by suggesting that the emergence of a number of new subjects for study can be seen as a superficial manifestation of the renewal of an ecopolitical paradigm for education: some are superficially practical (political), some are superficially holistic (ecological) and some may be attempting to be both. All of these subjects are products of both the older and the emerging (or re-emerging) world views and thus each reflects to some extent the contradictions and conflicts that accompany a major paradigm shift. For example, referring back to Harman's characterization of the 'global mind change' exemplified by the Green political movement, it is my experience that most environmental educators claim to hold 'ecological' and 'humane' values but that many are suspicious of 'spiritual' values and cling to the 'confident scientific materialism' of the recent past. This is particularly evident in formal courses of 'environmental science' or 'environmental studies', which usually embody, in uncritical ways, assumptions about the value of 'scientific method' in resolving practical problems of, say, natural-resource management or the assessment of new technologies.

Another contradiction is that many of the new studies preach a holistic perspective but still present themselves as separate entities and, worse, many have preserved the teaching practices and learning experiences that go with a fragmented world view. To use again the example of environmental education, my experience of it in schools is that it strongly resembles conventional science education in being dominated by the authority of teachers, textbooks and timetables and by the trivial pursuits of memorizing information and routinely performing technical tasks.

Nevertheless, despite these contradictions, the emergence of more practical and more holistic subject matters may be crude indicators of changes in the cultural norms, assumptions, values and myths underlying curriculum work. These changes are part of the 'deep ecopolitics'[4] of a paradigm shift in so far as they can be seen to be products of our pursuit of better understandings of the interrelationships that ought to prevail among humans and their environments and of the struggle to reconstruct our world views in the light of such understandings.

Ecopolitics and education 4: exemplary practices

There are several examples of well-developed educational practices which seem to be consistent with an ecopolitical paradigm but which also seem to have arisen more or less independently of each other and of the theoretic warrant for their effectiveness. These include de Bono's approaches to the teaching of thinking (e.g. de Bono 1979), an approach known as 'structural arithmetic' (Stern and Stern 1971) and a number of approaches to reading and writing (e.g. Gibson and Levin 1975, Hughes 1971), each of which is cited by Emery as an example of an educational application of ecological theories of perception.

My own inquiries have been concerned with the educational programmes and practices developed by the Institute for Earth Education, an international organization of volunteers committed to what would now be called environmental education. Earth Education originated in camp nature programmes in the USA and, according to one of its founders, was 'created partially out of frustration with the usual identifying–collecting–dissecting–testing approaches to nature' and to help learners 'build a sense of relationship – through both feeling and understandings – with the natural world' (Van Matre 1979: 5). Earth Education emphasizes direct sensory experience and the sharpening of learners' perceptions of their environments: 'Our aim is to help young people interact more directly with the fascinating array of living things around them' (Van Matre 1979: 6–7).

It will be apparent from the above quotations that the rhetoric of Earth Education suggests that it has many similarities with the 'education of attention' that follows from ecological theories of perception. Thus it is somewhat disappointing to find that the literature of Earth Education represents learning in ways which are more consistent with an epistemological paradigm. For example, many Earth Education teaching techniques are described as though they exemplified the principles of Skinnerian behaviourism (indeed, there is frequent reference to the 'mechanics' of learning). But I think it can be demonstrated that the effectiveness of the teaching techniques used in Earth Education is better explained by what Emery calls the 'recentering' of teaching on the learner–environment relationship, as is evident from their similarities with de Bono's techniques for the teaching of thinking. It is particularly significant that both approaches are characterized by the imaginative use of tools.

Table 1. Comparison of conventional schooling and Earth Education

	Concepts	Senses
Conventional schooling (Epistemological)	Differentiated (Focus is on smaller and smaller bits of knowledge)	Undifferentiated (Emphasis is more on thinking than perceiving)
Earth Education (Ecopolitical)	Undifferentiated (Emphasis is on the big picture)	Differentiated (Focus is on sharpening individual senses)

Source: Adapted from Van Matre 1979: 8.

De Bono provides learners with tools which deliberately block (or at least hinder) perceptual habits, such as taking cursory samples of sensory data, making snap judgements about these data and retreating rapidly into mental abstraction, classification and generalization. De Bono's tools are designed to prevent learners from making easy slips into perceptual error – they are reminders to look again. These tools encourage learners to attend to their own habits of perception and help them to sustain the perceptual work by which they can gain information directly from their environments – information that cannot be gained from analytical abstraction and logical inference (Emery 1981: 7). Thus de Bono's tools can be thought of as instruments of metaperception.[5]

The tools used to sharpen sensory awareness of natural environments in Earth Education programmes range from such simple devices as blindfolds and mirrors to more elaborate props and gimmicks, including the mental 'tools' of role play, theatre and fantasy. But in each instance of their use these tools function in much the same way as de Bono's, that is they sustain perceptual work as distinct from allowing the learner to retreat into abstraction. However, the heightening of learners' perceptual discriminations of natural environments is not the sole project of Earth Education; it also sets out to improve learners' understandings of a conceptual 'environment', specifically the small number of big ideas which currently encapsulate our theoretic knowledge of natural ecosystems (viz., energy flow, cycles, interrelationships and change). These key concepts are not taught by processes (nor presented as products) of analytic abstraction and logical inference. Rather, they are treated and used as further tools for perceiving and searching natural environments.

The Institute for Earth Education offers a critique of conventional schooling which parallels Emery's critique of education based on the assumptions of eighteenth-century epistemology. Van Matre asserts that conventional schooling 'differentiates conceptual learning and generalizes sensory awareness'; learners do not study their world as a whole but only progressively smaller bits of it in the form of generalizations, propositions, definitions and facts:

> Youngsters are encouraged to view the world as an infinite set of boxes – there are always smaller boxes inside the box before them. They study the world by studying smaller and smaller boxes . . . they gain knowledge by

adding to their stock of facts day by day. The senses, however, are lumped together in one homogenized mass. Kids are encouraged to believe that they should not trust their own perception and thus should do little to sharpen their senses. They are told to be objective, disregard emotions and feelings, and experiment to find the truth. Even when they 'take a look' at a problem, they rarely *see* anything; they merely talk it to death in the classroom (Van Matre 1979: 8)

In Earth Education, this approach is reversed: 'We differentiate in our sensory awareness and generalize in our conceptual understanding. We strive to strengthen individual senses, but opt for the big picture in understanding life' (Van Matre 1979: 8). The Earth Education approach, and its relationship to the paradigm shift I am attempting to characterize here, is summarized in table 1.

Earth Education programmes show that it is possible to create conditions for learning which encourage the simultaneous development of a holistic conceptual understanding *and* a highly differentiated sensory awareness of the learner's environments. It has also been my experience that, in the *social* conditions for learning they create, Earth Education programmes illustrate many of the virtues (and some of the vices) of the contemporary Green political movement. The main task of teaching in Earth Education is to contrive or to provide the materials, tools and settings which enable learners to search their environments with their own perceptual systems. The result is that teachers *and* learners tend to 'share and do' rather than to 'show and tell'. However, it is also patently obvious that these democratic conditions are usually achieved through very artful contrivance and a willingness to manipulate learners towards behaviours which are consistent with Earth Educators' moral purposes. The assertiveness with which Earth Educators display their sense of moral superiority can be irksome even to those of us who believe that their moral convictions are defensible.

This reservation notwithstanding, the approach to developing both conceptual understanding and perceptual discrimination that is exemplified by Earth Education, and generalized in table 1, provides a useful model of how practical bridges can be built between conventional schooling and practices more characteristic of an ecopolitical paradigm. This model does not suggest that either we or our learners should turn our backs on what Emery calls 'the storehouses of [theoretic] knowledge' (see above) but it does require us to be discriminating in making judgements about what we consider to be important in these storehouses. For example, my attention was drawn recently to a study guide for high school students called *Main Points in Chemistry* (Cook 1985). The author considers that students often have difficulty in retrieving the 'fundamental' ideas in chemistry because these ideas are 'lost in the narrative text' of lengthy school chemistry textbooks. Cook's approach, therefore, is to list what he believes to be the 'crucial points' in a 'simple, easily understood and useful fashion' (Cook 1985: v). The result is a short book (less than a hundred pages) that lists nearly 900 'facts, laws, rules, memory aids and ideas', presented in such a way that no discrimination is made between propositions and stipulative definitions, or between minutiae and key organizing concepts. The only discernible criteria for including items in this agglomeration seem to be their frequency of occurrence in conventional school chemistry textbooks and the ease with which they can be stated (and thus memorized and restated) without resort to compound sentences. By divorcing his 'main points' from any form of narrative (even the reductionist narratives of conventional school textbooks) Cook fails to acknowledge that ideas may be 'fundamental' or 'crucial' by virtue of their history and/or their hierarchical relationships with other concepts. Cook's approach has been used to produce equivalent books of 'main points' in physics and biology (the latter has around 1200 entries) and provides a kind of *reductio ad absurdum* of atomism in education. Such an approach to determining and representing what is important in a discipline – to providing, as it were, 'signposts' within our storehouses of socially constructed knowledge – is in sharp contrast to Earth Education, in which it was initially suggested that there were seven 'big ideas' in ecology – and these have since been reduced to four. As already noted, this is because Earth Education opts for the big picture in understanding life:

> The minutiae of life's workings are not of foremost importance for us; our goal is not pulling apart the insides of a frog, but understanding the frog inside the pond and the pond inside the water cycle. . .
>
> This does not mean that . . . we think the 'small picture' of life is unimportant, but only that such study should be self-motivated and should follow the individual's grasp of the big picture. (Van Matre 1979: 8)

In some Earth Education activities these 'big ideas' are used figuratively as 'keys' to the storehouse of socially constructed knowledge that we call 'ecology' and, after some initial practice in their use, learners are encouraged to use the keys in a self-motivated way. These key concepts are given further meaning by their use as tools for sustaining learners' peceptual work in natural environments. For example, in the environs of an Earth Education campsite, it would be most unusual to find trees or shrubs bearing conventional labels showing their botanical names or any other specific information about them. It would be more likely for one to find a piece of plumber's pipe bearing the word 'cycles' lashed to a tree; when curious observers look through the pipe, they could find themselves attending to what might otherwise have been a rather unobtrusive fungus decomposing a fallen branch. It is also more than likely that the pipe would have been placed there by a child rather than an Earth Education leader.

I have used the model illustrated in table 1 to design learning activities in a variety of pre-service and in-service teacher education programmes concerned with curriculum development and administration, evaluation in education, futures study in education, and so on. My experience satisfies me that strategies based on the model are effective in developing learners' perceptual skills and conceptual understandings regardless of the environments being searched. The strategies are as effective at developing perceptual discriminations in social environments, such as perceiving subtleties in the institutional arrangements through which schools administer the assessment of students' achievements, as they are in developing sensory awareness of biological diversity in natural environments.

In encouraging teachers to adopt ecopolitical practices, I have found Garth Boomer's notion of teaching as a kind of 'bushcraft' to be a useful and appropriate metaphor:

> In the ecology of the school 'bush' there is a bewildering array of texts, tests, assignments and artefacts. The teacher should be used to finding interesting and pertinent specimens and talking about their characteristics, habits and habitats. Students should be encouraged to familiarise themselves with funny creatures like science textbooks, learning how to tame them, remembering where dangers lurk . . .
>   Teachers should not drive students in a tourist bus through the school curriculum, encouraging the bland recital of tourist blurbs. Students should be obliged to savour the texture of life, wild and rich. (Boomer 1982: 119)

I would add that educational 'bushcraft' in socially constructed environments needs to be a form of *praxis* rather than mere 'craft' and, like more familiar forms of 'bushcraft', it is most likely to be learned from personal experience and by apprenticeship to someone who models it, such as a teacher who communicates his/her perceptual discriminations and conceptual understandings with genuine enthusiasm and flair. Part of the significance of the Green political movement for education may be that it supplies practitioners with some of the moral convictions that make such enthusiasms credible – and, perhaps, contagious.

For example, profound insights into the nature of socially constructed environments have emerged from the feminist critique of human history and culture. Feminist scholarship has helped us to perceive and understand the global and local destructiveness of patriarchy in its myriad forms – in industrialism, militarism, the exploitation of developing countries by transnational corporations, and the gross undervaluing of women's work in sustaining families, communities, societies and ecosystems. Thus feminism has functioned as a conceptual tool that has assisted us in searching personal and social environments and has done a great deal to aid the deconstruction and reconstruction of educational myths; it has helped us to perceive the ways in which the education systems in which we practice, and of which we are ourselves products, provide structured misrepresentations of reality, nature and human nature.

From epistemology to ecopolitics: a summary

The educational practices and experiences which characterize epistemological and ecopolitical paradigms for education are summarized and compared in table 2. It will be apparent to readers who are familiar with education in Australia that many of the practices and experiences listed as being 'ecopolitical' are supported by much of the Australian rhetoric of curriculum reform during the past two decades. Thus, for example, moves towards more school-based curriculum development, greater community involvement in school decision-making, increasingly non-competitive and co-operative assessment practices, more issues-based and inquiry-based learning, and so on, all seem to be consistent with an ecopolitical paradigm. However, this

Table 2. **Comparison of epistemological and ecopolitical paradigms**

	Epistemological	Ecopolitical
View of learning	Practice in cognition	Practice in perception
Purposes of learning	Transmission of existing knowledge Abstraction of generic concepts	Perception of invariants and discovery of serial concepts and generic concepts in learners' environments
Control of learning	Asymmetrical dependence (of learners on teachers) Competition between learners	Symmetrical dependence (all participants are co-learners) Co-operation between learners
Co-ordination of learning settings timing	Schools/classrooms Age-grading, school calendar and class timetable	Community settings Synchronized to, and negotiated with, community settings
View of knowledge	Socially structured (theoretic, technical)	Individually structured (practical, personal)
Learning materials	Textbooks and standardized procedures (e.g. laboratory exercises)	Reality-centred projects
Learning activities	Paying attention Rote performance Memorizing	Discrimination Differentiation Searching Creating
Teaching activities	Distribution of structured knowledge Lecturing and demonstrating	Creating and re-creating learning settings and tools which sustain learners' perceptual work

Source: Parts of this table are adapted from Emery 1981: 15.

consistency may be more illusory than real since each of these reforms may affect procedures and practices in a superficial way without altering deeper (and often tacit) purposes and values.

As already noted, the core of Western industrial society's educational world view lies in assumptions about how people gain *theoretic* knowledge. Many educators who pay lip-service to the practices and experiences which exemplify an ecopolitical paradigm, and many of those who attempt to make such practices and experiences a reality in schools, have not abandoned these assumptions. Thus they fail to achieve any significant change in the purposes of learning, and the disposition of the learner, because they are still orienting themselves and their learners towards the storehouses of theoretic knowledge. That is, practices which appear to be consistent with an ecopolitical paradigm can be used simply to provide a more attractive route to achieving the objects typical of an epistemological paradigm. The incorporation of 'reality-centred projects', 'community settings' and the 'co-operation of learners' into many educational programmes does not necessarily serve an 'education of attention', but rather may merely make the transmission of existing theoretic knowledge seem more palatable.

In conclusion, I wish to reaffirm that my convictions concerning the virtues and wisdom of bringing an ecopolitical perspective to curriculum work do not lead me to reject epistemological perspectives in their entirety nor to accept the polarization of world views implicit in much of the 'new paradigm' literature. However, I have no doubts about the relative importance of ecopolitics and epistemology in shaping my own practice as a curriculum scholar and teacher educator. To revert to metaphors employed earlier in this narrative, I find that an ecopolitical world view provides me with more interesting stories to tell (and hear), and more 'playful' work in which to engage. My students and I are more adventurous and, while the outcomes of learning are less predictable, I have become increasingly confident of being pleasantly surprised by our achievements.

Notes

1. Earlier versions of this story can be found in Gough (1987b, 1987c).
2. The 'longer-term . . . perspective' to which I refer also has a future dimension, though it is beyond the scope of this paper to explore it. The conceptual territory of futures in education is explored in considerable detail by Slaughter (1988). The implications of futures study for curriculum inquiry and curriculum design are outlined in Gough (1987a, 1988a).
3. The order in which these subjects are listed is alphabetical and the selection encompasses studies which have been legitimated in Australia's education systems by, for example, their representation in the curriculum policies and guidelines issued by Ministries of Education, their incorporation into Higher School Certificate and tertiary entrance courses, the development of specialist pre-service and/or in-service teacher education programmes supporting these studies, and the formation of relevant subject teacher associations.
4. 'Deep ecology' is a term used by certain ecophilosophers to describe the cultivation of a 'state of being . . . that sustains the widest (and deepest) possible *identification*' of oneself with one's environments (Fox 1986: 87). Deep ecology can be contrasted with the 'shallow ecology' of positivist empirical science which maintains clear distinctions between subject and object. I prefer to use 'deep ecopolitics' in this context to emphasize that the process of identifying oneself with one's environments – with *ecopolis* – is a matter of practical (i.e. political) choice, decision and action rather than the contemplation of a *logos*.
5. Awareness of one's thinking and control of it has recently come to be called 'metacognition' by a number of researchers who have studied children's 'cognitive strategies' (see, for example, Baird 1986, Brown 1980). However, if thinking is more a matter of perception than cognition, then 'metaperception' may be a more appropriate term.

References

Ash, M. (1980) *Green Politics: The New Paradigm* (London: The Green Alliance).

Baird, J. R. (1986) Improving learning through enhanced metacognition: a classroom study. *European Journal of Science Education*, 8: 263–282.

Boomer, G. (1982) Ten strategies for good teaching. In G. Boomer (ed.) *Negotiating the Curriculum* (Sydney: Ashton Scholastic), 119–121.

Briggs, J. and Peat, D. (1984) *Looking Glass Universe* (New York: Simon and Schuster).

Brown, A. L. (1980) Metacognitive development and reading. In R. J. Spiro, B. C. Bruce and W. F. Brewer (eds) *Theoretical Issues in Reading Comprehension : Perspectives from Cognitive Psychology, Linguistics, Artificial Intelligence and Education* (Hillsdale, NJ: Erlbaum).

Capra, F. (1983) *The Turning Point* (London: Fontana).

Cook, A. (1985) *Main Points in Chemistry* (Milton, Queensland: The Jacaranda Press).

De Bono, E. (1979) *Learning to Think* (London: Penguin).

DOLL, W. (1986) Prigogine: a new sense of order, a new curriculum. *Theory into Practice*, 25 (1): 10–16.

DOLL, W. (1988) Curriculum beyond stability. In W. Pinar (ed.) *Contemporary Curriculum Discourses* (Scottsdale, AZ: Gorsuch Scarisbrick).

DUFTY, D. and DUFTY, H. (eds) (1988) Thinking whole: the quest for a new educational paradigm. Readings and resources prepared for the conference of the Social Education Association of Australia (University of Sydney, NSW).

EMERY, F. (1981) Educational paradigms. *Human Futures*, Spring.

FOX, W. (1986) *Approaching Deep Ecology: A Response to Richard Sylvan's Critique of Deep Ecology*. Environmental Studies Occasional Paper No. 20 (Hobart: Centre for Environmental Studies, University of Tasmania).

GIBSON, E. and LEVIN, H. (1975) *The Psychology of Reading* (Cambridge, MA: MIT Press).

GIBSON, J. J. (1979) *The Ecological Approach to Visual Perception* (Boston: Houghton Mifflin).

GOUGH, N. (1987a) Forecasting curriculum futures: arts of anticipation in curriculum inquiry. Paper presented at the annual meeting of the American Educational Research Association (Victoria College, Victoria, Australia).

GOUGH, N. (1987b) Greening education. In D. Hutton (ed.) *Green Politics in Australia* (Sydney: Angus and Robertson), 173–202.

GOUGH, N. (1987c) Learning with environments: towards an ecological paradigm for education. In I. Robottom (ed.) *Environmental Education: Practice and Possibility* (Geelong, Victoria: Deakin University Press), 49–67.

GOUGH, N. (1988a) Children's images of the future: their meaning and their implications for school curriculum. *Curriculum Concerns*, 5 (2): 6–10.

GOUGH, N. (1988b) Review of S. Grundy (1987) *Curriculum: Product or Praxis?* (London: The Falmer Press). *Curriculum Perspectives*, 8 (2): 94–96.

HARMAN, W. W. (1985) Colour the future green? The uncertain significance of global Green politics. *Futures*, 17 (4): 318–330.

HARMAN, W. W. (1988) *Global Mind Change: The Promise of the Last Years of the Twentieth Century* (Indianapolis, IN: Knowledge Systems).

HEIDER, F. (1959) *On Perception and Event Structure and the Psychological Environment: Selected Papers* (New York: International Universities Press).

HOLT, M. (1987) *Judgment, Planning and Educational Change* (London: Harper and Row).

HUGHES, F. (1971) *Reading and Writing before School* (London: Pan).

LOBKOWICZ, N. (1967) *Theory and Practice* (Notre Dame: University of Notre Dame Press).

McCUSKER, B., and McCUSKER, C. (1988) The modern scientific view of the universe. In D. Dufty and H. Dufty (eds) Thinking whole: the quest for a new educational paradigm. Readings and resources prepared for the conference of the Social Education Association of Australia (University of Sydney, NSW), 77–81.

McKEON, R. (1977) Person and community: metaphysical and political. *Ethics*, 88: 207–217.

MICHAEL, D. N. and ANDERSON, W. T. (1986) Norms in conflict and confusion. In H. Didsbury (ed.) *Challenges and Opportunities: From Now to 2001* (Washington: World Future Society), 114–124.

PAGANO, J. (1988) Review of M. Holt (1987) *Judgment, Planning and Educational Change* (London: Harper and Row). *Journal of Curriculum Studies*, 20 (3): 284–288.

PIAGET, J. and INHELDER, B. (1956) *The Child's Conception of Space* (London: Routledge & Kegan Paul).

PBIGOGINE, I. (1980) *From Being to Becoming* (San Francisco: W. W. Freeman).

PRIGOGINE, I. and STENGERS, E. (1984) *Order out of Chaos* (New York: Bantam Books).

REID, W. A. (1981) The deliberative approach to the study of the curriculum and its relation to critical pluralism. In M. Lawn and L. Barton (eds) *Rethinking Curriculum Studies* (London: Croom Helm), 160–187.

SCHWAB, J. J. (1969) The practical: a language for curriculum. *School Review*, 78 (1): 1–24.

SCHWAB, J. J. (1971) The practical: arts of eclectic. *School Review*, 79 (4): 493–542.

SCHWAB, J. J. (1973) The practical 3: translation into curriculum. *School Review*, 81 (4): 501–522.

SHAW, R. and PITTENGER, J. (1977) Perceiving the face of change in changing faces. In R. Shaw and J. Bransford (eds) *Perceiving Action and Knowing: Toward an Ecological Psychology* (New York: Wiley).

SLAUGHTER, R. A. (1988) *Recovering the Future* (Clayton, Victoria: Monash University Graduate School of Environmental Science).

SPRETNAK, C. and CAPRA, F. (1984) *Green Politics: The Global Promise* (London: Hutchinson).

STERN, C. and STERN, M. (1971) *Children Discover Arithmetic* (New York: Harper and Row).

VAN MATRE, S. (1979) *Sunship Earth* (Martinsville, IN: American Camping Association).

WAUGH, P. (1984) *Metafiction: The Theory and Practice of Self-Conscious Fiction* (London: Methuen).

WLLBER, K. (1983) *Eye to Eye: The Quest for the New Paradigm* (New York: Anchor Books).

WILSON, N. (1987) The state of the planet and of young people's minds. *Ethos 87: Journal of the Victorian Association of Social Studies Teachers*, 9—13.

Sustainability and the learning virtues

JOHN FOSTER

Learning is important to sustainability—but how? On the dominant sustainable development picture, various kinds of learning are seen as instrumental to one's behaving responsibly towards future generations, within a framework of present actions and ecological consequences. This whole picture of future-oriented responsibility is radically flawed, fundamentally misrepresenting our creative engagement in change. It grossly exaggerates our powers to predict and control and licenses an endemic bad faith in the construction of sustainability goals supposedly derived from obligations to the future. This process is not only the opposite of genuine learning, but is very likely to ensure practical failure. In contrast, a model of ecological responsibility that might work will have learning not as a subsidiary and instrumental feature, but right at its core. The only way in which one really comes up against the constraint of the future is by acknowledging the demands of active learning—critical self-awareness, exploratory-creative commitment, and a robust tolerance for uncertainty—as *virtues*. The paper develops this account of the learning virtues in detail, and shows how embodying their practice across all our institutions and activities constitutes the only kind of responsibility to the future which we can genuinely exercise.

We must learn to understand nature from ourselves, not ourselves from nature. (Schopenhauer, *The World as Will and Representation*, ii, XVIII)

Nobody questions that education in general, and thus more specifically the curriculum and its development, are vitally important for the relations between humans and their natural environment. Learning, we can be sure, matters crucially for sustainability. However, there *is* a real question (albeit rarely raised) about *how* it matters. That is what I explore in this paper. I want to suggest that we have got our understanding of the relation between sustainability and learning wrong end on. The requirements of sustainability don't set us a curriculum, or tell us what and how to learn—rather, the nature of learning and its proper virtues tell us how sustainability must be conceived and how we must pursue it (including the business of educating for it) if we are really to take it seriously.

It may well seem that, for a piece in a journal concerned with the curriculum, this discussion takes rather a long time to get round to learning as such. That was inevitable, since a mistaken conceptual model, deeply embedded, had to be cleared out of the way first. However, implicit throughout the preparatory under-labouring constituting the first half of the paper are the ideas to which I subsequently appeal—those of the human meaning of learning and its relevant strengths of character.

A failing model: sustainable development

Standardly, discussions of the relation between learning and sustainability start with the latter and move to the former. Given sustainability as a requirement, a desirable condition to which society has to find its way, we then work back to the kinds of learning which that implies—the things that need to be learnt, who needs to learn them, and the ways in which they need to do so. Learning is framed instrumentally as a means (albeit a central and indispensable means) to the end of putting society on a sustainable pathway.

Thus, on the now-dominant 'sustainable development' picture of sustainability policy and practice, various kinds of learning are seen as instrumental to our behaving responsibly towards future generations, within the nexus of present actions and their scientifically-predictable ecological consequences. Characteristically, this is the way in which the current United Nations Decade of Education for Sustainable Development has been framed. Integrating 'the principles, values, and practices of sustainable development into all aspects of education and learning' will, according to UNESCO (2005) 'encourage changes in behaviour that will create ... environmental integrity, economic viability, and a just society for present and future generations'. We learn about technologies, lifestyles, trade patterns, and so on, and about their ecological effects, and we acquire relevant new skills and adaptabilities, in order to pursue an ethically-warranted trajectory towards this identified future state.

However, despite its having become something like the political default mode in this area, the whole sustainable development paradigm of what is involved in behaving responsibly towards the future is radically flawed. As a result, it is already revealing itself (could we only admit to ourselves what we really see) as delusory and ineffectual in practice. Correspondingly, acting on this paradigm doesn't actually exercise our capacities for genuine learning, but rather our capacities for something like its opposite: the intricate, tacit practice of self-deception and bad faith.

I have devoted a good deal of a recently-published book (Foster 2008) to arguing along these lines, and clearly have no scope to reprise that argument in any detail for this paper. It can, however, be summarized fairly brutally under three heads. First, the 'journey' model of our progress into the future ('getting to *there* from *here*') fundamentally misrepresents our creative engagement in emergent change. Secondly, in the environmental domain in particular, this misrepresentation depends on gross scientist exaggeration of our powers to predict and control. And, thirdly, it does so in the service of supposed obligations to future people which actually are no more than pseudo-obligations. The upshot of all this is to supply us with a battery of floating standards—that is, sustainability benchmarks and targets which quietly flex, under pressure, according to how comfortable or uncomfortable we find it to be constrained by them.

I will try to say enough under each of these heads to establish that there is a case to be answered, apologizing in advance for any slight appearance of dogmatism in the necessary compression of what follows. (I can only point the reader annoyed by this to the book, where the case is made in full.)

The 'journey' model

What *is* the future, really? It is only the advancing edge of the present—it exists just in virtue of the fact that the edge *is* always 'advancing'. The past, by contrast, shaping the conditions we now inhabit, has a comparative substantiality and reality—'another country, where they do things differently' makes a kind of sense as a metaphor for the past which simply doesn't transfer to the future. Of course, that doesn't stop us projecting the future in imagination well beyond the breaking crest of now. We can 'think long-term'—for we know that there *will* come a point, viewed from which the present will be as firmly shaping a past as the already-past is now, and we can consider how it will be to be present at that point. Similarly, we can want it to be, or feel that it ought to be, a certain way; and then we can ask, what must we do now to bring that about? 'Where do we want to be', we wonder, 'in a month's time ... a year's time ... in 2050 ...?' The very tempting model, already implicit, is then of a journey ahead of us from here to there, a trek plotted out across a region of time towards an envisaged destination. Typically, this seems to involve us in prediction-stages—we plan the journey as far ahead as we think we can reliably see, and then extrapolate. (Compare: 'If we can get on to that col, there's a route off it to the gill-course, and that will surely lead us down towards ...'.)

However, this model is dangerously misleading if taken too literally—as we much too readily slip into taking it. We are betrayed here by the pervading scientism of the culture, by a sense of the future as real and graspable, which is largely constituted by over-confidence in our powers of prediction. (A related cultural factor may be the popularity of televisual and cinematic science-fictions, which depict fantasy futures for a mass audience with all the falsifying realism of those media.) Actually, however, rather than following any route towards an identifiable destination, what we have always to deal with in human agency are paths that change as we go along them, shift their direction indeed just because we choose to go along them, and ramify unguessably according to how fast or slowly we pursue them. Moving into the future, in other words, is really *nothing* like negotiating a topography. Instead, in the words of the sociologist John Urry (2005: 3), it is 'walking through a maze whose walls rearrange themselves as one walks through: new footsteps have to be taken in order to adjust to the walls of the maze that are adapting to each movement made ...'. The future is always, and essentially, under construction, to a plan which is itself being perpetually reconstructed. We also, it is true, 'construct' or reconstruct the past, in memory and imagination—we have no other cognitive access to it, after all—but the radical difference is that these constructions are made true or false (whether we know it or not) by what did actually happen. Thus, they parallel the way in which a map can represent, more or less accurately, a mapped terrain. The distinction between the actual past and our construction of it is then one which we can put to good use (in trying to confirm our account of past events by finding out more about what *really* happened, for instance). However, the distinction between what will happen, and what we think will happen (or, crucially, *want* to think will happen) can be of no similar avail.

Scientism

Mistaking the future for a destination is, as I say, inherent in scientism—that is, the expectation of more from science than it could possibly deliver, in the mistaken belief that it would be unscientific to expect less. Where our environmental relations and policies are concerned, that mistake is nothing short of deadly.

In practice, we live with the uncertainty, malleability, and fundamental indeterminacy of the future all the time—it is the permanent sub-text of human agency that the best-laid plans are liable to go pear-shaped and one never really knows what's going to happen. However, embedded in the formation of sustainable development is the idea that 'rational' science-based planning can somehow transcend these inherent conditions on our knowledge and action. Hence 'strategies' which envisage specific percentage cuts in CO_2 emissions by specific dates 50 (or 20) years ahead, as a way of keeping below specific global warming thresholds, and which assume that adopting such a planning horizon is an act of sober and responsible administrative realism—when what it really represents is hubristic disdain for the glaring fact that all the ecological synergies and feedbacks involved are infinitely too complex to guess at beyond the very short-term, never mind predict and try to plan for.

Of course this doesn't mean that we can't predict environmental futures *at all*. However, our prediction must be quantitatively modest. We must seek to predict in terms of directions, trends, broad magnitudes, and possible scenarios, rather than anything more numerically specific. This is an essential function, and one for which we rightly turn to science. At a sufficiently low resolution, there are things in prospect about which we can be fairly confident. We know, for example, that world population is likely to go on increasing, and (bar apocalyptic accident) to pass the nine billion mark before stabilizing, though we don't know just when or by how much. Similarly, global mobility and the pressures of techno-agriculture are likely to produce more pandemics of the swine-flu variety, though each particular outbreak is likely to take us as much by surprise as that one did. And there is now a huge expert consensus behind the conviction that greenhouse gases generated by human activity are indeed warming up the world.

We should not have known this last, vitally important truth without the science. However, there are so many interlocking ifs along the way to computing its consequences—*if* the Arctic ice disappears, *if* the Gulf Stream alters direction, the Amazonian tree-cover is reduced by vast forest fires, the capacity of the oceans to pump down carbon dioxide collapses ... and so much unavoidable contingency in conditioning for them, that any precise prediction will have the uncertainty of effective indeterminacy. The point is, however, that on the mainstream sustainable development approach our predictions do have to be fairly precise, because they have to be translated back into specific targets and standards for present action. And no scientific prediction of ecological developments extending that far out into the future can be anything like robust enough to warrant one specific target, standard, or action *rather than another*.

It is also important to acknowledge that with at least part of our minds we know this to be the case all the time. Everyone actually understands quite well the uncertainties inherent in this kind of science, whether or not our policy discourse has a language in which to admit to them. For instance, research on the GM issue has shown that 'lay' people take uncertainty about the long-term consequences of GMOs for granted. Representatively, focus group participants in a landmark study took the view that nobody knows *and nobody can know* the full impacts of GMOs in the long-term. Unintended effects (both harmful and beneficial) will necessarily occur, but these will only become apparent later, when consequences 'which had not been imagined' become apparent (Marris *et al.* 2001)

The trouble here is two-fold. On the one hand, the part of our minds which recognizes all this (and not just about GMOs) is systematically disconnected, by the pervading scientistic culture, from the part of our minds which tries to grapple with the policy implications of a hugely complex techno-civilization. On the other hand, that disconnect doesn't prevent a tacit awareness of indeterminacy from undermining the robustness of 'scientifically' quantified sustainability targets—whenever, without admitting it to ourselves, we actually *want* these putative constraints on present Western lifestyles to be undermined.

Shadow stewardship

Such processes of tacit adjustment are greatly aided by a defective picture of the kind of responsibility which is incumbent on us in this connection—that is, the 'stewardship' model. It is clear enough in general terms that we now have the power to disrupt the ecological workings of the world in far-reaching and long-term ways. Correspondingly, it might seem, we have a clear moral obligation not to do so, but rather to deal justly by our successors, stewarding the earth's resources and ecological capacities in our time so that future people can rely on them as we have been able to do. This picture of the present human generation as stewards of the planet has had a powerful presence in environmental thinking from its beginnings. A *locus classicus* for its impact on mainstream political thinking is the former UK government's 1990 White Paper *This Common Inheritance*, which states that

> the ethical imperative of stewardship ... must underlie all environmental poli-
> cies ... It is mankind's duty to look after our world prudently and conscien-
> tiously ... we do not hold a freehold on our world, but only a full repairing
> lease. We have a moral duty to look after our planet and hand it on in good

order to future generations. That is what experts mean when they talk of 'sustainable development'. (HMSO 1990: 10)

However, this whole idea of an ethical standard of fairness or impartiality towards future people is an illusion. The concept of stewardship obligation simply will not make the kind of sense nor do the kind of work for which it is here being enlisted. For reasons closely connected with the nature of the future as discussed in the previous section, we cannot treat future people unproblematically as others, existents standing independently over against us to whom we can be accountable for the consequences of what we do. We are only genuinely accountable where we can at least in principle be *held* to account—and all that can ever happen when we make a decision with implications for longer-term futurity is that we try to stand in for successor generations and hold ourselves to account on their behalf—try to govern our actions by some sense of what they could reasonably expect from us. This can certainly be done, in a serious attempt to take, as it were, a back-bearing from the future on present conduct and try to see such matters from their perspective. However, we are, too, necessarily the only judges of how far we have succeeded. And, holding ourselves to account in this way isn't being held to account, any more than locking oneself in a room and pocketing the key is imprisoning oneself, however firmly one intends to stay put. There is no way in which the basic judgements of the human present can be challenged here, except by present humans—so that, indefeasibly, we have only our own best judgement to go on. However, if that is the case, there can be no genuinely constraining accountability in the frame, which means that we can't actually deal fairly (nor, of course, unfairly) with the future at all.

What sustainable development tries to offer instead is an *ersatz* fairness constituted by the numbers: the practice of counting going proxy for accountability. If only we can accurately predict what quanta of what forms of natural capital will be available at what future dates as a result of what action taken or refrained from by us now, we can have in effect a mode of unforced, timeless agreement with the people who will be alive then. Any such quantum is, and will be, the one quantum it is, for them as for us. It thus provides a standard for fair dealing to which both we and they must subscribe, without the need to maintain it in being through all the collaborative dialogical processes which are here impossible. If less than 2°C of global temperature rise will enable the climate to go on providing for people after 2050, or 2100, more or less the quantified levels of service which it provides for us now, and we can determine with sufficient precision what reductions in CO_2 emissions volumes that requires immediately, we can act fairly towards futurity by making those reductions.

As we have seen, however, this proxy impartiality is unachievable. The numbers in this whole domain, being essentially indeterminate, in fact depend on our present judgements, including our judgements of value in relation to future consequences. To take their objectivity as independently warranted (by 'scientific reality') and then use it as a substitute standard for the impartiality of our judgements is really no more than to go in a circle. And, once we acknowledge this, we can see that the sustainable development paradigm provides, at bottom, no robust way to mark the difference between preparing to meet our obligations to the future, and construing the future as putting us under obligations which we are prepared to meet. What is to *count* as handing on an 'equivalent level' of biospheric capacity? Does significantly reduced biodiversity together with significantly improved techniques for manipulating the residual gene pool so count, for instance? Or what about an infrastructure increasingly dependent on electricity and all the hills covered with wind turbines? The answers to such questions are not simply a matter of 'doing the science right'; they involve *choices* about what attitudes we are willing to adopt towards the whole human-natural context in which future people will have to meet their needs. And it is clear that present humans will make both individual and collective choices on such issues not impartially at all, but with an eye very much to their own present interests and convenience.

Bad faith

That present humans will do this is not only clear—we also need to face up to the crucial sense in which it is entirely natural and right. Here again, sustainable development's false picture of our relations with the future tends to muddle our thinking. If future generations genuinely subsisted over against us, we could indeed betray or short-change them, and that is what we should certainly be doing by deliberately preferring unabated globalized consumerism, in all its meretricious vacuity, over conservation of the ecological resources to which they would then appear to be entitled. As it is, however, and as far as the human world goes, *there is never really anyone there but us*—that is, presently-existing people. At the same time, the will-to-life, as it drives human action, is the will to life *now*. It requires us, irresistibly, to make the best of present opportunity: we breed, create, delight in movement or beauty or music, all in the present and only under its conditions. Those conditions may be shabby, demeaning, culturally impoverished—they may, as currently, quite sharply restrict the scope of humane possibility—but they can't be avoided if we would really live in the only time we have. Contemporary civilization is the only terrain on which human beings driven by the overwhelming force of the life-urge can operate. The emphasis on material 'success' (as, essentially, distinction in the breeding game), the associated consumerism which now infects our pursuit even of life's genuine goods, the urban clutter and restless mechanical motion, the ever-ramifying technological mediation of human exchange—all this constitutes the diminished, damaged, but nevertheless unevadable, present circumstances under which our life-energy has no option but to express itself. And such life-energy, in itself, isn't concerned with the future at all. Its concern is self-realization, with the future entering only as the forward-directedness of its creative and reproductive urges. Any concern for longer-term consequence, introduced at the conscious intellectual level of perceived ecological upshots and calculated 'obligation', will simply struggle to be heard if it tries to countervail that fierce focus on the will's present opportunities. (Still less is this basic life-impulse in humans likely to be inhibited by worries over the effects of its expression on the life-chances of other species.)

Perhaps the worst feature of the ruling sustainable development paradigm, indeed, is the way it blandly glosses over these uncomfortable but not unobvious facts—affecting to suppose (what no-one, surely, can really believe) that if individuals and society collectively only summoned enough moral fibre, we could actually subdue our drive to present gratification in the interests of the longer-term future. This is self-deluding and self-serving, and the louder and more often we say so the better.

None of that *excuses* us from concern for consequences. We are the creatures who inhabit the conceptual sense which we consciously make of the world, and an uneasy awareness of our new ecological reach and destructiveness now forms, rightly, an increasingly important part of that sense. What the sustainable development paradigm represents in sum, however, is a radical misunderstanding of how that awareness of underlying ecological responsibility can be brought effectually to bear on changing our present ways of going on. If we take it, on that paradigm, as recognition of an obligation to future people to pursue, or refrain from, specific present actions determined according to their calculable longer-term ecological effects, all we really do is issue ourselves a covert licence for bad faith—a warrant for continuing to accord an instinctually-driven priority to present satisfactions, while pretending not to. Sustainable development is a hook, the real purpose of which is to let ourselves continuously off it. The trick works because, knowing science-supported long-term targets to be essentially indeterminate, we can always let them turn slippery under pressure. They can then yield floating standards of 'obligation to the future', standards and benchmarks which will never seriously constrain our present inclinations. Making the changes we find we can make without unacceptable present pain is warranted by the in-principle-calculable beneficial consequences (the quanta as determinate). Not making those we find too uncomfortable is warranted by the recognition that any particular calculation will lack robustness (the quanta as inherently uncertain). The sustainable development model of ecological responsibility can never substantiate the charge that to have gone only *thus* far, rather than somewhat further, wasn't meeting our obligations—and, knowing this, we settle very naturally in practice for roughly what we find we can manage. That is why, for instance, a UK government which introduced the most (theoretically) stringent set of public sustainable development commitments in the world, commitments which *blatantly* require, among other things, a massive reduction in recreational air travel, could at the same time plan to build yet another runway at Heathrow. The sustainable development model of environmental responsibility spurs us to go just far enough not to make any difference that hurts. That is not a corrigible shortcoming in the model, but its tacit point.

We are not talking here, it must be emphasized, about any of the various forms of denial which social psychologists have identified as operating in the environmental arena. The denier's tactic is to ignore unpleasant facts, or else to challenge their status as facts ('There's no proof that climate change is really happening'), plead powerlessness ('Nothing I do can make a difference'), or divert onto the perennial issue of casting the first stone ('All these environmentalists use electricity and drive around ...'). Such denial can indeed be in bad faith if it is a refusal to acknowledge truths which, underneath, one really knows. However, my point is that, beyond these tactics of denial, bad faith is still always soliciting even those who are genuinely trying to confront the issues, because it is inherent in the sustainable development picture of *how* to confront them.

The US historian of science Theodore Porter provides a brilliant cameo case-study which is directly relevant. 'A congressional mandate', he notes, 'permits the United States Forest Service to cut no more lumber than is renewed by annual growth. Since that law was put into effect, growth rates have been greatly enhanced, at least in the Forest Service accounts, by new herbicides, pesticides and tree varieties' (Porter, 1995: 44). Here we can see very clearly in miniature just how the characteristic move is made. The illicit real dilemma—'We have to comply with the mandate, but we really do need the income from going on cutting the lumber'—becomes the licit aspiration to have it both ways by boosting compensatory growth rates well above previous levels. This in turn depends on finding means to quantify the forecast effects of new interventions so that they show this to be what, over the longer term, is actually being done. The nature of scientific prediction of the quantified consequences of human interaction with ecological systems over the medium- to longer-term, with its inherent indeterminacy arising both from the socially constructed nature of such numbers and the methodological limitations on mathematical modelling of natural systems, ensures that opportunities for—in effect—equivocating with the figures in this kind of situation, will abound.

A better bet: the learning model of sustainability

That cameo of Porter's should also make plain how the inherent dynamic of sustainable development travesties any genuine learning. Bad faith is, rather, the essential negation of such learning. Its structure is not *finding out*, but 'finding' what we have put there because we want it to be found; it involves tacitly closing off the various ways in which emergent reality might subvert what we have already decided is to be the case. We can of course, at superficial levels, learn lots of things about and in pursuit of sustainable development goals. There is also learning of an all-too-familiar kind going on when we register the essentially floating nature of standards in this domain, and acquire the facility in floating them where necessary which enables us to make an affordable peace, individually and collectively, with our sense of ecological unease. However, none of this alters the fact that sustainable development's deep demand on human learning is that it should abdicate its fundamental responsibilities and strengths.

By the same token, we can begin to appreciate how genuine learning must be at the heart of a genuinely sustainable relation between the human present and the human future. Our real responsibilities in regard to this future, as well as any likelihood of our being able to meet them, are a matter of insisting in all our practices and attitudes on precisely the kind of openness to the emergent which sustainable development tacitly closes off. Our only grappling with the future, the only kind of activity where we feel its grit under our fingernails, as it were, happens when we take a learning stance at the forward edge of the present. At the advancing frontier between the known and the emergent, the essence of responsibility is keeping open our communications directly with possibility, which is the condition of the future—adapting to the way the possible is perpetually hardening into the actual. So here we get a touchstone for how genuinely responsible engagement with the future might feel, what the lived texture of it might be.

However, it should also be clear that we need to set aside the sustainable development picture of scientifically-determinable obligations to future people, in order to grasp these real responsibilities. This will not be easy, given the established dominance of the mainstream model. It is now quite a challenge to think of what environmentally-concerned or ecologically-responsible people are being responsible about, what the green turn entails, except in terms of this 'equity vis-à-vis the future' idea. Still, if I am right, that picture simply cannot provide a viable model of the relation between learning and sustainability.

We need to start again from the beginning. We are all intimately familiar with the *feel* of learning, of that kind of future-oriented engagement. What can be got out of a direct phenomenological confrontation with that very familiar domain of experience and activity that might help us with thinking about sustainability? The right approach to this question, I suggest, is to enquire after what I call the 'learning virtues'.

Let me re-emphasize the key issue here. For responsibility, we must change what we do now in the light of the future—but *we cast* that light when it is the medium- to longer-term which is in question. This envisaged future is almost inevitably, in the image which I use to organize my book, a mirage—delusory because the extent of our involvement in its construction is concealed by the mode of its appearance. So where does 'deciding what we do by the light of the future' represent a real constraint? The basic thought is that it does so where we are finding out how to make sense, listening for a sense still forming, 'feeling our way forward'. I want to look at how this stance is at the core of learning, and how it expresses a centrally characteristic set of 'learning virtues'—and then see what that can tell us about what any genuinely hopeful pursuit of sustainability must involve.

Virtue, life-form, and learning

Classically, the virtues are grounded in the life-activity of the sort of creature that human beings are. In broad terms, they are dispositions needed for living successfully as that particular sort of creature, just as the right amount of sunshine or water for a particular plant is needed for its living successfully, that is its thriving, as that sort of plant. Consider for example *courage*. Humans are naturally gifted with reason and foresight, which means among other things that we can anticipate everything that might go wrong, across a vastly wider range of possibility than any other species. (Humans are the only creatures subject to Sod's Law, and just because they are the only ones capable of formulating it.) Correspondingly, our life-activity would tend to be paralysed by this all-round risk-awareness, without the disposition to hazard ourselves, where necessary, in action of which we can't see the safe issue beforehand. That disposition, which has to be established and strengthened by training and deployed appropriately situation-by-situation, is courage. Its value is its enabling us to negotiate the generalized human life-condition of having consciously to go on chancing our arms. Its exercise might often promote the individual or general happiness, or brace us up to our duty, but *qua* virtue it is a specific human good prior to that and notwithstanding. (This is why it can be displayed in 'bad' actions as well as morally sound ones, although typically—or so moralists hope—such actions will tend to fail by the tests of other virtues.)

Again, take *honesty*. Humans have language, and so the capacity to mislead one another. (Voltaire says cynically somewhere that that's *why* we have language.) We also have imagination and wishful thinking, and so the capacity, uniquely among terrestrial species, to deceive ourselves. A settled disposition to honesty helps us both to acknowledge things for what they are and to trust one another—both essential for a highly-intelligent, curious, technologically ambitious, and radically social species endowed with those powerful resources for illusion. Or again, *patience*: the capacity to defer gratification makes the most of our species gifts for foresight, communication, and collaborative planning. It might be said to be one of the main reasons why we have hitherto been such a 'successful' species, at least in terms of extending our ecological niche from the ancient African savannahs to the entire planet.

The virtues generally, that is, represent what we might call life-strengths: acquired dispositions, the exercise of which avails us for living the kind of life belonging naturally to our form of being. Now one of the fundamental ways in which the human form of being is lived is our engagement in active and perpetual *learning*. We learn when experience brings us to augment or revise the resources of memory and concept-formation on which we can then draw for negotiating new experience. This can happen involuntarily through mere habit-acquisition, which is the sense in which other kinds of animals learn (to recall salient associations, pathways, and so on), and indeed are genetically programmed to do so. However, its distinctively human manifestation is reached in its becoming a conscious and deliberate process beyond any programming, an intentional interpretive scrutiny of whatever we encounter in order to understand our way forward. As such, learning is about as centrally human a business as one could hope to find—it starts very early in infancy, ceases in many cases only on our deathbeds, and is at the heart of our specific rational distinctiveness.

This centrally human activity operates at a number of levels, which have been described and differentiated by learning theorists in various ways. The key distinction, however, is one in terms of degrees of reflectiveness, first made decisively 40 years ago by Argyris and Schon (1974) in their account of single- and double-loop learning. It is brought out very clearly in a helpful characterization of three 'orders' of learning by the environmental educationalist Stephen Sterling (2001), who draws here on the seminal work of Gregory Bateson (1980). First-order learning is what might be called adaptive learning, and takes place within accepted boundaries, leaving basic values and beliefs unchanged. Second-order learning is more critically reflective, and leads us to examine the assumptions according to which we proceed unquestioningly in first-order learning. Third-order learning occurs when this reflective examination of our working assumptions undermines our established core beliefs and leads to a transformative perspective-shift.

(This analytical framework will no doubt be familiar to readers, and the terms in which I go on to rehearse its significance may seem merely to rework old ground. However, I would ask again for patience. The relation which I am proposing between all this and sustainability as future-oriented responsibility will be found to involve a radical inversion of conventional understanding which should slake the most raging thirst for novelty.)

We can re-express this distinction in terms of the conceptual framework which we always and necessarily bring to the understanding of any experience. In first-order learning we take this conceptual framework for granted: we acquire new information under the concepts we already have, or new concepts which fit comfortably onto the structure of the existing. In second-order learning, however, this conceptual framework itself comes to be at issue, so that what we acquire from experience is a new organization of the concepts which we bring to it, or a new understanding of key nodes within that organization, usually together with some new concepts which only cohere within that revised framework. The key difference between these two levels is that, at the first, we are making sense *within* our existing interpretive framework, while at the second we are also making potentially new sense *of* that framework. Then at the most challenging, third-order level, the framework is not just reconfigured (something which we could at least contemplate out of its present configuration, though we might not yet see a need for it), but essentially replaced—in ways which we necessarily can't now imagine, since, if we could, they would be graspable from within our existing framework. Of course, 'essentially replaced' can't mean that *everything* goes into the melting-pot, or else we should lose our cognitive bearings totally. However, elements in a structure of sense are constituted by their place within that structure, and very significant shifts in our understanding of central concepts and their relations can occur even though, unforeseeably, some of them will afterwards be recognized as continuous with what went before.

The same analytical pattern emerges if we think from the perspective of personal development, about learning our way onwards in pursuit of our various life-projects. I can learn how better to pursue my needs in a given life-situation; and I can also go further, to question how far that situation actually meets my needs, and learn how to change it. However, the third-order possibility is that I come to see both needs and situation—and thus to understand myself—in wholly new ways which I can't yet properly imagine, still less substantively grasp.

Given this hierarchy of levels, it is important to recognize the inherently second- and third-order potential in *all* learning. In even the simplest learning processes at first-order level, a unique kind of inner distancing goes on which we might try to capture with the idea of the self standing back from itself. This is already evident in learning by rote, as when we commit something (say, a number sequence) to memory. Here we move beyond simply attending to the sequence, to deliberately rehearsing our attention until it can summon its object unseen, as it were: a process, that is, of inner attention to our attention. In such a process, we are both an active conscious self and a further active awareness of that consciousness. The same kind of thing can be observed in the ordinary business of learning a skill. Think for instance about learning to drive, where the pragmatic paradox involved in acquiring any skill (you learn by doing something you *can't* do) is particularly evident. You learn to co-ordinate hands, feet and eyes in putting the car into motion and controlling its starting speed, by 'letting it come', that is, by letting attention step back from the explicit instruction you have received. As you feel for the right moment to declutch, let off the handbrake and slightly accelerate, the key thing is again this kind of inner duality, in which the co-ordination of what you are doing is entrusted to a competence you haven't yet acquired. We practice ourselves, as it were, in a poised expectant submission to the emerging embodied know-how we are trying to gain.

It is this characteristic stepping-back of self from itself which provides the essential phenomenological continuity between first-order learning, where we can be acquiring new information or abilities merely additively, and second-order learning where what is added changes our wider understanding of what it is added *to*. Second-order learning is a matter not just of *reflexive* attention (the self working in one way or another on itself) but of critically *reflective* attention by the self to the assumptions and concepts constituting its self-framing. Its reconstructive work on these, in turn, always foregrounds the contingency of its frame of self-understanding—the fact that it is thus-and-so, but could have been different: and so its deep provisionality in the face of new experience. Thus it always, too, contains the possibility of a radical reconstruction, in the course of which we will step not just back from, but out of and beyond, our current self-understanding, into something unforeseeable.

The learning virtues: negative capability

Learning, as all this demonstrates, is always potentially an existentially challenging process. Correspondingly, we can see that it always (and increasingly, as it moves up through the three levels) tends to demand a certain distinctive kind of life-strength or virtue in the learning self. In view of what has been said, we might be inclined to describe that strength in deliberately paradoxical terms—as the self's strength for contemplating, or confronting, being supplanted by something *not yet itself*. Genuine learning, that is, always somehow requires the self not to seek to protect its position, but to be ready to expose itself—whether to the possible failure of rehearsed attention or pre-competent activity, or to the mounting insecurity of recognized intellectual contingency and provisionality.

This sense of paradoxical power recalls Keats' famous formulation of 'what quality [goes] to form a Man of Achievement especially in literature':

> Negative Capability, that is, when a man is capable of being in uncertainties, mysteries, doubts without any irritable reaching after facts and reason ...[1]

Keats has particularly in mind a contrast that has struck him between Shakespeare and Coleridge, in terms of how the impulse to get everything clear in explicit knowledge can inhibit creative poetic insight—but he also offers, it seems to me, a very important clue to the capabilities essential for effective learning.

Think about what learning involves in any complex area: learning how to understand an unfamiliar topic (a new subject area, a new job), or learning new life-skills, the sort of self-development that goes on as one moves through life and encounters new kinds of situation and life-problem. In all these activities we are most successfully attentive, and most effective as learners, when we are able to project possibilities, act despite uncertainty, and discover from what happens which path we are actually on. An awful lot of life is taken up with this kind of exploratory 'learning on the job', especially in key transitions—becoming a pupil, a student, a lover, a parent, a mentor, an elder. As La Rochefoucauld (1967: 95) observes, 'we come quite fresh to the different stages of life, and in each of them we usually lack experience however old we are'. However, at least the rules of thumb are pretty similar in all cases: make as few working assumptions as you can, find your way forward by trying things out to see what runs, be aware of reactions, be ready to shift your weight, rebalance, move responsively with what emerges. We learn on the job at any level, indeed, by acting (as far as we can) in ways that open up options and deliberately don't decide prematurely between them. So, in developing a course for a new group, you sketch out the first couple of sessions in detail and the remainder much more roughly, with different possible lines of enquiry and main emphases depending on how the initial phase goes (and you are then alert to how it does go, rather than simply ploughing on regardless). In writing, you rework not just your material but your plans and expectations, as you constantly review and revise what you actually produce. In developing any practical enterprise you 'don't put all your eggs in the one basket', but diversify and risk different simultaneous bets on what will run, while always being prepared to junk what isn't working.

It would make obvious sense to think of the relevant dispositions and character traits here as virtues—that is, as life-strengths in the exercise of which we thrive. However, they cannot just be applied forms of the standardly-recognized virtues, honesty, patience, persistence, and so on, though obviously these are all very relevant both to learning itself and to bringing learning to bear in action. The standard virtues, however, are aspects of (as it were) *positive* capability: they are characteristics of a self which readily confronts, and by asserting its own powers reliably masters, circumstance—or at any rate survives it relatively unscathed. The learning virtues, on the other hand, must exhibit the kind of paradoxical inversion which we have been noting: they must be strengths for robust hesitancy, poised insecurity, winning by surrender ...—strengths, that is, for negotiating the permanent existential challenge of seeking out something unknown, something one does not yet know how to recognize.[2]

What might such virtues look like? Here are three candidates. (They are evidently complementary, and may indeed turn out to be all aspects of just the one virtue.)

In the first place, *criticality*. This is the clear-sightedness of appraisal that passes naturally into self-appraisal and self-critique. It means being able not just to stand back within oneself and rehearse oneself in some new fact or ability, but as part of the same movement of mind to stand back from what one believes and *is*, and recognize that these may be lacking, inadequate, needing to be changed and developed. (And of course they always are, because whenever we try to rest in stasis, life has moved on beyond us.[3]) Criticality thus understood is not, as it might appear, a virtue only of second- and third-order learning. It involves the disposition always to be alert to what is going on in any seemingly first-order learning, and to recognize the point where we must pass beyond that level. It responds to our perpetual temptation to shirk the emergent, to accommodate the new to the familiar, to domesticate potentially challenging experience under concepts with which we have grown comfortable. There is always the danger that this will lead us subtly to misrepresent experience, tweaking its actuality here and there in order to fit it to the frame of the already known.[4] Bad faith expands alarmingly from these beginnings, but criticality is the life-strength to resist it—to be even *too* watchfully ready to challenge our taken-for-granted assumptions, if only to ensure for them that permanent instability without which they so easily solidify and jam up our being.

Closely associated with criticality is, in the second place, *option-fecundity*—a robust persistence in exploring and testing possible interpretive frameworks of assumption alternative to the currently taken-for-granted, even before one is moved to adopt any of them in its place. This multiplication of available cognitive options increases the epistemic value of one's already-learnt understanding, in something of the way in which building real options into a capital asset (the technical possibility of an engine's switching between different fuels, for instance, according to changing market price) increases present, rather than discounted future, asset value (Foster 2005). The epistemic equivalent of such present value might be a reduced temptation to insulate and preserve one's current learnt understanding in the face of emerging experience which it can't really accommodate.

The third learning virtue might be called (perhaps a bit provocatively) *epistemic mischievousness*—the readiness and resolve to mix it sometimes, just to see what happens, when things get too clear and straightforward. This involves not merely enduring uncertainty when interpretation is difficult or ambiguous (as in Keats' formulation), but actively courting it as a potentially creative matrix for deeper insight. A person evinces this virtue, who insists on complicating speculatively, positively valuing and welcoming the uncertainty and open-endedness which is after all endemic in many a life-situation— who exults, indeed, in having to trust to intuition and make his or her own luck.

Together these virtues or virtue-aspects define a characteristic disposition towards the future—a kind of focused attentiveness in which we deliberately let the unanticipated inform us out of experience, a self-possessed and resourceful waiting on what transpires, by way of connections that will make themselves and insights that will be vouchsafed. It is the capacity to restrain as far as possible any deliberate configuring of what is to emerge in terms of who we are or what we want *now*—which would be an imposition of the already-present on the future, driven by a need to secure ourselves in what we already grasp. Rather, learning is letting new sense take shape and shape us. It is, precisely, *negative* capability—the capability *to not be* what we are: or better, not to *be*, but always to *become*.

Reconfiguring sustainability

If human beings thrive, in general, by approaching the future with this radically learning stance, what can we conclude about the requirements for paying *ecologically-responsible* regard to the future in our present actions?

Certainly, this regard will be a very different matter from the sustainable development approach of 'predict and evaluate'. For it will involve understanding, not just that all longer-term prediction is irredeemably 'soft', but that we can only act responsibly in respect of longer-term consequences by explicitly recognizing this. The crucial practical challenge is to get an institutional and political framework flexible and adaptive enough to be capable of the continuous exploratory social learning which this concept of responsibility implies.

Of course, acting responsibly in any field always involves being prepared to learn. However, the crucial factor in shaping appropriate learning here is uncertainty. Acting responsibly when we are basically not uncertain—when we have reason to suppose that our methods of predicting outcomes are comparatively robust and our evaluative criteria comparatively reliable—involves what has been aptly called *single-loop* learning (Argyris and Schon 1974). In such cases, we review alternative possible actions, predict their respective outcomes, compare these predictions evaluatively, and learn from this process how we should proceed. When real uncertainty comes into the frame, however, what we need instead is *double-loop* learning. We make some assumptions, and act on them in the expectation that certain results will follow. Almost invariably, rather different results actually follow, in the light of which we don't just trim our course of action, but also revisit and maybe somewhat revise the assumptions on which we then continue to act. Note that this is a short-term process—albeit iteratively so: it doesn't usually take very much time, if we are paying attention, before the upshots of our assumptions start differing visibly from our expectations.

Double-loop learning, however, can still be what I call *closed-loop*. This essentially defensive model of learning is the one which underlies precautionary behaviour. We acknowledge uncertainty, accept provisionality, and explore forwards, but still with a notion of ourselves as pursuing a desired ideal state in which we shall know what we are doing and where we are going. *Open-loop* learning, in contrast, is responsibility under uncertainty which tries to get the best out of our standardly, characteristically, and permanently *not* really knowing what will happen. We do this not just by constant attention to how things are actually turning out, but by building in options— moves we will be able to make, but are not required to make, when the appropriate time comes. Thus, we maximize our room to respond creatively to the emergent. The aim of open-loop learning is not to minimize uncertainty, but to ride its advancing wave with as few tumbles as we can.

This is the context in which sustainability policy can work with the grain of capitalism—and in which capitalism can be seen, not grudgingly and *faute de mieux* as the only political-economic context we are likely to have for pursuing sustainability, but as (despite its manifold flaws) the context we actually need. Capitalism at its best is open-loop learning applied to the ongoing production, distribution, and exchange of goods and services. The clearest insight here comes from von Hayek (1960), who is a much more important thinker in relation to these issues than his caricature as a free-market ideologue suggests. His argument, historically mounted against the bureaucratic ambitions of centralized socialist economies, draws on conceptual resources which make it powerfully relevant to a radical learning model of sustainability:

> If we are to advance, we must leave room for a continuous revision of our present conceptions and ideals which will be necessitated by further experience ... Though we must always strive for the achievement of our present aims, we must also leave room for new experiences and future events to decide which of these aims will be achieved ... the advance and even the preservation of civilization are dependent on a maximum of opportunity for accidents to happen .... (von Hayek 1960: 23–24, 29)

The role of government in this advance is to make dispositions which free up a society's capacity for creative mistakes and illuminating accidents and to facilitate (though not to organize or direct) the processes of ongoing learning from them, and of adaptive accommodation to their upshots. Genuine sustainability requires a rich texture of optionality in its productive arrangements. We cannot be proactively learning our way into the open future if there is nothing we can *do* with what we ongoingly find out about the assumptions to which we had provisionally committed ourselves—if we haven't left ourselves scope for our learning to make a real difference. Capitalism, insofar as it is not hypertrophied monopolism but flexible, light-footed, and creatively entrepreneurial, represents the condition of that optionality.

That doesn't mean the end of planning. A society operating on this kind of rationale for sustainability will seek to configure policy within a learning framework as an absolutely basic requirement, but it will still need to be structured by plans and targets. In that learning spirit, however, we will take a different attitude to the figures and the science which supplies them, essential as these will remain. Quantified targets will be seen as heuristics, exploratory approximations to be ongoingly revised, the more so the further ahead they are cast. What we do with them will reflect that recognition. A key application here is to sustainability targets—which of course we are going to go on needing. A quantified target on the sustainable development model—say, for an 80% carbon emissions cut as against 1990 levels—is meant to be the figure it is, because it is meant to be the inter-generationally equitable balancing point. (Smaller cuts damage the future, larger ones penalize the present unnecessarily.) Bad faith then has its chance, because we know, tacitly, that if the target presses too hard on us, it needn't actually be the figure it is, but can slide or float—that is, be treated as just sufficiently imprecise to allow us the leeway we need. However, such a quantified target set and treated heuristically is an acknowledged place-holder for *whatever amount of cut it will turn out that it is going to take* to do the job—here, the job of trying to ensure that climate change remains such as can be adaptively coped with, ongoingly. We can fully admit, on this understanding, that the figure is very unlikely to be a place-holder for itself (that is, to be the figure it is)—we can only hope to learn more about what it should be, or whether it is even the right quantum to be targeting, as we go on. However, we can still mark it as a line in the sand, and insist on it as a working target.[5]

... and the curriculum?

This paper is not the place to explore in greater detail the large implications of such a reconfigured sustainability for economic, fiscal, industrial, and social policy, for democratic governance and community empowerment, nor for the broader patterns of living and working in our kind of society. (I make a start, though hardly more than that, in my book.) However, it is appropriate to conclude with at least a pointer to what it would involve for our understanding of the 'sustainability curriculum'.

As I began by noting, education for sustainability as currently advocated and practised is really education *about* the sustainable development model, and *in* the knowledge and skills needed for making a contribution, as worker, householder, or citizen, towards the pursuit of sustainability conceived on that model. However, what we have been doing in the foregoing considerations is essentially to re-conceive sustainability as the continuous exploratory pursuit, through open-ended learning, of ways to ensure that life goes on. That learning mode of living has ultimately nothing to depend on except our collective social intelligence. However, social institutions and procedures of whatever sort are finally only going to be as intelligent as the individual men and women who constitute and operate them. Sustainability really consists, therefore, in the life-effort of people whose education has equipped them with enough knowledge, sensitivity, emotional range, and moral imagination to act together as a genuinely learning community in modern conditions of ecological responsibility.

Education seen (or rather, recognized anew) in that light will be a matter of the whole curriculum, not of some corner—even some quite substantial corner—allocated to issues of ecology and the human environment. It will require us to re-conceive the purpose of both schooling and tertiary studies as the empowering of integrated, flexible, creative, and well-informed human beings, rather than the production of suitably skilled employees and consumers for the global economy. It will, that is, be a very long way from the current spirit and practice of the curriculum (and not just the sustainability curriculum) in UK schools.

This is a matter for a whole book in itself, and suggestive links are all I have room for here. At the heart of education in genuine sustainability must be developing the capacity for critical understanding. This involves recognizing that there is nothing given, by way of information, which does not have to be interpretively applied in making any kind of knowledge-claim in any field. By the same token, interpretive paradigms will be the upshot or precipitate of past judgements, arising within and expressing the living activity of the culture. From this perspective it is evident that none of them can be taken as definitive; the judgements underwriting any knowledge-claim can only be assessed from the standpoint of other judgements having only the same ultimate status, so our judgement as a whole comes to be understood as necessarily self-dependent and self-warranting. The relation of education so conceived to the inculcation and development of the learning virtues is clear.

It is a mistake to think of induction into such criticality as something which has to wait until Sixth Form or even the university. The study of history offers a very clear example of how critical understanding can be developed. Attention to mere events passes inevitably into attention to how varying accounts of them are recorded and transmitted, in ways which students can begin to appreciate from the early secondary years. Not all fields of study display their inherent criticality as readily as this—notably, study of the natural sciences doesn't. However, we should now have no difficulty in seeing the analogy between scientific and historical paradigms, and their role in relation to the judgements which underpin 'normal-scientific' practice. Again, understanding this is something which can well be begun in the secondary school—indeed the study of environmental issues as such offers an excellent opportunity. Meanwhile, critical thinking skills—of reflectiveness, open dialogue, and attention, as well as logical sharpness— can be practised from a surprisingly young age in the primary school, as various experiments in adapting philosophical enquiry for these age groups demonstrate (see for instance Lipman 1988).

The key point is to see the real sustainability curriculum as, again, a matter of widening our options—that is, our options to understand our situation, imagine possibilities and create the emergent future as we go on. This is to rediscover the thought about education in general which the humanist psychologist Carl Rogers was expressing when he wrote in the 1960s (well before the 'sustainability' discourse, but very presciently) that:

> ... the goal of education, if we are to survive, is the *facilitation of change and learning*. The only man who is educated is the man who has learned how to learn; the man who has learned how to adapt and change; the man who has realized that no knowledge is secure, that only the process of *seeking* knowledge gives a basis for security. (Rogers 1969: 104)

This thought, too, brings us to the final point to be made in connection with bad faith. As it is just mere honesty to admit, the will to life—the urgencies of present satisfaction—will always tend to prevail over merely conceptual considerations about the future. At best, the rational force of these considerations will only be strong enough to produce scientifically-fudged anxieties dignified with the earnest moral portentousness of shadow-stewardship. However, a crucial part of what we also need, here and now, in order to find human life liveable, is a different order of *present* fulfilment—satisfaction of the will to *life-as-sense*. We need the sense of meaningfulness which comes with the recognition of what we fundamentally are: and that is, essentially learners, creatures that thrive in exploratory openness and creativity.

There is of course no guarantee. The tendency to deceive ourselves, so richly and copiously illustrated in our recent dealings with 'sustainable development', remains a permanent pathology of the human condition, whatever may happen to that particular policy framework. A model of ecological sustainability built on the learning virtues is, however, at least directly opposed to, rather than covertly in league with, this tendency. The vice of hiding things from oneself to protect oneself is directly challenged, and may sometimes even be overcome, when we explicitly recognize the virtues of acknowledging the genuinely emergent even at the expense of one's present existential comfort. Sustainability, like the security of which Rogers writes, is ultimately not about playing fair with the future—it is about not playing false with ourselves.

Notes

1. *Letters*: To George and Thomas Keats, 21st December 1817.
2. Cf. Socrates: 'We shall be better, braver and more active men if we believe it right to look for what we don't know'—*Meno*, 86b.
3. As we find out, always, in trying to put experience into words—since
   one has only learnt to get the better of words
   For the thing one no longer has to say, or the way in which
   One is no longer disposed to say it. (T.S. Eliot, *Four Quartets* ('East Coker')).
4. Eliot again:

   There is, it seems to us
   At best, only a limited value
   In the knowledge derived from experience.
   The knowledge imposes a pattern, and falsifies,
   For the pattern is new in every moment
   And every moment is a new and shocking
   Valuation of all we have been. (*ibid*)

5. Lines in the sand are drawn to stop oneself fudging, as much as to stop whatever it is one wants to stop fudging about.

References

Argyris, C. and Schon, D. (1974) *Theory in practice: increasing professional effectiveness* (San Francisco: Jossey-Bass).

Bateson, G. (1980) *Mind and Nature: a Necessary Unity* (New York: Bantam Books).

Eliot T.S. (1963) *Collected Poems 1909–1962* (London: Faber).

Foster J. (2005) Options, sustainability policy and the spontaneous order. In J. Foster and S. Gough (eds), *Learning, natural capital and sustainable development: options for an uncertain world* (London: Routledge), 111–131.

Foster J. (2008) *The Sustainability Mirage* (London: Earthscan).

HMSO (1990) *This common inheritance: Britain's environmental strategy* (London: Her Majesty's Stationery Office).

La Rochefoucauld, F. (1967) *Maximes,* no.405 (Paris: Editions Garniers Freres) (my translation).

Lipman, M. (1988) *Philosophy goes to school* (Philadelphia: Temple University Press).

Marris, C., Wynne, B., Simmons, P. and Weldon, S. (2001) *Public perceptions of agricultural biotechnologies in Europe: final report of the PABE research project* (Lancaster University: CSEC).

Porter, T. (1995) *Trust in numbers: the pursuit of objectivity in science and public life* (Princeton, NJ: Princeton University Press), 44.

Rogers C. (1969) *Freedom to learn* (Columbus, OH: Merrill).

Schopenhauer, A. (1818/1966 trans.) E.F.J. Payne, 2 vols. (New York: Dover).

Sterling S. (2001) Sustainable education: revisioning learning and change (Totnes: Green Books), 15–16.

UNESCO (2005) Available online at: http://portal.unesco.org/education/en/ev.php-URL_ID=23295&URL_DO=DO_TOPIC&URL_SECTION=201.html, accessed 8 September 2009.

Urry J. (2005) The complexity turn. *Theory, Culture & Society*, 22(5), 1–14.

von Hayek, F. (1960) *The Constitution of Liberty* (London: Routledge and Kegan Paul).

Ideology, political education and teacher education: matching paradigms and models

JOHN FIEN

In April 1989, ten broad goals for a national curriculum were adopted by the Australian Education Council, the 'peak' educational policy group in Australia. Two of these goals are relevant to the theme of this paper. These focused on:

1. the development of 'knowledge, skills, attitudes, and values which will enable students to participate as active and informed citizens in our democratic society';
2. the promotion of 'judgment in matters of morality, ethics, and social justice'.

On one level, these two goals might be seen to parallel Giroux's (1980: 357) call for education to promote civic courage and a sense of justice:

> If citizenship education is to be emancipatory, it must begin with the assumption that its major aim is not 'to fit' students into the existing society; instead, its primary purpose must be to stimulate their passions, imaginations, and intellects so that they will be moved to challenge the social, political, and economic forces that weigh so heavily upon their lives. In other words, students should be educated to display civic courage, i.e., the willingness to act as if they were living in a democratic society. At its core, this form of education is political, and its goal is a genuine democratic society, one that is responsive to the needs of all and not just of a privileged few.

On another level, however, such goals beg questions about what is meant by politics, democracy, and citizen participation, and the form that education for political understanding and citizenship should take. Such an analysis is especially important, given the use of such concepts by politicians and the use of the language of democracy and active citizenship by educators from widely divergent political perspectives. Much educational research and theorizing in recent years has sought to clarify central issues in the debate about the purposes and impact of schooling in the area of citizenship or political education (Kickbusch 1985). However, this debate usually has not been linked to the role played by different approaches to teacher education in political education, and the different conceptions of political education that teacher education might serve. In fact, explicit efforts to provide teachers and student teachers with the political understandings and curriculum insights and skills to make a reality of (any form of) the rhetoric of political education in schools, let alone debate, are reported as being quite rare in teacher education (Curriculum Review Unit Initial Training Panel (CRUIT) 1983).

The role of teacher education for citizenship education was first addressed publicly in Australia at a conference on political education initiatives in teacher education in October 1989. This conference was one of the first outcomes of an enquiry by the Australian Senate Standing Committee on Employment, Education and Training (ASSC) (1989) into 'education for active citizenship in Australian schools and youth organizations'. The all-party Committee received 135 submissions, including oral evidence from 68 witnesses, in the second half of 1988 and issued its report, *Education for Active Citizenship*, in February 1989. This report describes in detail a general failure in Australian schools successfully to teach factual (names of politicians, etc.), conceptual ('democracy', 'preferential voting', etc.), and functional (how and when to vote, etc.) knowledge, or to develop a commitment in young people to democratic values and involvement. One conclusion drawn from these findings was that 'the quality of Australian democracy . . . [is] under threat' because 'hand in hand with ignorance goes apathy and a sense of powerlessness . . . indifference and an unwillingness to become involved right down to the local level' (ASSC 1989: 8, 13). Many submissions sought to account for this situation, with the major explanations invoking (1) the low status given to citizenship education in schools, (2) a lack of adequate teaching and learning resources, and (3) major shortcomings in preservice and inservice teacher education in political education. A variety of plans to address these three problems was at the centre of the report's recommendations.

This paper is concerned with a recommendation of the ASSC (1989) which called on the Australian federal government to:

> . . . ask all higher education institutions with responsibility for teacher education to ensure that education faculties recognize the importance of education for active citizenship and make provision for it as a component in preservice courses, particularly for those teacher education students who are likely to teach in social studies and related areas of the curriculum.

This recommendation was implemented by the Parliamentary Education Office which convened a three-day conference on 'parliamentary' (not political or citizenship) education for teacher-educators in October 1989, and subsequently wrote to all education faculties in Australia informing them of the above recommendation. However, the general curricular autonomy of higher-education institutions in Australia means that action on this request will depend upon the priorities of individual schools of education, the micropolitics of course-planning committees, and the political skills of interested staff in these institutions. Thus, as most decisions about educational change are political decisions and reflect changing patterns of social priorities and power, decisions about appropriate curriculum responses in teacher education also will be political decisions. They are political in at least three senses:

1. choice of model or interpretation of political education to be adopted;
2. decisions about the style of course(s) and pedagogy to be employed in promoting active citizenship in teacher education;
3. power relationships in course-development committees and the role, status, and time they allocate to the promotion of active citizenship in the preservice (and continuing) education of teachers.

Planning appropriate teacher education studies to reflect the second issue, and developing strategies for addressing the third, are the responsibilities of individual institutions and their education faculties. As a preliminary to such deliberations, this paper explores the first sense in which promoting active citizenship in teacher education is political. Hence, it raises issues and questions about the nature of the political education advocated in *Education for Active Citizenship*, especially its interpretations of politics and education, and the ideology underpinning the model of political education it advocates. The central concepts, findings, and recommendations of the ASSC report are analysed in relation to different models or 'paradigms' of teacher education and political education in order to problematize the ways in which the discourses of democracy and political education are used in the report. The reason for attempting to clarify these issues is to uncover the form(s) of teacher education that might most appropriately complement an emancipatory form of political education. Thus, in sequence, this paper explores the following questions:

1. What paradigm(s) of teacher education can promote active citizenship?
2. What model(s) of political education is/are advocated in *Education for Active Citizenship*?
3. What directions for future action can be developed by matching paradigms of teacher education with models of political literacy?

Alternative paradigms in teacher education

Writing of the political role of schooling, Woods (1984: 220) describes a 'paradox . . . central to the educational enterprise in a democracy':

> On one hand, the state strives to obtain legitimacy and stability for the existing order. On the other hand, it embraces a creed that stipulates the right of citizens to alter any existing social arrangement. Children are caught in the middle of this ideological tug-of-war through the educational process. On one hand, schooling is to make children fit the system, on the other it is to help children remake the system to fit them. Even this paradox can be overcome by a democratic culture if that culture provides for an equal, just, and plentiful social order. Yet when that existing order begins to deny the basic material and social goods to certain of its members, schooling is forced into deciding between the social order or the interests of each of its members.

Connell *et al.* (1982: 208) describe the socially-critical type of approach, as advocated by Woods, as 'the only education worth the name [education]'. However, if education is to serve emancipatory interests, the political assumptions in alternative approaches to education – and to teacher education – need to be clarified. As Giroux and McLaren (1986: 224; my emphasis) argue:

> At issue here is whether schools of education are to serve and reproduce the existing society or to adopt the more critical role of challenging the social order so as to develop and advance its democratic imperatives. Also at issue is *developing a rationale for defining teacher education programmes in political*

terms that make explicit a particular view of the relationship between public schools and the social order, a view based on defending the imperatives of a democratic society.

In choosing an approach to teacher education, especially to promote active citizenship, teacher educators need to be mindful that, like schooling, all approaches to teacher education are based upon, and in turn support, particular ideologies. As Spodek (1974: 89) argues:

> All teacher education is a form of ideology. Each programme is related to the educational ideology held by a particular teacher educator or teacher education institution, even though the relationship may not be explicit. There is no such thing as a value-free teacher education just as there is no such thing as a value-free education for children.

Models of teacher education

A number of ideologically oriented models of teacher education can be identified. Kirk (1986), for example, has identified three dominant paradigms in teacher education, namely the traditionalist, the rationalist, and the critical or radical.

Kirk's first paradigm, traditionalism, sees teaching as a technical craft and teachers as craftpersons who develop a body of practical knowledge largely from trial and error in their work. The central role of teacher education from this perspective is to provide the knowledge and skills that comprise good practice. Student teachers are seen as 'trainees' whose major studies comprise methods courses related to prolonged periods of teaching practice. Kirk (1986: 158) characterizes this approach to teacher 'training' as the 'sitting with Nellie' approach in which 'Nellie is the factory worker (or "master teacher" or "methods lecturer") who has been doing the job for years and to whom new recruits are attached while they learn the job'.

Kirk sees this traditionalist approach to teacher education as inherently conservative in its social orientation and the interests it serves. Liston and Zeichner (1988: 22) describe its social impact this way:

> Prospective teachers are viewed largely as passive recipients . . . and play little role in determining the substance and direction of their preparation for teaching. The political thrust of this approach is essentially conservative. Preparing prospective teachers to fit smoothly into existing teaching roles encourages the continued acceptance of the educational, political, and social contexts in which teaching now occurs.

Giroux (1981: 156) describes the resultant 'paradox of teacher education' as one in which teacher education works to reproduce and legitimate social inequality through the influence of a hegemonic ideology that sees all curriculum and teaching problems as technical ones. Thus, teacher-education programmes based upon the traditionalist paradigm are very likely to result in the 'production of uncritical teachers' (Kirk 1986: 156) who are unable to stand back from their experience of teaching 'to make problematic the basic beliefs, values, and structural socio-economic arrangements of . . . society' (Giroux 1981: 152).

Kirk's second paradigm of teacher education is based upon rationalism. A rationalist paradigm of teacher education sees teaching as an 'applied science' based upon the proven 'best' current practices in education. Whereas traditionalist teacher education involves a relatively haphazard socialization into the skills and the work context of teaching, rationalist programmes of teacher education are based upon principles and practices derived from 'scientific' research on teaching. Microskills programmes (see, for example, Turney et al. 1983) and certain approaches to reflective teaching (see, for example, Cruickshank 1985, Cruickshank and Armaline 1986) might be seen as examples of this approach. Like the traditionalist paradigm, this approach to teacher education sees teaching in technical rather than moral terms. Decisions about educational purposes and curriculum content, and the interests served by such decisions are not part of the rationalist approach. Thus, in terms of teacher education for political education, the rationalist paradigm is also an essentially conservative activity.

Kirk's third paradigm of teacher education is a radical or critical approach which encourages student teachers to reflect critically upon the craft of teaching, the social context of education and the interests served by alternative educational practices. The aim of such reflection is the promotion of critical and emancipatory practices in schools. As Connell et al. (1982: 208) have argued:

> Education has fundamental connections with the idea of human emancipation . . . In a society disfigured by class exploitation, sexual and racial oppression, and in chronic danger of war and environmental destruction, the only education worth the name (education) is one that forms people capable of taking part in their own liberation. The business of the school is not propaganda. It is equipping people with the knowledge and skills and concepts relevant to remaking a dangerous and disordered world. In the most basic sense, the process of education and the process of liberation are the same.

Preparing prospective teachers to promote this emancipatory style of education requires both reflection and action to be used in an analysis of the purposes, craft, contexts, and effects of teaching. While the term 'reflective' in 'reflective teaching' and 'reflective teacher-education' has been appropriated to refer to non-critical reflection on some occasions, a critically reflective approach to teacher education has the potential to provide prospective teachers with learning that can prepare them as both role models and reflective guides to active citizenship for their students. Such teachers have been described as 'transformative intellectuals' (Giroux 1983, 1988) who see 'education as a virtue-laden social practice and as an exercise in reflective inquiry' (Liston and Zeichner 1988: 8).

The purpose of this brief summary of three paradigms of teacher education is to outline the importance of analysing the ideology underlying alternative educational programmes in teacher-education institutions (or schools). If Kirk's analysis of teacher-education paradigms is valid, then the reports in *Education for Active Citizenship* of the low and relatively unsuccessful impact of teacher education in promoting active citizenship should not be unexpected. The dominance of traditionalist and rationalist approaches to teacher education in Australia have prevented the education of a profession skilled in, and committed to, promoting active citizenship participation in schools. These dominant paradigms of teacher education give rise to models of political education in schools which, at best, promote the rhetoric of participation – within existing social and political frameworks – and, at worst, disenfranchise most students from the political process. This is because their academic and institutional orientations are unrelated to students' everyday experiences of politics; or as the Queensland Department of Education's submission to the ASSC (1989: 28) lamented:

> Citizenship learnings tend to be descriptive and functional, often trivial, usually bookish and removed from everyday experience . . . confined to the classroom and school environment, with limited application to community life.

However, the thoughtful adoption and implementation of the political education learnings and forms of learning experiences for preservice teachers recommended in *Education for Active Citizenship* requires more than a critique of existing practices or an analysis of ideological paradigms in teacher education. It also requires an analysis of alternative approaches to political education.

What model of political education is advocated in *Education for Active Citizenship*?

In the UK in the early 1980s, an interuniversity panel was established to improve the provision of political education experiences in teacher-education courses. Its report, *Teaching Political Literacy: Implications for Teacher Training and Curriculum Planning* (CRUIT 1983), canvassed a range of models of political education and recommended objectives, key concepts and issues, and readings for institutions to adapt to their own needs in foundation and specialist curriculum-studies courses. A brief analysis of the five models of political education identified in this report helps to clarify the more general contest over meanings in political education and is informative in analysing the approach to political education advocated in *Education for Active Citizenship*. The five models are:

1. A conservative model which derives its aims from a belief in the need to maintain existing social, economic, and political structures. This model sees providing knowledge of how the present political system is believed to function, and of the duties and responsibilities of citizens, as the main task of political education. This model promotes a values commitment to supporting the existing system, often under a guise of 'values neutrality'.
2. A liberal model which derives its aims from the perceived political needs of individuals and a belief in the rights of the individual. The liberal model sees empowering individuals with the knowledge and participatory skills to achieve their goals and rights within the system as the main task of political education. This model promotes a

commitment to liberal-democratic beliefs such as majority rule, minority rights and freedom of speech, as well as more traditionally conservative values such as rule of law, respect for authority, the sanctity of private property, etc. It also promotes a commitment to the procedural values of tolerating all viewpoints, and carefully considering all views and supporing evidence before acting.

3. An apolitical model which derives its aims from the perceived educational needs of individuals and a belief that knowledge drawn from the publicly acknowledged disciplines or forms of knowledge can develop the mind. A well developed mind is thus seen as the hallmark of a well educated person who, with knowledge of how the system works and how the citizen can act within it, also can operate effectively as a good citizen. The values underlying this approach are said to be 'educational' rather than 'political', but are derived from a commitment to conservative beliefs in existing social arrangements and liberal views of the needs of individuals.

4. A reformist or reconstructionist model which derives its aims from the perceived political needs of individuals, especially as members of particular groups based upon class or race. The reformist model seeks to achieve the knowledge, values, and participatory objectives of the liberal model but adds to it a concern to empower those with limited access to power with the means and self-confidence to adopt a more active political role.

5. A radical or socially critical model which derives its aims from a critical analysis of the power relations in current social arrangements and the need to challenge them if justice and fairness for all is to be achieved. This model promotes similar knowledge, skill, and participation objectives to the liberal and reformist models but adds to them a pedagogical dimension which promotes skills for reflective action and learning by doing (through social action rather than analysis of, and reflection on, action), and skills for collective rather than individual participation.

These five models represent differing interpretations of the nature of politics, the location of the political arena, the nature of political actors, and the role of education. Nevertheless, they are not five distinct models in that several share similar concerns, with many differences between them reflecting degrees of emphasis rather than different emphases. As a result, Williamson-Fien (1987) suggests that the various models might be reduced to three with little loss of meaning.[1] Table 1 outlines the main features of Williamson-Fien's three models, as well as their relation to paradigms of teacher education. She describes the first model as conservative; it emphasizes what students need to know about politics and how the political system operates. She describes the second model as liberal-pluralist; it principally addresses what students need to be able to do in order to play an active and effective part in civic life. Because Williamson-Fien's paper was specifically about the political education of women and girls, she describes the third model as a feminist approach to political education. This model addresses the knowledge, skills, and values that students need in order to recognize, evaluate, and challenge

Table 1. Three models of political education (after Williamson-Fien 1987: 61).

Political arena	Politics	Political actors	Political education	Teacher education paradigm
Conservative				
The public sphere with particular emphasis on formal institutions	Knowledge about government institutions, formal procedures, the rights and duties of the citizen	Individuals as unitary abstractions, formal groups – e.g., political parties	Civics and citizenship courses featuring institutional knowledge; traditional structures and the individual's place in them	Traditionalist
Liberal/pluralist				
The public sphere with more emphasis on informal activities, lobbying etc.	Conflict resolution via active participation (conscious and deliberate) in political processes – elections, lobbying etc. Concern to influence decision-making at all levels in public life	Individuals as unitary abstractions, formal groups, informal lobby groups	Political literacy approach – issues based, promoting skills for active participation in public life	Traditionalist – rationalist
Socially critical (feminist)				
Social life as a whole. Distinctions between public and private realms essentially false	The conscious and unconscious articulation of relationships in pre-existing power structures – notably patriarchy and capitalism. The importance of the personal as political	Everyone: men, women, children, groups, families etc., in any social context	The development of skills to recognize, evaluate, and challenge the dynamics of existing power structures and their implications for girls (and boys). The promotion of autonomy – self definition for all irrespective of gender, race, and class	Emancipatory

the dynamics of existing power structures (especially as they affect women and girls) through collective reflective action. While Williamson-Fien and other feminist educators might insist that the special concerns of the feminist model dictate that it stand alone as a distinct approach to political literacy, the conceptions of politics, the political arena, political actors, and education in the feminist approach does make it very similar to the socially critical model of political literacy as described by CRUIT (1983).

Where does *Education for Active Citizenship* fit in this pattern of models of political education? In its first paragraph the report (ASSC 1989: 7) outlines its view on the nature of active citizenship and the knowledge, skill, and values objectives of political education:

> What is meant by the term 'active citizenship' and why was it chosen for this inquiry? The Committee does not equate active citizenship with knowledge about politics. An active citizen is not someone who has simply accumulated a store of facts about the workings of the political system, someone who is able to perform well in a political quiz. An understanding of how the social and political systems work is an essential element, but equally important is the motivation and the capacity to put that knowledge to good use. Essentially, it is a question of active commitment to democracy. An active citizen in the Committee's view is someone who not only believes in the concept of a democratic society but who is willing and able to translate that belief into action. Active citizenship is a compound of knowledge, skills and attitudes: knowledge about how society works; the skills needed to participate effectively; and a conviction that active participation is the right of all citizens.

On the surface, this statement appears to rule out the conservative and socially critical models and to locate the report and its recommendations in the liberal-pluralist category. A liberal-pluralist orientation in the report may also be found in several of its other aspects: for example, its concern that education be restructured to help students overcome personal apathy and powerlessness, its appeal to the vocational utility of political education, and a number of cautions about bias in, and possible negative effects of, social action.

Thus, on one level the ASSC report warrants support because it lends authority to those educators who have decried the conservative 'knowledge only' approach to social education, not just political education, for so long. The report also highlights the important need for teacher-education institutions to redress the self-perpetuating inadequacies in their courses and, in so doing, lends authority to those who have not been successful in raising the profile of political education (or the related areas of global, development, and environmental education) in their own institutions, ASSC's recommendations for action across all educational sectors, not just teacher education or formal schooling, is also significant.

Nevertheless, it is within the spirit of this positive assessment of some aspects of the ASSC report that I suggest that its underlying model of political education should be analysed in detail. My conclusion is that, while the rhetoric of the report is liberal-pluralist, the message or practical ideology of its hidden curriculum is, essentially, conservative. There are three reasons for this claim: (1) *Education for Active Citizenship* is based on a narrow definition of politics; (2) it uses a narrow research base to justify its conclusions; and (3) it fails to address questions of power in Australian society.

A narrow view of the nature of politics: Education for Active Citizenship eschews the conservative definition of the political arena as the preserve of political parties and, instead, advocates the liberal-pluralist notion that the political arena is for everyone and that individuals can play a role in influencing political decisions. However, the report is silent on what the ASSC understands by the nature of politics and democracy. The report talks about participation, but participation in what type of politics? Planning a political-education (and teacher-education) curriculum to foster participation in personal politics and community politics is a fundamentally different task from planning a curriculum to promote participation in party politics and representative democracy. There is a similar difference between a curriculum for participation in a participatory democracy and in a representative democracy. It might be said that the intelligent reader of the ASSC report should be able to deduce the conceptions of politics and participation upon which it is based. However, it is as one tries to make such a deduction that mixed, but essentially conservative, messages about the nature of its view of politics emerges. For example, the report's endorsement of a view of Australia as one of 'the most enduring continuous democracies' in which governments 'have come to power by the free vote of the people and which at all times held themselves effectively accountable to the people' promotes a conception of politics as a form of accountability by governing élites to the wishes of majorities of the voting public. This conservative view contrasts with the conflict model of politics implicit in the liberal-pluralist concept of politics (*see* Table 1) which sees politics as a process of conflict resolution according to democratic principles.

However, in whatever sense the conception of politics in *Education for Active Citizenship* is read, the report endorses a conservative institutional view of politics in which democratically legitimate groups make decisions on behalf of a broad constituency. This is an unnecessarily restricted view of politics for, as Gould (1988: 1) argues, the concept of politics has to be broadened so that 'democratic decision-making should not only apply to politics but be extended to economic and social life as well'. And as Aronowitz and Giroux (1985: 140) argue, there is a need 'to rescue the concept of politics from its conservative and liberal advocates by arguing that its meaning should include the entire way we organize social life along with the power relations that inform its underlying social practices'. Such alternatives to an institutional view of politics are provided by political economists and feminists. For example, Leftwich (1983: 11–12) suggests that:

> Politics consists of all the activities of cooperation and conflict, within and between societies, whereby the human species goes about obtaining, using, producing and distributing resources in the course of the production and reproduction of its social and biological life. These activities are not isolated from other features of social life. They everywhere influence, and are influenced by, the distribution of power and decision-making, the systems of social organization, culture, and ideology in a society, as well as its relations with the natural environment and other societies. Politics is therefore a defining

characteristic of all human groups, and always has been. The way people use and distribute resources – their politics – also helps to explain the problems which occur within or between societies, institutions, or groups, whether it be unemployment, war, famine, disease, overcrowding, or various forms of conflict.

Feminists add to such a political economist's conception of politics the principle of dialectical relationships between power, social arrangements, and personal life. As Williamson-Fien (1987: 57) writes:

> ... in general feminists argue that the political arena is synonymous with social life as a whole and that politics represents the conscious and unconscious articulation of power relationships in any social context. Politics, therefore, is not simply a matter of who occupies *The Lodge*, or what issues are attracting public lobbying activity, but who (for example) decides and who accepts responsibility for the household chores and why one particular type of household 'agreement' on these matters is common. Personal life is undoubtedly political and any attempt to deny this must be seen as a political act in itself.

None of these views of politics and the related approaches to active citizenship is reflected in *Education for Active Citizenship*.

A narrow research base: Arguments to justify new emphases or directions in education generally focus on the social benefits likely to accrue from addressing particular deficiencies among students. Such individualistic explanations are typically superficial and cloud the social, political, and economic problems that give rise to the perceived deficiencies. In blaming the victim, little attention is paid to wider questions that might challenge the status quo. The research base in the ASSC report is comprised chiefly of such individualistic explanations. This is not the place to challenge the reported evidence of low levels of student knowledge of political personalities and institutions, or their low level of interest in institutional politics, and feelings of powerlessness (though questions do need to be asked about sample size and the comprehensiveness and validity of several of the reported studies). Rather, I argue that the evidence submitted to the ASSC to justify the need for political education does not constitute a (1) relevant, (2) convincing, or (3) comprehensive case because of the narrow research base upon which it draws and the analytic difficulties it poses.

1. *Irrelevant evidence.* In itself, data on low levels of political knowledge do not constitute valid evidence to justify a liberal-pluralist approach to political education. Evidence which supports a conservative approach to political education (low knowledge levels) is not relevant to an argument for a liberal-pluralist approach. The liberal-pluralist argument requires evidence of low levels of political interest and an aversion to involvement in political matters.

2. *Unconvincing evidence.* Evidence of disinterest in political matters is provided in *Education for Active Citizenship*. However, it relates only to low student interest in the dominant institutional and adult party-political conceptions of politics at the local, state, and national levels. Evidence which suggests that young people are interested in personal-political or

political-economic problems paints quite a different picture of their political awareness and interest (*see*, for example, Australian National Opinion Polls (ANOP) 1984, Youth Council of Australia 1984, Finlayson *et al.* 1987, Eckersley 1988). Issues in which young people are very interested and which they might perceive as relevant areas in which to assess their political knowledge and interests include: personal relationships, environmental problems, nuclear war, racism and sexism in everyday life, drug problems, education, and alternative job-creation schemes. Unfortunately, no evidence related to the possibility of active citizenship in these areas is canvassed in *Education for Active Citizenship*. The Youth Council of Australia (1984) study confirms the findings of other studies of youth alienation from institutional politics, but it also indicates that young people are politically motivated and active, and are more likely than adults to participate in political activities such as joining a political party, signing a petition, participating in boycotts and strikes, and so on.

Evidence on the political socialization of young Australians dating back nearly 20 years (Connell 1971: 228–240) also indicates that children and adolescents are 'active, enterprising beings' who 'selectively appropriate' from the range of political issues. Connell urged caution about research which 'distorts' accounts of political interest by failing to 'recognize and account for the conscious creative activity of the children themselves in the development of their own beliefs'.

3. *An uncomprehensive case*. Much of the research evidence submitted to the Senate Standing Committee was related to student knowledge and interest in conservative views of politics and citizenship. This evidence was decontextualized from the lifeworlds of Australian young people and failed to take account of the social structures and perceptions that govern their lives. As I have already noted, there is a growing body of literature within (Youth Affairs Council of Australia 1984) and outside (Wilson and Arnold 1986) mainstream life in Australia which draws one to the conclusion that *Education for Active Citizenship* might really be about citizenship in a 'world' which young people do not recognize as their own.

Education for Active Citizenship is the product of the evidence submitted to the ASSC and should not be criticized for not addressing matters that were not submitted as evidence. However, when judged against this wider body of research and a wider conception of politics, the ASSC report must be seen as advocating a liberal-pluralist model of political education with strong overtones of conservatism.

Neglect of power relationships in Australia: Australia is not an equal society. It is a society in which many groups suffer from marked inequalities of access to income, services, and political power (Hollingsworth 1983, Bottomley and de Lepervanche 1984). This is a result of the institutionalization of patriarchy, racism, and class differences in Australian society which results in differentiated access to the means of production, distribution, and consumption of economic, cultural, and political resources. There are sharp distinctions between those who enjoy the profits from economic development and the control of resources and those who pay the costs of environmental and social disruption. *Education for Active Citizenship* does not consider these power relationships and argues idealistically, and perhaps naïvely, that education can help all young people to participate effectively in active citizenship.

A similar neglect of social power relations was found in the UK Programme for Political Education (Crick and Porter 1978), and Vincent (1978: 14) has suggested that the programme 'did not propose to repeal the iron law of oligarchy' but assumed that 'the élite, acting in the name of equality' would share access to political power with the rest of society. He argues that this view of political education gives 'no serious answer to the question of what a society of politically literate people all taking part in making arrangements would involve'; unless power relations are challenged, all that political education can achieve is to educate the majority of citizens with 'the vocabulary and responses of an élite which they cannot in principle join'.[2] Similarly, Williamson-Fien (1988: 144) argues that liberal-pluralist approaches to empowering students, devoid of a structural analysis of power relations, 'run the risk of creating idealistic and individualistic people who become alienated, insecure, or power hungry' and who are merely inducted 'into existing structures of power and powerlessness'.

Unfortunately, these criticisms might be applied equally to *Education for Active Citizenship*. The conservative underpinnings of its liberal-pluralist model of education explain how a report which set out to encourage active citizenship can serve a limited conception of politics and political education. As Woods (1984: 228) writes, this brings us:

> . . . face to face with the paradoxical nature of education's social role: in attempting to serve students, the overriding logic of the state directs educators to work at fitting students to the pre-existing roles for them in the cultural, political, and economic matrix of postindustrial capitalism. Of course, to adopt such roles may, on the surface, make a great deal of sense. But in the long run the roles and actions endorsed and embraced limit and perhaps destroy the abilities, hopes, potentialities, and dreams students have for a better world. Thus, civic education works to reproduce the established social order through limiting the sphere of democratic operations and further refusing to develop the skills needed by democratic citizens critically to examine claims to objective truth, to challenge the opinions of experts, and to utilize their own histories in both opposing the dominant order and building a new one.

Such an analysis matches Bernstein's (1986) model of pedagogic discourse which denies that the language of government reports, policies, and syllabuses is context-free. Instead, Bernstein sees 'official pedagogic discourse' as located within national and global social, economic, and political contexts which generate a 'pedagogic discourse of *reproduction*'.

Political education and teacher education

I have argued two points so far. First, the emancipatory or critical paradigm of teacher education is the one most suited to raising the political consciousness and skills of prospective teachers. Other models tend to serve the technocratic interests of schooling as an agency for social reproduction. Second, the conservative and liberal-pluralist models of political education also serve conservative interests. *Education for Active Citizenship* can be interpreted as conservative, even though this may not have been its intent. Two conclusions may be drawn from these arguments. First, *Education for Active Citizenship* contains a number of flaws and contradictions, and may be judged more favourably for the assistance it lends to arguments for a greater emphasis on political education in teacher education than for the direction it provides for critical, emancipatory political education. Second, this conclusion leads to an argument for matching the critical paradigm of teacher education with the socially critical model of political education. This involves, in Giroux and McLaren's (1986: 227) terms, 'reconstructing teacher education programmes around a new vision of democratic schooling and teaching for critical citizenship'.

In commenting on the failure of teacher-education courses to address the needs of education for active citizenship, the ASSC (1989: 47–48) noted that:

> At present the problem is self-perpetuating. Students entering teacher education institutions lack knowledge and interest in these [citizenship and participation] topics because they have passed through a school system in which this area of the curriculum is often badly taught. At the conclusion of their training, large numbers of teachers remain ill-equipped because their teacher education courses have provided, at best, a patchy coverage. Inadequately trained teachers then continue to offer inadequate active citizenship education in schools. It is essential that this cycle be broken by decisive action to improve provisions for teacher education.

This cycle cannot be broken by reforms in curriculum-studies courses in political science, Australian history, or social education – or even by providing some opportunities for active citizenship participation – in existing teacher-education programmes. Such reforms, even if they sponsor a socially critical approach to political education, fail to address the conceptions of education and the nature of teachers' work that are part of the hidden curriculum of teacher education. And, course-development projects such as the CRUIT 1983 report, *Teaching Political Literacy*, which outlined a minimum preservice teacher-education curriculum for political education for both educational foundations and applied curriculum-studies ('methods') courses must also be seen as premature. *A wider restructuring of teacher education is needed.* Thus, as Zeichner (1983: 8) argues, debate in teacher education must first address:

> . . . the question of which educational, moral, and political commitments ought to guide our work in the field rather than. . .merely dwelling on which procedures and organizational arrangements will most effectively help us realize tacit and often unexplained ends. Only after we have begun to resolve some of these necessarily prior questions related to ends should we concentrate on the resolution of more instrumental issues related to effectively accomplishing our goal.

Giroux and McLaren (1986: 222) argue that, in the development of a critical emancipatory framework for teacher education, 'the fundamental concerns of democracy and critical citizenship should be central to any discussion of the purpose of teacher education'. One starting point, they argue, should be a recognition of the importance of educating student teachers in the languages of critique and possibility by:

> ... providing teachers with the critical terminology and conceptual apparatus that will allow them not only to critically analyse the democratic and political shortcomings of schools, but also to develop the knowledge and skills that will advance the possibilities for generating curricula, classroom social practices, and organizational arrangements based on and cultivating a deep respect for a democratic and ethically based community. In effect, this means that the relationship of teacher education to public schooling will be self-consciously guided by political and moral considerations.

A number of teacher-education approaches and courses based upon this critical paradigm have been described (for example, Goodman 1984, 1986 a, b, Giroux and McLaren 1986, Popkewitz 1987, Robottom 1987, Schön 1987, Shor 1986, 1987, Liston and Zeichner 1988, Ferguson 1989, Mercer and Abbott 1989 a, b, Ross 1989, Smyth 1989). Liston and Zeichner (1988) have summarized the common feature of many of these programmes as a study in 'cultural politics' that provides prospective teachers with concepts and skills for 'making the everyday problematic' in their own lives and for their pupils. Such programmes are designed to broaden prospective teachers' conceptions of the role of teacher and to help them to examine the political, social, and ethical issues that determine the goals, content, and methods of teaching. Such programmes model the style of learning they are promoting for schools by constructing the teacher-education process as a model of democratic and equitable ways of learning and self-empowerment. The critical approaches explored by Liston and Zeichner include: action research, practical ethnography, journal writing for critical reflection, experience in curriculum negotiation and curriculum development, and a restructuring of field-based experiences to enhance occasions for reflection and critical learning. A further characteristic of the critical paradigm of teacher education recommended by Liston and Zeichner (1988: 33) relates to the personal-political lives of the staff involved in that they:

> ... must serve as living examples of the very kind of critically oriented pedagogical practices that they seek to have their students adopt. This means that teacher educators need to reflect critically and act strategically upon the nature of their own pedagogical practices and the institutional contexts in which they work.

Promoting school-level reform in teaching for active citizenship demands reform at the teacher-education level. There is a challenge for all teacher-educators in Liston and Zeichner's proposals, and in their description of types of teacher-education programmes that can empower teachers to be transformative intellectuals who can promote the concepts, skills, and attitudes of active citizenship. However, that possibility depends upon the ethos of the whole teacher-education curriculum – not on the piecemeal reforms that committed individuals may make. Discussions about a low level of political knowledge and interest on the part of student teachers or the general failure of teacher-education institutions to provide adequate studies in political science or political education are diversionary. Such discussions address the symptoms, not the causes – which reside in the ideology uderpinning the courses. Developing new curriculum materials for political education or new ways of planning and teaching particular subjects in existing programmes, without an engagement in the critique of ideology, only contributes to the maintenance of the reproductive functions of teacher education.

Conclusion

On one level, this paper might be seen as pessimistic, and nihilistic. It could be read as a criticism of *Education for Active Citizenship* and of the efforts of many teacher-educators who work actively to promote among their students the political and pedagogical skills needed for active citizenship. However, nihilism is not my intention. Like such colleagues, I despair at the lack of political (and social, and economic, and environmental) knowledge and interests of many of my students. With like-minded colleagues, I plan alternative approaches and courses, and I celebrate when I notice the buds of conscientization in some students and hope that they will blossom into critically reflective transformative intellectuals in their personal lives and teaching careers. I struggle in my own institution with the institutional politics of course-development committees in a bid for course structures and processes which could be supportive of critically reflective approaches to teacher education, and I despair at the chasms between the rhetoric of course rationales and their day-by-day implementation.

This paper stands as an exercise in clarifying my own thinking on the sort of teacher-education programme that might support work in promoting active citizenship through teacher education. As Zeichner (1983: 8) wrote, we must first 'resolve some of those necessarily prior questions related to ends' before 'we concentrate on the resolution of more instrumental issues related to effectively accomplishing our goal'.

Acknowledgement

I with to acknowledge the advice of Noel Gough and two reviewers for assistance with this paper.

Notes

1. The common two-category classification of models of political literacy developed by Osborne (1984: 4041) and Land (1987: 64) was rejected for the purpose of this analysis as it does not provide for an understanding of an approach to political education based upon notions of a critical social science.
2. Gilbert (1987: 17) is critical of Vincent's arguments on the grounds that (1) Vincent's focus is on participation in national politics, and (2) those power relations do not need to be addressed so much in community and personal politics. While Gilbert's first point is valid, his second requires a more careful evaluation.

References

ARONOWITZ, S. and GIROUX, H. (1985) *Education under Siege: The Conservative, Liberal and Radical Debate over Schooling* (London: Routledge and Kegan Paul).

AUSTRALIAN NATIONAL OPINION POLLS (ANOP) (1984) *Young Australians Today: The New Traditionalists* (Canberra: AGPS).

AUSTRALIAN SENATE STANDING COMMITTEE ON EMPLOYMENT, EDUCATION AND TRAINING (ASSC) (1989) *Education for Active Citizenship* (Canberra: AGPS).

BERNSTEIN, B. (1986) On pedagogic discourse (revised). *Core*, 12 (1): 1–8.

BOTTOMLEY, G. and DE LEPERVANCHE, M. (eds) (1984) *Ethnicity, Class and Gender in Australia* (Sydney: Allen and Unwin).

CONNELL, R. (1971) *The Child's Construction of Politics* (Melbourne: Melbourne University Press).

CONNELL, R. W., ASHENDON, D. J., KESSLER, S. and DOWSETT, G. W. (1982) *Making the Difference: Schools, Families and Social Divisions* (Sydney: Allen and Unwin).

CRICK, B. and PORTER, A. (1978) *Political Education and Political Literacy* (London: Longman).

CRUICKSHANK, D. R. (1985) Uses and benefits of reflective teaching. *Phi Delta Kappan*, 66 (10): 704–706.

CRUICKSHANK, D. R. and AMMALINE, J. H. (1986) Field experiences in teacher education: considerations and recommendations. *Journal of Teacher Education*, 37 (3): 34–40.

CURRICULUM REVIEW UNIT INITIAL TRAINING PANEL (CRUIT) (1983) *Teaching Political Literacy*. Bedford Way Papers 16 (London: University of London Institute of Education).

ECKERSLEY, R. (1988) *Casualties of Change: The Predicament of Youth in Australia* (Melbourne: Commission for the Future).

FERGUSON, P. (1989) A reflective approach to the methods practicum. *Journal of Teacher Education*, 40 (2): 36–41.

FINLAYSON, P., REYNOLDS, I., ROB, M. and MUIR, C. (1987) *Adolescents: Their Views, Problems and Needs* (Sydney: Hornsby and Kuring-gai Area Health Service).

GILBERT, R. (1987) A cautionary note: Problems with the 'political literacy' approach to political education. In L. Lewis (ed) *Social Education and Political Literacy*. Monograph No. 3 (Melbourne: Social Education Association of Australia).

GIROUX, H. (1980) Critical theory and rationality in citizenship education. *Curriculum Inquiry*, 10 (4): 329–366.

GIROUX, H. (1981) *Ideology, Culture and the Process of Schooling* (Lewes: The Falmer Press).

GIROUX, H. (1983) *Theory and Resistance in Education* (Massachusetts: Bergin and Garvey).

GIROUX. H. (1988) *Schooling for Democracy: Critical Pedagogy in the Modern Age* (London: Routledge and Kegan Paul).

GIROUX, H. and McLAREN, P. (1986) Teacher education and the politics of engagement: the case for democratic schooling. *Harvard Educational Review*, 56 (3): 213–233.

GOODMAN, J. (1984) Social studies curriculum design: a critical approach. Paper presented at the annual meeting of the National Council for the Social Studies, Washington DC. ERIC ED 251 373.

GOODMAN, J. (1986 a) Making early field experiences meaningful: a critical approach. *Journal of Education for Teaching*, 12 (2): 109–125.

GOODMAN, J. (1986 b) Teaching preservice teachers a critical approach to curriculum design. *Curriculum Inquiry*, 16 (2): 181–201.

GORE, J. (1987) Reflecting on reflective teaching. *Journal of Teacher Education*, 38 (2): 33–39.

GORE, J. and BARTLETT, V. L. (1988) Pathways and barriers to reflective teaching in an initial teacher education program. *Research Grant Series*, No. 4 (Brisbane: Board of Teacher Education), 19–31.

GOULD, C. C. (1988) *Rethinking Democracy: Freedom and Social Co-operation in Politics, Economy and Society* (Cambridge: Cambridge University Press).

HOLLINGSWORTH, P. (1983) *Poverty in Australia*, 3rd edition (Melbourne: Nelson).

KICKBUSCH, K. W. (1985) Ideological innocence and dialogue: a critical perspective on discourse in the social studies. *Theory and Practice in Social Education*, 13 (3): 45–56.

KIRK, D. (1986) Beyond the limits of theoretical discourse in teacher education: towards a critical pedagogy. *Teaching and Teacher Education*, 2 (2): 155–167.

LAND, R. (1987) *Education for Democracy in Australian Schools*. Unpublished paper (Brisbane: Queensland Department of Education).

LEFTWICH, A. (1983) *Redefining Politics: People, Resources and Power* (London: Methuen).

LEWIS, L. (ed) (1987) *Social Education and Political Literacy*, Monograph No. 3 (Melbourne: Social Education Association of Australia).

LISTON, D. P. and ZEICHNER, K. M. (1988) Critical pedagogy and teacher education. Paper presented at the annual meeting of the American Educational Research Association, New Orleans, ERIC ED 295 937.

MERCER, D. and ABBOTT, I. (1989a) Democratic learning in teacher education: a comparative study. *Educational Review*, 41 (1): 3–8.

MERCER, D. and ABBOTT, I. (1989b) Democratic learning in teacher education: partnership supervision in the teaching practice. *Journal of Education for Teaching*, 15 (2): 141–148.

OSBORNE, K. (1984) *Working Papers in Political Education*. Monographs in Education XII (University of Manitoba).

POPKEWITZ, T. (ed) (1987) *Critical Studies in Teacher Education: Its Folklore, Theory and Practice* (Lewes: The Falmer Press).

POPKEWITZ, T., TABACHNICK, R. and ZEICHNER, K. M. (1979) Dulling the senses: research in teacher education. *Journal of Teacher Education*, 30: 52–60.

ROBOTTOM, I. (1987) The dual challenge for teacher education in environmental education. In Department of Arts, Heritage, and Environment, *Environmental Education: Past, Present and Future* (Canberra: AGPS), 72–83.

ROSS, D. D. (1989) First steps in developing a reflective approach. *Journal of Teacher Education*, 40 (2): 22–30.

SCHÖN, D. (1987) *Educating the Reflective Practioner* (San Francisco: Jossey-Bass).

SHOR, I. (1986) Equality is excellence: transforming teacher education and the learning experience. *Harvard Educational Review*, 56 (4): 406–426.

SHOR, I. (1987) Educating the educators: a Freirian approach to the crisis in teacher education. In I. Shor (ed) *Freire for the Classroom: A Sourcebook for the Classroom* (Portsmouth NH: Boynton/Cook), 7–32.

SMYTH, J. (1989) Developing and sustaining critical reflection in teacher education. *Journal of Teacher Education*, 40 (2): 2–9.

SPODEK, B. (1974) *Teacher Evaluation: Of the Teacher, by the Teacher, for the Child* (Washington, DC: National Association for the Education of Young Children).

TURNEY, C., ELTIS, K., HATTON, N., OWENS, L. C., TOWER, J. and WRIGHT, R. (1983) *Sydney Micro Skills Redeveloped* (Sydney: Sydney University Press).

VINCENT, J. (1978) In the country of the blind. *Times Educational Supplement*, 21 July.

WILLIAMSON-FIEN, J. (1987) Women, politics and the political education of girls. In L. Lewis (ed) *Social Education and Political Literacy*, Monograph Series No. 3 (Melbourne: Social Education Association of Australia), 51–64.

WILLIAMSON-FIEN, J. (1988) Power. In D. Hicks (ed) *Education for Peace: Issues, Principles and Practice in the Classroom* (London: Routledge and Kegan Paul), 143–167.

WILSON, P. and ARNOLD, J. (1986) *Street Kids* (Melbourne: Collins Dove).

WOODS, G. H. (1984) Schooling in a democracy: Transformation or reproduction? *Educational Theory*, 34 (3): 219–239.

YOUTH COUNCIL OF AUSTRALIA (1984) *Australian Values: Do Young People Share Them?* (Melbourne: Youth Affairs Council of Australia).

ZEICHNER, K. M. (1983) Alternative paradigms of teacher education. *Journal of Teacher Education*, 34 (3): 3–9.

ZEICHNER, K. M. (1986) Social and ethical dimensions of reform in teacher education. In J. Hoffman and S. Edwards (eds) *Clinical Teacher Education* (New York: Random House), 87–108.

Ecological consciousness and curriculum

MARLA MORRIS

This paper explores competing stories around consciousness, ecology and education, with particular reference to conceptual refinement of the idea of an 'ecological consciousness'. Phenomenological and functional models of consciousness are examined in terms of their implications for developing ecological consciousness in and for education. Ecological consciousness has significant implications for education as it resituates us as human creatures dwelling in the world—as interrelated ecological identities embedded in a many-voiced landscape.

Consciousness is a key to understanding personhood (Varela *et al.* 1996) and a good starting-place for reconceptualizing cognition. Whatever cognition is, it must take account of consciousness, because it is this mysterious something that allows us as human beings to exist. Whatever consciousness is, it is that something which integrates us into the very 'web of life' (Capra 1996). Consciousness enables us to be open to others and the world. It is that very something that connects us to nature and the earth. It does not make sense, therefore, to talk of consciousness as an isolated event or phenomenon. Whitehead (1951: 423) notes that 'there is no possibility of a detached, self-contained local existence. The environment enters into the nature of each thing.' Consciousness is not separated from the environment; the subject is immersed in nature.

An ecological paradigm recognizes our integration with the world and others. Many scholars in the field of education (Marsden 1971, Hewson 1988, Gough 1989a, b, 1991, 1993, 1994, Bowers and Flinders 1990, Eisner 1991, Lydon 1992, Bowers 1993, 1995, Greenall Gough and Robottom 1993, Gabbard 1994, Prakash 1994, Block 1995, Jardine 1998) have grappled with rethinking education ecologically. Although Gough (1989b) questions the rhetoric of paradigmatic shifts, he suggests that a shift toward thinking more ecopolitically becomes key when rethinking educational work. He contends that rethinking understandings of perception and their relation to ecology has significant implications for a curriculum that embraces calls to action. For Gough, it is not enough to talk about an ecologically sustainable world, one must also engage in ecopolitically defensible educational work. Gough emphasizes that when educators re-imagine what an ecologically sustainable education might be, they must keep in mind that there are many stories to tell, not one. Building on Gough's (1994: 193) work, I embrace a 'multistoried' and interdisciplinary framework in order to get outside the frame of implicitly or explicitly (or even unconsciously) offering a 'final' truth on educating ecologically. In other words, I do not offer a final truth on what might move educators to think more ecologically. Rather, this paper reflects my own struggles in grappling with competing stories around consciousness, ecology and education.

In Gough's (1989b: 225) ground-breaking paper, 'From epistemology to ecopolitics: renewing a paradigm for curriculum', he observes that since the 1980s there has been a call for educators to think about what it might mean to embrace an 'ecological consciousness'. It is this challenge that I take up here; it is this paradigm that I would like to flesh out when talking of consciousness. An ecological consciousness 'fuses' (Whitehead 1951) human beings back with nature, fuses mind back with body. An ecological consciousness has important implications for education as it re-situates us as human creatures dwelling-in-the-world. Learning about who we are as we dwell in the world may save our ecosphere. As Prakash (1994) points out, it was novelist and poet Wendell Berry who, long ago, called for a relationship of dwelling with the earth, for it is dwelling that keeps us rooted in a soulful way to our surroundings. Like Berry, Heidegger (1971) called for a poetic dwelling with the earth in order to re-situate being. Dwelling can become a potentially significant experience because it makes us slow down, take walks, and take in psychologically the very earth on which and in which we are situated. As Wittgenstein (cited in Theater of Voices 1996) once said, 'If you want to go down deep you do not need to travel far'. Travelling requires re-locating. Dwelling requires sitting in place, being where we already are.

When thinking about re-conceptualizing cognition as ecological consciousness, one needs to stop and dwell on where we as human beings already are, to dwell on our consciousness—to dwell on the ways in which human beings are conscious. Thus, I will examine phenomenological and functional models of consciousness. From these I will build an ecological paradigm of consciousness. Finally, I will discuss implications of reconceptualizing cognition as ecological consciousness for education.

Consciousness: the murky waters

Consciousness defies language and is in some ways beyond representation because of its ineffableness. It is mysterious, and yet it is that something that people can almost name, and it is that unnameable stuff of everyday experience. Although philosophers have been grappling with the mysteries of consciousness for centuries, cognitive scientists and neuroscientists, if they are to grapple with it at all, are newcomers to the adventures of consciousness. 'It is widely argued that consciousness is an intractable, or irrelevant, problem for cognitive science' (Newman *et al.* 1997: 393). However, as Güzeldere (1998a) points out, this position may be changing, especially with the work of Francis Crick and Christof Koch in neurobiology. Güzeldere (pp. xi–xii) explains:

> In neuroscience and psychology, perhaps the biggest positive impact in favour of consciousness research was due to Francis Crick and Christof Koch. Their 1990 article, 'Towards a Neurobiological Theory of Consciousness', carrying the secure stamp of approval by a Nobel Laureate, helped shatter the current psychological and professional barrier against the credibility of studying consciousness in science.

Crick and Koch's (1998: 69) neurobiological theory of consciousness turns on the assumption that consciousness is caused by 'oscillations' in the 'lower layers' of the cerebral neocortex. These oscillations of neurons register at 40-hertz. Crick and Koch maintain that it is this 40-hertz oscillation that causes consciousness to emerge. However, Chalmers (1996: 116) raises an interesting and important question: 'Why should [40-hertz oscillation] be accompanied by conscious experience?' Neurobiological explanations of what may cause consciousness still cannot explain the more difficult question of why it arises at all.

Whatever consciousness is, and whatever disagreements cognitive scientists may have about this most elusive something that allows human beings to experience life, agreement abounds that most lives are experienced unconsciously. Freud was right after all. Cognitive scientists and neuroscientists, however, depart from Freud because they do not hold that unconscious experience is due to repression. Many cognitive scientists suggest that unconscious processes operate in ways that scientists simply have no access to (Newman *et al.* 1997, Crick and Koch 1998, Flanagan 1998, Jackendoff cited in Crick and Koch 1998: 64, Lycan 1998, Nørretranders 1998, Lakoff and Johnson 1999). Lakoff and Johnson (1999: 13) remark that: 'Conscious thought is the tip of an enormous iceberg. It is the rule of thumb among cognitive scientists that unconscious thought is 95% of all thought—and that may be a serious underestimate.'

The complexities of mind are far too vast for our understanding. What scholars know is very little indeed. Some cognitive scientists and neuroscientists, however, may believe that with enough study and with perhaps the advancement of technology we may someday be able to figure out these conscious and unconscious processes. However, I suspect that studies on consciousness will always, at the end of the day, remain a mystery. Nørretranders (1998: 220) claims that it is the unconscious that actually motivates us to do anything.

> Our actions begin unconsciously! Even when we think we make a conscious decision to act, our brain starts half a second before we do so! Our consciousness is not the initiator—unconscious processes are!

The unconscious muddies the waters. Clarity and certainty around the notion of consciousness will not win the day. Murkiness becomes key. If people are willing to step into the murky waters of studies of consciousness, two paths seem relevant.

The study of consciousness is divided roughly into two realms: one realm focuses on phenomenological aspects of conscious experience; the other focuses on functional aspects. First, I turn to some of the questions raised by phenomenological inquiry.

Phenomenological questions

Phenomenological questions around consciousness concern what it *feels* like to be conscious, what it *is* to be conscious. Chalmers (1996) and Nagel (1998) note that if people are to understand consciousness at all, scholars must argue from the phenomenological perspective. Chalmers (1996: 4) points out that when we ask the question about what it is 'like to be a cognitive agent' we have entered the realm of the phenomenological. 'We can say that a being is conscious if there is *something it is like* to be that being. . . . if it has a *qualitative feel*'. What is it like to experience something? The notion of experience, for Chalmers (pp. 3–4) is 'central to consciousness'. However, what exactly experience is, of course, is quite complicated. Experience is that something that eludes description in a way. As soon as I think I can explain my experience, it seems to slip beyond me.

Like Chalmers, Nagel (1998: 520)—another proponent of phenomenological explanations of consciousness—asks: What is it like to be a bat? '[T]hat bats have experience is that there is something that it is like to be a bat'. This 'something that it is to be' is the experience of consciousness. However, this experience is so intensely personal that it becomes difficult for one person to translate this experience in a language that makes sense to another person. There is something strange and idiosyncratic about consciousness. In some way, it is untranslatable. It is this untranslatable quality of consciousness that makes some people uncomfortable. Nagel suggests that the agenda of what he terms 'neo-behaviorists' is ultimately to 'reduce' consciousness to what can be explained and, in essence, to explain consciousness away (p. 524). Indeed, as Güzeldere (1998b: 5) points out, naturalists (those who believe that scholars can explain consciousness like any other physical property of the body) have been accused of 'trying to do away with consciousness for the sake of explaining it'. Chalmers (1996) and Nagel (1998) maintain, and I agree, that no matter how much explaining one does, consciousness will always express an ineffable something that defies explanation. Consciousness will always remain a stranger. Explanation leads to more explanation but still does not and cannot answer the fundamental questions: what is it like to be conscious and why are agents conscious at all? Chalmers (1996: 118) declares that:

> Newtonian physics involves an Euclidean space-time; relativity theory invokes a non-Euclidean differential manifold; quantum theory invokes a Hilbert space for wave functions. And different theories invoke different kinds of dynamics within those structures: Newton's laws, the principles of relativity, the wave equations of quantum mechanics. But from structure and dynamics, we can only get more structure and dynamics. . . . No set of facts about physical structure and dynamics can add up to a fact about phenomenology.

Chalmers argues, then, that no matter how much explaining and theorizing one does about how consciousness is structured, arguments of structure do not lead one to answer the hard question of why it is that agents are conscious at all and it cannot answer the question of what it feels like to be conscious.

Functional questions

Questions around the structure of consciousness are questions of function. Functional explanations attempt to grapple with how consciousness works. There are many kinds of functional explanations: neurobiological approaches (Edelman 1992, Crick and Koch 1998), emergentist approaches (Clark 1997), biomolecular approaches (Narayanan 1997), evolutionary approaches (Edelman 1992), cultural approaches (Bruner 1986, 1990, Ó Nualláin 1997), information-processing approaches (Dennett 1996, 1998), imaginative approaches (Ellis 1995) and embodied approaches (Edelman 1992, Clark 1997, Ó Nualláin 1997, Lakoff and Johnson 1999). Van Gulick (1998) and Güzeldere (1998a, b) straddle the phenomenological and functional approaches to consciousness. Although I am drawn more to phenomenological questions, I suspect that a position that balances phenomenology and functionalism might be more defensible. However, unlike functionalists, who maintain that humankind will one day be able to explain consciousness, I concur with phenomenologists who contend that, no matter how much explaining we do, the ineffable quality of consciousness is most likely to prevail. However, to focus exclusively on phenomenological questions is problematic because these questions seem to lapse into a solipsistic or subjectivist puzzle. How does one consciousness translate experience to another? How do people translate the ways in which consciousness gets enfleshed in the world and the ways in which conscious beings are interconnected with the world? How do scholars explain our difference and alterity as well as interconnection and unity in the ecosphere? A more adequate route in our inquiry would be to combine phenomenological and functional questions.

I turn now to some questions raised by functional arguments. Edelman (1992: 27) points out that when people examine the structure of the brain in neurobiological terms, they soon discover that all brains are not the same. 'Unidentifiable (and not necessarily repeatable) patterns of synapses [create] a crisis for those who believe that the nervous system is precise and "hardwired"...'. Neurobiological systems may not produce consciousness, but they are in some way related to consciousness and whatever conscious experience emerges from these patterns of synaptic activity, they are idiosyncratic. Whatever consciousness is, it expresses an alterity that cannot be reduced. One of the difficulties of taking up a phenomenological approach is that the experience of consciousness (what it is like to be a bat, for example) seems to suggest continuity. Flanagan (1998: 89) suggests that consciousness, although it may feel continuous (like a 'stream' of 'consciousness', as William James argued long ago), is actually 'gappy'. Certainly, the contention that there is a half-second delay between conscious and unconscious processes (Nørretranders 1998) supports this notion of gappiness. Furthermore, Nørretranders contends that one of the operations of mind is getting rid of, or erasing, irrelevant or unusable material. Gappiness could be caused by this erasure. Nørretranders (1998: 125) points out that:

> Every single second, everyone of us discards millions of bits [a bit enables people to make distinctions between objects] in order to arrive at the special state known as consciousness. ... Consciousness involves information that is not present; information that has disappeared along the way.

Human beings are conscious of usable material, but most of our mental activity goes unnoticed and slips away, perhaps into the unconscious. Freud would argue that nothing is really erased, it just gets repressed. He suggests that people never forget anything. Cognitive scientists might not agree but, nevertheless, much of the workings of the mind are inaccessible and puzzling. If people were conscious of everything they would probably become mad.

Dennett and Kinsbourne (1998: 144–145) propose that consciousness is heavily edited. Consciousness produces what they term 'multiple drafts', which are like 'parallel stream[s] of conflicting and continuously revised' texts. Consciousness is not one text but many. The content of these texts can change at a moment's notice. Like memory, consciousness is selective. Neither do we as human beings remember everything, nor are we conscious of everything.

Functional approaches to consciousness vary. Yet, despite their disagreements, functionalists agree, for the most part, that whatever consciousness is, it does not operate from any single point (Clark 1997, Newman *et al.* 1997, Dennett and Kinsbourne 1998, Lycan 1998). Most philosophers and cognitive scientists have dismissed the notion that there is a single homunculus operating the site of consciousness. As Edelman (1992: 82) points out, the homunculus (the 'little man' or woman inside who controls conscious experience) is not a valid idea:

> The homunculus is the little man that one must postulate 'at the top of the mind', acting as an interpreter of signals and symbols in any instructive theory of mind. If information from the world is processed by rules in a computerlike brain, his existence seems to be obliged. But then another homunculus is required in *his* head and so on. . . .

The infinite regress of the homunculus makes little sense when people understand that processes of consciousness are probably widely distributed over the brain. There is no single site that controls all brain activity.

Computer-based models of consciousness and mind abound. Computer-based models are termed 'cognitivism' (Edelman 1992: 14). Dennett (1996), perhaps one of the best-known proponents of cognitivism, maintains that consciousness is computer-like. At different points in his writings, Dennett uses computer metaphors to describe consciousness. Consciousness, Dennett (1998) declares is a 'virtual machine'; it is like a 'computer program' (p. 100), 'a virtual serial machine (which we might call the Joycean machine, since it is the stream of consciousness we are modelling)' (p. 108). 'All brains', he suggests, 'are, in essence, "anticipation machines"' (p. 101). Because consciousness is like a computer, Dennett (1996: 68) contends that human beings can replicate consciousness in non-human things: 'once we figure out what minds do ... we ought to be able to make minds ... out of alternative materials...'. For Dennett (1996: 16), 'a conscious robot is possible in principle'. Bruner (1990: 6) points out that computer metaphors for consciousness are not new: 'Computers and computational theory had by the early 1950s become the root metaphor for information processing. ... Very soon, computing became the model of mind'.

However, human beings are not computers and consciousness does not operate as computers do. Lakoff and Johnson (1999: 6) claim that 'there is no such thing as a computational person, whose mind is like computer software ... whose mind somehow derives meaning from taking mean-ingless symbols as input, manipulating them by rule, and giving mean-ingless symbols as output'. Edelman (1992: 14) declares that there is 'a large body of evidence that undermines the view that the brain is a kind of computer'.

Finally, information processing is not an adequate way in which to describe consciousness (Nørretranders 1998) because there is little infor-mation in consciousness in the first place. Mostly, at a pre-conscious state, information is thrown out, only traces remain. Dreyfus and Dreyfus (1986: xiv) point out that 'computers are certainly more precise and more predictable than we, but precision and predictability are not what human intelligence is about'. Computers lack emotion, imagination and motiva-tion. Ellis (1995) claims that these qualities are the prime movers of conscious experience.

Mechanistic metaphors for human consciousness are problematic, then, for many reasons. Human beings are not machines and to mechanize human beings is a way to devalue the very stuff that makes us human. To turn humans into machines is, in many ways, obfuscating and even immoral. To dehumanize is to desacralize. Machines do not have souls. Robots are not conscious. Broadly speaking, what is termed classical cognitive science and Artificial Intelligence (AI) (theories modelling mind on machine) have been roundly criticized. Narayanan (1997: 21) explains that:

> rule-following by itself is not sufficient (and may not even be necessary) for intelligence, awareness and consciousness ... that because CCS [classical cognitive science] and AI are anti-materialist and perhaps anti-reductionist in nature they cannot explain how brain gives rise to mind ... that CCS and AI, because they succumb to the same formal limits that apply to computa-tion and algorithms, cannot account for certain types of mental processes....

Although some speak about computers having a kind of intuition, com-puters are not intuitive in the same way that human beings are. To intuit that a close friend might be ill, or to intuit that something will come to pass when the evidence of such an event is not forthcoming is something that computers cannot do. Computers do not feel, they do not suffer and they do not mourn. They are on again, off again, user-friendly or not.

A more adequate functional explanation of consciousness is the neuro-biological. Edelman's (1992) neurobiological model is termed the theory of neuronal group selection. This position is counter to Dennett's, or to any cognitive position that models mind on machine. Edelman claims that 'The theory of neuronal group selection ... will enable us to come to grips with the daunting problem of consciousness' (pp. 82–83). Perhaps he promises too much. Yet, he does offer an important theoretical model that has explanatory power. He points out that consciousness is selective and that 'the unit of selection' is a 'neuronal group' (pp. 85–86). Edelman (1992: 89) suggests that these neuronal groups ignite consciousness by what he terms 're-entry'. The 'interaction' of 'brain maps' composed of neuronal groups 'emerge through successive and recursive re-entry'. He contends that these re-entry systems are dependent upon 'evolutionary morphology' (p. 29). He claims that 'all cognition and all conscious experience rest solely on processes and orderings occurring in the physical world' (p. 82). Consciousness emerges in relation to the world, it is not cut off from the world. These brain maps and re-entry sites are 'a self-organizing system' (p. 25). No two self-organizing systems are alike. This accounts for the radical alterity of consciousness. The way my brain system allows consciousness to emerge at those re-entry sites may differ radically from someone else's. There seems to be much uncertainty, too, around the ways in which these neuronal groups interact. Brain 'maps are connected by massively parallel and reciprocal connections' (p. 85).

Like Edelman, Clark (1997) offers a biological approach to consciousness. As against classical cognitive science and much of AI, Clark contends that it may be misleading to image the brain as a manipulator of data and information. Clark wants to re-situate cognitive science back in the world and to re-connect 'real brains, bodies, and environments' (pp. 1–2). His position is termed 'emergentist', because he suggests that when people talk about the brain (or consciousness or mind) it must be situated in the body and in the world, because consciousness emerges with the world, not as an isolated thing.

Clark's (1997) is a more ecological position and one to which I adhere. However, Clark and I part ways because he still uses mechanistic metaphors to describe human cognition. Clark also compares the brain to a computer: 'human reasoners are truly *distributed* cognitive engines: we call on external resources to perform specific computational tasks, much as a networked computer may call on other networked computers ...' (pp. 68–69). In spite of Clark's mechanistic metaphors, his work is important because of the emergentist nature of consciousness. Consciousness emerges together with the world and in response to others because consciousness is embodied and connected to the world. Clark remarks, interestingly, that 'the true engine of reason ... is bounded neither by skin nor skull' (p. 69). Neither is consciousness. Although I maintain that consciousness is embodied, it seeps outward, it is intentional, and it is always about something and comes into existence in response to surroundings. As Lakoff and Johnson (1999: 37) explain:

> The claim that the mind is embodied is, therefore, far more than the simple-minded claim that the body is needed if we are to think. ... [T]he very properties of concepts [I would say consciousness] are created as a result of the way the brain and body are structured and the way they function in interpersonal relations and in the physical world.

Furthermore, consciousness is immersed in the world, not separate from it. It is not solipsistic, but develops in relation to others and the world in which it is immersed. As Edelman (1992: 122) points out, 'consciousness ... is efficacious and likely to enhance evolutionary fitness'. Ó Nualláin (1997: 43) claims that we are 'enmeshed in a Life-world'. He declares that 'mind (and cognition) is manifest in the co-adaptation of species and environment over time' (p. 43). As Capra (1996: 305) says in relation to the Santiago theory (which has most notably been discussed by Francisco Varela and Humberto Maturana), consciousness 'brings forth a world'.

Ecological paradigm of consciousness: educational implications

Whatever consciousness is, it is that unnameable something which integrates us as human beings into our environment, into our world. An ecological paradigm webs consciousness back into nature. Consciousness situates us into the ecosystem, embodied creatures in continuum with the world. No strict boundaries or divisions separate conscious creatures from the ecosystem. Haeuber (1996: 6) notes that the ecosystem has no 'rigid boundary demarcation. Ecosystems are dynamic, constantly changing, and vary continuously along gradients in space and time.' Ecosystems are highly complex processes that are mysterious. However, one thing is certain: human beings are part of these ecosystems, not separate from them. Consciousness and environment are entangled, confused, co-related, co-dependent. There is no way to separate consciousness from environment without doing violence to the very ecosystem that sustains us.

If consciousness emerges in the world and is in continuum with the world, it is also, to a certain extent, 'co-emergent' (Davis and Sumara 1997) and co-extensive with others. Consciousness arises in what I call co-conscious inter-relations with others and the world. We are co-conscious creatures. We are entangled in a web of consciousness. As Gell-Mann (1994: xviii) concludes, there is a 'chain of relationships linking the quark to the jaguar and to human beings'. And there is a chain linking human beings together, because co-consciousnesses emerge in interactions and inter-relations.

Co-consciousness, however, does not mean we merge into some oceanic unity of oneness or sameness. As Krall (1994: 242) points out, 'connection to place ... is not some sublime, undifferentiated oneness, not a merging with Nature...'. Clearly, the jaguar is not the same as a human being, and human beings are different from one another. Co-consciousness, then, arises out of differences. Gabbard (1994: 181), drawing on the work of Ivan Illich, suggests that discussions around ecology, and our complex interrelations to and with nature, might embrace a threefold 'communion', that emerges with others, with nature, and with ourselves. This is a communion, not of sameness or oneness, but of alterity and otherness. Gabbard notes that for Illich what becomes important is recognizing our otherness with others, the otherness of nature, and the otherness within the self. The strangeness of the world and our enmeshment within the world must never be taken for granted. The jaguar, the cat, the tree and the starfish are all wildly different from each other, even though all are webbed into the same weave of the world. And that place we (as human beings) call the world is wildly different from us as well. Even though I belong to a landscape, it can seem alien to me. Although the marshes of Georgia are home to me, they are forbidding places with alligators and snakes, which crawl and sometimes land in backyards or on the interstate highway.

When talking of connections to our ecoscape and the inter-connections of creatures with us as human beings, I am not advocating a taming of nature. An ecological consciousness hopefully re-integrates us back into the wilderness. As Gabbard (1994: 173) indicates, scholars must engage in a 'wild discourse' when talking about developing a more ecologically sensitive education.

An ecological model of consciousness has significant implications for education. 'Curriculum is a[n] extraordinarily complicated conversation' (Pinar *et al.* 1995: 848), and this complicated conversation has everything to do with our inter-relationships with each other and the world. If learning happens as a co-conscious event, that is, if learning happens in community, it is that process of inter-relation and interaction between students, teachers and texts that becomes crucial. Bruner (1986: 68) contends that 'most of our approaches to the world are mediated through negotiation with others'. Learning, like consciousness, then, is not an isolated event because human beings are shaped by complex inter-relations with one another. Learning is a living, breathing 'holomovement' (Bohm 1978) that changes us at each moment.

Co-consciousness arises out of differences, and much learning happens when we recognize this. In fact, Bateson (1991) claims that it is the notion of difference that makes us recognize and perceive. Bateson's 'ecology of mind', the notion that cognitive processes are situated in a larger ecosphere, is similar to Maturana and Varela's Santiago theory.[1] Here, it is thought that a 'living organism brings forth a world by making distinctions. Cognition results from a pattern of distinctions, and distinctions are perceptions of difference' (Capra 1996: 305). We come to know by learning what we do not yet understand; we come to know who we are by learning who we are not. We come to understand race, class and gender through our differences with one another. We come to know in the wilderness. And yet, co-consciousness connects us to one another as well.

Re-immersing ourselves into the ecosystem as conscious creatures calls for the eradication of views that reproduce violence. Cognition-as-machine may justify all sorts of violence as it separates minds from bodies and bodies from the world. A vigilant wisdom is necessary to overcome these oppressions. Curriculum workers who understand the sort of vigilance I am talking about understand that re-conceptualizing cognition is more than rethinking an idea. Our very survival depends on the way we think about thinking. Freyfogle (1996: 647) points out that 'the teaching of ecologists, the observations of ethicists, the sophisticated discussions about sustainability and ecosystem health—all have fallen on deaf or disinterested ears'. It takes a certain vigilance to awaken the deaf. Schools are, for the most part, deaf to the sound of a dying planet. The call for an ecological consciousness, then, is prophetic. It raises the noise, sounds the alarm.

An ecological consciousness thrusts humankind back into the world and down into the earth. Perhaps a new Copernican Revolution is needed for us to understand that we are not separate from our environment. A move, therefore, from anthropocentric to ecocentric thinking becomes key (Bowers 1993, 1995). Ecocentric thinking shifts the focus from human-centredness to earth-centredness. Ecocentric thinking shifts the focus from students and teachers in an isolated schoolroom, to students and teachers in society-in-the-world. Ecocentric thinking is a more integrative way to think about ourselves as creatures living in an ecosphere.

It will be difficult for many people to embrace ecocentric thinking because schooling reproduces ways of thinking that are completely antithetical to it. Anthropocentrism is a product of modernism and Enlightenment, and we as human beings are, in a sense, children of the Enlightenment. It is very difficult, therefore, to move away from thinking that is couched in Enlightenment discourse. It is hard not to think that we are not the centre of everything. It is very difficult to move away from thinking that the earth is merely a tool for us to exploit. To conceptualize the earth in mechanistic terms is to do violence to it. Benton (1996: 2156) indicates that modernism teaches that 'the environment [becomes] a utility for progress, an object to be studied and manipulated'. Using and exploiting the land, without regard for her, without respect for her, amounts to rape.

An ecological view of consciousness, therefore, is incompatible with violence and violent metaphors. If anything, it embraces humility. As Fox (1983: 59) points out, humility is an earthy word:

> Meister Eckart, who repeats this message dozens of times in his writings, points out that the word 'humility' comes from the word *humus* or earth ... to be humble means to be in touch with the earth. ...

To be in touch with the earth means that we, as human beings, have to move out of our way, move out of our heads and stop thinking about thought as an isolated event. And we have to stop thinking about ourselves as isolated creatures. Freyfogle (1996: 651) emphasizes that environmental thinking is a 'profound challenge to ... individualism'. Enlightenment mentality is that of the lone ranger, the lone individual who conquers the land. However, the lone ranger forgets that he is not alone. And he forgets, too, that conquering the land means killing the land. Let us, then, give up lone ranger mentality.

However, this is not easy. Lone ranger mentality is deeply inscribed in Western culture. As Gough (1994: 193) has demonstrated, many who do educational theory still perpetuate the myth of the lone ranger:

> This narrative detachment of human culture from the earth that sustains it is manifested in educational theory by stories that construct the 'cultivated' subject—the 'educated' person—as an individual consciousness 'dislocated' (positioned otherwise) from nature.

The lone ranger, this individually 'dislocated' conqueror, is the symbol of 'westward-ho' expansion. And, as Freyfogle (1996: 644) comments, 'the frontier ethic [is] the focus on man as the locus and measure of value'. If value has everything to do with human life, but nothing to do with animals and non-human life, it becomes clear that the earth and her animals become devalued. Furthermore, for human beings, value is associated with economic well-being. Consumer-capitalist mentality values human beings based on economic worth. Consumer capitalism, therefore, has much to do with what is considered valuable and what is not. Freyfogle (p. 643) points out that 'our national infatuation with economic growth, measured in a distorted way that looks at market transactions . . . ignores degradation of the land'.

Consumer capitalism reproduces dangerous value systems that serve to exploit and degrade earth. It is not surprising, then, that corporate USA is uninterested, for the most part, in saving the planet. Little funding goes into saving rivers, lakes, forests, the ozone layer, and endangered species. Gandy (1996: 27) reports that environmental problems, according to neo-marxists, seem to stem from 'structural features in the economy'. Complex issues like ecological degradation have economic and historical causes. Thus, ecological degradation needs to be tackled at the structural level if real change is to occur. It is not enough to change individual attitudes and beliefs about our place in the world, about our failing ecosystems, about our dying planet. More needs to be done. Like institutionalized racism, sexism and homophobia, structural inequalities need to be dismantled if justice is to prevail, if human beings are to make this earth a more equitable place. Teachers know that changing beliefs and attitudes of students is not enough to combat racism, homophobia and sexism. These problems are structural, these problems are cultural, these problems are embedded and encoded in the very discourses inherited from schooling. And the problems of ecological degradation, too, are primarily cultural.

A more collective and integrative way of thinking about our relationship with the ecosphere is needed. To be in touch with the earth is to also be in touch with the sacred, for we are born from the earth and will return to her as dust. We live in what Bateson (1979: 19) terms 'a sacred unity' with the larger ecosphere. The earth is our home.[2]

Cognitive arguments that perpetuate the idea that thought thinks thought (after the fashion of Hegel) are simply unethical. Thinking minds separated from bodies and the world foster all sorts of oppressions. However, it seems that cognitive studies, for the most part, remain stuck in this Hegelian paradigm. However, there are alternative models of cognition that argue for our situatedness in the larger cultural community (Bruner 1986, 1990, Kincheloe and Steinberg 1993, Davis and Sumara 1997, Kirshner and Whitson 1997, Lemke 1997, St. Julien 1997, Walkerdine 1997, Steinberg et al. 1999). If cognition deals simply with disembodied minds, the current ecological crisis will continue. Many ecologists warn that if value systems do not change, if attitudes and institutionalized belief systems about the planet are not altered, extinction looms near (Zimmerman 1994, Hayward 1995, Thomashow 1995).

Callicott (1986) recommends an ecocentric ethic that demands changes in the way human beings think about thinking. Instead of valuing only the thinking creature, an ecocentric ethic values non-human creatures and the earth as well. Education as an ethical enterprise might teach young people to value their world and the very ecosystem that sustains life. For the sake of children, animals and the earth, teachers must educate to heal what Callicott (1986: 417) terms the 'bruised and tattered planet'. However, these hopes will be dashed if we continue to think that what counts is in the head only. What knowledge is of most worth? The knowledge that will save the planet from complete destruction is of most worth.

I have argued for what I call an ecological consciousness because I think scholars need to begin re-conceptualizing the ways in which human creatures are embedded in the world. As co-conscious creatures we must take responsibility for our actions and interactions with both human and non-human creatures. Abram (1996: ix) suggests that Americans are completely wrapped up in 'human-made technologies' and have forgotten the 'age-old reciprocity with the many-voiced landscape'. As ecological consciousnesses, we are related and embedded in this many voiced-landscape. Abram (p. ix) contends that:

> Humans are tuned for relationship. The eyes, the skin, the tongue, ears, and nostrils—all are gates where our body receives the nourishment of otherness. This landscape of shadowed voices, these feathered bodies and antlers and tumbling streams—these breathing shapes are our family, the beings with whom we are engaged, with whom we struggle and suffer and celebrate.

Curriculum as relationship might embody Abram's teaching, because the tumbling streams and feathered birds have everything to do with lived experience, with life-in-schools. Schools are dead places without these birds and streams, these shadowed voices. Curriculum as a running stream pours out toward the wilderness as wolverines and walruses might enliven students' lives-in-schools.

Educating for an ecological consciousness is also educating for an ecological identity. Thomashow (1995) suggests that developing an ecological identity is more than thinking about thoughts and how thoughts are constructed in relation to nature. He maintains that 'each person's path to ecological identity reflects his or her cognitive, intuitive, and affective perceptions of ecological relationships' (p. 3). Identities are constructed around reflections and relations; identities are invented and imagined around others and the world. Ecological consciousness co-constructs identities with others (human and non-human). In fact, the animals and plants are not absolutely exterior to us, but are part of us.

However, identities are also produced through social discourse, language and institutions. And, if the language and discourse of schooling perpetuates isolationist, individualist, lone ranger ways of thinking about who we are, ecological identities become very difficult to embrace. Ecological identities might be vigilant ones that fight to dismantle harmful metaphors. Ecological co-consciousnesses offer a promise to tear down the walls between schooling and society, teachers and students, texts and world, animals and human beings, human beings and non-human creatures.

Cognition as ecological consciousness is a broader and wiser knowing. According to Bateson (1979: 5), cognition might embrace a *'wider knowing* which is the glue holding together the starfishes and sea anemones and redwood forests and human committees'. This wider knowing, however, is a paradoxical one, too. We as human beings are interconnected to the wilderness, but we feel alienated from it. We feel inter-connected, at some level, to the jaguar; yet we feel estranged from it as well. We feel inter-connected to one another, but we know that our differences matter most; our differences, in fact, create what I call co-consciousness. This broader and wiser knowing is a complex, paradoxical knowing.

Curriculum workers can change the way educators think about thinking. Working across differences, educators can learn to care about jaguars and trees, quarks and cats. Cognition as ecological consciousness is a path toward vigilant wisdom through the wilderness we call life.

Acknowledgement

I thank Brent Davis for his careful and thoughtful suggestions.

Notes

1. In the Santiago theory, 'the brain is not necessary for the mind to exist. A bacterium, or a plant, has no brain but has a mind' (Capra 1996: 174–175).
2. Russell (1997) notes that the word 'ecology' comes from the Greek word for home.

References

ABRAM, D. (1996) *The Spell of the Sensuous: Perception and Language in a More-Than-Human World* (New York: Pantheon).
BATESON, G. (1979) *Mind and Nature: A Necessary Unity* (New York: Dutton).
BATESON, G. (1991) Mind/environment. In R. E. Donaldson (ed.), *A Sacred Unity: Further Steps to an Ecology of Mind* (New York: HarperCollins), 161–173.
BENTON, L. M. (1996) The greening of world trade? the debate about the North American Free Trade Agreement (NAFTA) and the environment. *Environment and Planning A*, 28 (12), 2155–2177.
BLOCK, A. A. (1995) *Occupied Reading: Critical Foundations for an Ecological Theory* (New York: Garland).
BOHM, D. (1978) The implicate order: a new order for physics. *Process Studies*, 8 (2), 73–102.
BOWERS, C. A. (1993) *Education, Cultural Myths, and the Ecological Crisis: Toward Deep Changes* (Albany, NY: State University of New York Press).
BOWERS, C. A. (1995) *Educating for an Ecologically Sustainable Culture: Rethinking Moral Education, Creativity, Intelligence, and Other Modern Orthodoxies* (Albany, NY: State University of New York Press).

BOWERS, C. A. and FLINDERS, D. J. (1990) *Responsive Teaching: An Ecological Approach to Classroom Patterns of Language, Culture, and Thought* (New York: Teachers College Press).

BRUNER, J. S. (1986) *Actual Minds, Possible Worlds* (Cambridge, MA: Harvard University Press).

BRUNER, J. S. (1990) *Acts of Meaning* (Cambridge, MA: Harvard University Press).

CALLICOTT, J. B. (1986) The search for an environmental ethic. In T. Regan (ed.), *Matters of Life and Death: New Introductory Essays in Moral Philosophy*, 2nd edn (New York: Random House), 416–422.

CAPRA, F. (1996) *The Web of Life: A New Scientific Understanding of Living Systems* (New York: Anchor Doubleday).

CHALMERS, D. J. (1996) *The Conscious Mind: In Search of a Fundamental Theory* (New York: Oxford University Press).

CLARK, A. (1997) *Being There: Putting Brain, Body, and World Together Again* (Cambridge, MA: MIT Press).

CRICK, F. and KOCH, C. (1998) The problem of consciousness. In J. Toribio and A. Clark (eds), *Consciousness and Emotion in Cognitive Science: Conceptual and Empirical Issues* (New York: Garland), 63–69.

DAVIS, B. and SUMARA, D. J. (1997) Cognition, complexity, and teacher education. *Harvard Educational Review*, 67 (1), 105–125.

DENNETT, D. C. (1996) *Kinds of Minds: Toward an Understanding of Consciousness* (New York: BasicBooks).

DENNETT, D. C. (1998) The evolution of consciousness. In J. Toribio and A. Clark (eds), *Consciousness and Emotion in Cognitive Science: Conceptual and Empirical Issues* (New York: Garland), 99–120.

DENNETT, D. C. and KINSBOURNE, M. (1998) Time and the observer: the where and when of consciousness in the brain. In N. Block, O. Flanagan and G. Güzeldere (eds), *The Nature of Consciousness: Philosophical Debates* (Cambridge, MA: MIT Press), 141–174.

DREYFUS, H. L. and DREYFUS, S. E. (1986) *Mind over Machine: The Power of Human Intuition and Expertise in the Era of the Computer* (New York: Free Press).

EDELMAN, G. M. (1992) *Bright Air, Brilliant Fire: On the Matter of the Mind* (New York: Basic Books).

EISNER, E. W. (1991) *The Enlightened Eye: Qualitative Inquiry and the Enhancement of Educational Practice* (New York: Macmillan).

ELLIS, R. D. (1995) *Questioning Consciousness: The Interplay of Imagery, Cognition, and Emotion in the Human Brain* (Amsterdam, The Netherlands: John Benjamins).

FLANAGAN, O. (1998) The robust phenomenology of the stream of consciousness. In N. Block, O. Flanagan and G. Güzeldere (eds), *The Nature of Consciousness: Philosophical Debates* (Cambridge, MA: MIT Press), 89–93.

FOX, M. (1983) *Original Blessing: A Primer in Creation Spirituality* (Santa Fe, NM: Bear).

FREYFOGLE, E. T. (1996) Ethics, community, and private land. *Ecology Law Quarterly*, 23 (4), 631–661.

GABBARD, D. A. (1994) Ivan Illich, postmodernism, and the eco-crisis: reintroducing a 'wild' discourse. *Educational Theory*, 44 (2), 173–187.

GANDY, M. (1996) Crumbling land: the postmodern debate and the analysis of environmental problems. *Progress in Human Geography*, 20 (1), 23–40.

GELL-MANN, M. (1994) *The Quark and the Jaguar: Adventures in the Simple and the Complex* (New York: Freeman).

GOUGH, N. (1989a) Becoming ecopolitical: some mythic links in curriculum renewal. Paper presented at the annual meeting of the American Educational Research Association, San Francisco, CA (Faculty of Education, Deakin University, Victoria, Australia).

GOUGH, N. (1989b) From epistemology to ecopolitics: renewing a paradigm for curriculum. *Journal of Curriculum Studies*, 21 (3), 225–241.

GOUGH, N. (1991) Narrative and nature: unsustainable fictions in environmental education. *Australian Journal of Environmental Education*, 7, 31–42.

GOUGH, N. (1993) Environmental education, narrative complexity and postmodern science/fiction. *International Journal of Science Education*, 15 (5), 607–625.

Gough, N. (1994) Playing at catastrophe: ecopolitical education after poststructuralism. *Educational Theory*, 44 (2), 189–210.

Greenall Gough, A. and Robottom, I. (1993) Towards a socially critical environmental education: water quality studies in a coastal school. *Journal of Curriculum Studies*, 25 (4), 301–316.

Güzeldere, G. (1998a) Preface and acknowledgments. In N. Block, O. Flanagan and G. Güzeldere (eds), *The Nature of Consciousness: Philosophical Debates* (Cambridge, MA: MIT Press), xi–xxvi.

Güzeldere, G. (1998b) Introduction: the many faces of consciousness: a field guide. In N. Block, O. Flanagan and G. Güzeldere (eds), *The Nature of Consciousness: Philosophical Debates* (Cambridge, MA: MIT Press), 1–67.

Haeuber, R. (1996) Setting the environment policy agenda: the case of ecosystem management. *Natural Resources Journal*, 36 (1), 1–28.

Hayward, T. (1995) *Ecological Thought: An Introduction* (Cambridge: Polity Press).

Heidegger, M. (1971) *Poetry, Language, Thought*, trans. A. Hofstadter (New York: Harper & Row).

Hewson, M. G. A'B. (1988) The ecological context of knowledge: implications of learning science in developing countries. *Journal of Curriculum Studies*, 20 (4), 317–326.

Jardine, D. W. (1998) *To Dwell with a Boundless Heart: Essays in Curriculum Theory, Hermeneutics, and the Ecological Imagination* (New York: Peter Lang).

Kincheloe, J. and Steinberg, S. R. (1993) A tentative description of post-formal thinking: the critical confrontation with cognitive theory. *Harvard Educational Review*, 63 (3), 296–320.

Kirshner, D. and Whitson, J. A. (eds) (1997) *Situated Cognition: Social, Semiotic, and Psychological Perspectives* (Mahwah, NJ: Erlbaum).

Krall, F. R. (1994) *Ecotone: Wayfaring on the Margins* (Albany, NY: State University of New York Press).

Lakoff, G. and Johnson, M. (1999) *Philosophy in the Flesh: The Embodied Mind and its Challenge to Western Thought* (New York: Basic Books).

Lemke, J. L. (1997) Cognition, context and learning: a social semiotic perspective. In D. Kirshner and J. A. Whitson (eds), *Situated Cognition: Social, Semiotic, and Psychological Perspectives* (Mahwah, NJ: Erlbaum), 37–55.

Lycan, W. G. (1998) Consciousness as internal monitoring, I. In J. Toribio and A. Clark (eds), *Consciousness and Emotion in Cognitive Science: Conceptual and Empirical Issues* (New York: Garland), 49–62.

Lydon, A. T. (1992) Cosmology and curriculum: a vision for an ecozoic age. Doctoral dissertation, Louisiana State University, Baton Rouge, LA, USA.

Marsden, W. E. (1971) Environmental studies courses in colleges in education. *Journal of Curriculum Studies*, 3 (2), 163–178.

Nagel, T. (1998) What is it like to be a bat? In N. Block, O. Flanagan and G. Güzeldere (eds), *The Nature of Consciousness: Philosophical Debates* (Cambridge, MA: MIT Press), 519–527.

Narayanan, A. (1997) Biomolecular cognitive science. In S. Ó Nualláin, P. McKevitt and E. Mac Aogáin (eds), *Two Sciences of Mind: Readings in Cognitive Science and Consciousness* (Amsterdam, The Netherlands: John Benjamins), 21–36.

Newman, J., Baars, B. J. and Cho, S.B. (1997) A neurocognitive model for consciousness and attention. In S. Ó Nualláin, P. McKevitt and E. Mac Aogáin (eds), *Two Sciences of Mind: Readings in Cognitive Science and Consciousness* (Amsterdam, The Netherlands: John Benjamins), 393–417.

Nørretranders, T. (1998) *The User Illusion: Cutting Consciousness down to Size*, trans. J. Sydenham (New York: Viking).

Ó Nualláin, S. (1997) The search for mind: a new foundation for cognitive science. In S. Ó Nualláin, P McKevitt and E. Mac Aogáin (eds), *Two Sciences of Mind: Readings in Cognitive Science* (Amsterdam, The Netherlands: John Benjamins), 37–49.

Pinar, W. F., Reynolds, W. M., Slattery, P. and Taubman, P. M. (1995) *Understanding Curriculum: An Introduction to the Study of Historical and Contemporary Curriculum Discourses* (New York: Peter Lang).

PRAKASH, M. S. (1994) What are people for? Wendell Berry on education, ecology, and culture. *Educational Theory*, 44 (2), 135–157.

RUSSELL, E. P. (1997) Cost among the parts per billion: ecological protection at the United States Environmental Protection Agency 1970–1993. *Environmental History*, 2 (1), 29–51.

ST. JULIEN, J. (1997) Explaining learning: the research trajectory of situated cognition and the implications of connectionism. In D. Kirshner and J. A. Whitson (eds), *Situated Cognition: Social, Semiotic, and Psychological Perspectives* (Mahwah, NJ: Erlbaum), 261–279.

STEINBERG, S. R., KINCHELOE, J. L. and HINCHEY, P. H. (eds) (1999) *The Post-formal Reader: Cognition and Education* (New York: Falmer).

THEATER OF VOICES (1996) 'Proverb' [Song]. In Theater of Voices and Steve Reich Ensemble, *On Proverb, Nagoya Marimbas, City Life* [CD recording and #79430-2] (New York: Nonesuch Records, Warner Music Group).

THOMASHOW, M. (1995) *Ecological Identity: Becoming a Reflective Environmentalist* (Cambridge, MA: MIT Press).

VAN GULICK, R. (1998) Understanding the phenomenal mind: are we all just armadillos? Part I: phenomenal knowledge and explanatory gaps. In N. Block, O. Flanagan and G. Güzeldere (eds), *The Nature of Consciousness: Philosophical Debates* (Cambridge, MA: MIT Press), 559–566.

VARELA, F. J., THOMPSON, E. and ROSCH, E. (1996) *The Embodied Mind: Cognitive Science and Human Experience* (Cambridge, MA: MIT Press).

WALKERDINE, V. (1997) Redefining the subject in situated cognition theory. In D. Kirshner and J. A. Whitson (eds), *Situated Cognition: Social, Semiotic, and Psychological Perspectives* (Mahwah, NJ: Erlbaum), 57–70.

WHITEHEAD, A. N. (1951) Nature and life. In M. H. Fisch (ed.), *Classic American Philosophers: Peirce, James, Royce, Santayana, Dewey, Whitehead* (New York: Appleton-Century-Crofts), 418–437.

ZIMMERMAN, M. E. (1994) *Contesting Earth's Future: Radical Ecology and Postmodernity* (Berkeley, CA: University of California Press).

Environmental education and the secondary school curriculum

C. G. GAYFORD

Introduction

Environmental education in secondary schools in Britain has had a somewhat uneven and disappointing existence to date. From its emergence as a recognized area of the curriculum nearly twenty years ago, consistent efforts have been made to introduce environmental education into the formal curriculum of schools by a small and strongly motivated group of people.[1] Almost twenty years later with various changes in thinking within the educational system, the situation is still far from encouraging. This is especially disappointing when it is considered that many people generally subscribe to the idea and importance of environmental education. Indeed, an extremely strong case has already been made for the importance of environmental education in terms of human survival.[2]

The limited interest in environmental education has also been reflected in the small number of published studies relating to this part of the curriculum. Richmond and Morgan[3] carried out a study into the attitudes and knowledge of pupils in the fifth forms of secondary schools to environmental matters. Carson[4] edited a series of useful discussions concerned with teaching environmental studies, and since that time major published items have been the HMI discussion document, *Curriculum 11–16*,[5] the DES publication, *Environmental Education: A Review*,[6] and an extensive survey and discussion of environmental subjects in the curriculum by Goodson.[7]

It is the purpose of this paper to explore some important aspects of the relationship between environmental education and the curriculum, to suggest some of the likely reasons for the lack of enthusiasm for it in educational circles and to indicate some possible ways forward. A good deal of the evidence on which this paper is based is from a series of surveys carried out over the last four years and involving a large proportion of the centres taking GCE examinations in environmental studies or environmental science.[8]

Approaches to environmental education

Part of the problem of environmental education lies in initial confusion, even among advocates of environmental education, over its nature and identity and how it should relate to the curriculum. Some of the reasons for this stem from the variety of sources from which environmental education has been developed. A strong historical link exists with rural studies and also with geography and biology and the tension between these subject areas and environmental education has been documented and discussed extensively by Goodson.[9]

There have been two main approaches to the inclusion of environmental education into the formal curriculum. One has involved the development of environmental studies or environmental science as separate areas of the timetabled curriculum with public examinations set by GCE and CSE boards. The other approach has involved attempts to identify environmental components within established subject areas in the curriculum.

The problems that relate to the first approach are often concerned with the perceived status of the subject in the estimation of teachers, parents, pupils, and others. Environmental subjects in this context are seen as minority subjects, which at the same time adopt an integrated approach; both of these features probably contribute to the lowering of the status of the subjects. Also, it is notoriously difficult for new subjects to become established in the formal curriculum.[10] An added problem is that in many seconday schools where environmental studies are offered, they are taken by the lower ability pupils and the more able ones continue to study the traditional academic disciplines. Thus, 'Environmental Studies are seen as the preserve of those whom the school finds it difficult to involve.'[11]

The second approach requires considerable planning across the curriculum, which probably means that senior staff need to be both involved and enthusiastic. The effects of subject chauvinism should also be recognized, as should the frequent unwillingness on the part of specialist subject teachers to incorporate new areas into their teaching, particularly if these areas have a strongly affective component. Perhaps it is this important aspect relating to the affective nature of much of environmental education, which is essential to this type of education and at the same time creates problems for those teaching within the formal curriculum where cognitive areas are of prime concern as a result of examination pressures. However, this approach, if it can be achieved, does have the advantages, outlined by Goodson,[12] that this results in more children being offered environmental education and that children of all abilities become involved.

The development of GCE examinations

The development of environmental studies and environmental science as separate subjects on the timetable has been accompanied by attempts to establish GCE examinations in these subjects. The existence of public examinations in a subject area, particularly examinations at GCE A-level, raises the status of that subject in the eyes of pupils and parents alike.[13] The Schools Council[14] cited rural studies as an example of a subject which suffers because it lacks the hallmark possessed by many other disciplines through the provision of a GCE A-level examination. It is important to note that, particularly at the end of the 1960s and early 1970s, most of the impetus for a GCE A-level examination in environmental studies came from certain leading rural studies specialists.[15]

GCE A-level examinations have existed for environmental studies and environmental science for more than a decade now (London University

Schools Examinations Department, Associated Examining Board, Northern Universities Joint Matriculation Board), but the subjects do not seem to have been accorded the appropriate status by pupils, teachers or parents. The combined entry of candidates for all examination boards for A-levels in these subjects is still only a few hundred and growth to this number has been extremely slow. This has been the situation during a period when growth in A-level entries in many other subjects has been rapid. The problems related to this phenomenon have been complex, but one of the most notable contributory factors has been the perceived status of environmental subjects as A-level studies. Part of the reason for this aspect of the problem is the attitude shown by the universities to A-level qualifications in these subject areas, and this will be discussed below.

The situation regarding environmental education and the relatively low status given to it in schools is perhaps rather surprising since the adoption of subjects within the academic curricula in the UK involves assumpitons that some kinds and areas of knowledge are more worthwhile than others.[16] On these grounds, environmental subjects would seem to have a strong claim on a place in the curriculum.

At this point it is appropriate to consider the stages in the evolution of a subject in the secondary school curriculum that are suggested by Layton.[17] This offers important insights into the particular problems of environmental education. Layton's model seems to have considerable relevance for environmental subjects when they are considered as part of the formal curriculum in schools. The model identifies three stages in the development of a subject. The first stage involves:

> justifying its presence on grounds such as pertinence and utility. During this stage learners are attracted to the subject because of its bearing on matters of concern to them. The teachers are rarely trained specialists, but bring the missionary enthusiasm of pioneers to the task. The dominating criterion for selection of subject matter is relevance to the needs and interests of the learners.[18]

In the second stage:

> the internal logic discipline of the subject becomes increasingly influential in the selection and organisation of subject matter.[19]

In this stage also, scholarly work in the subject begins to emerge and this is accompanied by the production of trained specialists from whom teachers may be recruited. The attractiom of the subject to pupils at this stage is as much by the reputation and increasing academic status of the subject as by the relevance that it has to matters that concern them.

In the third state there is:

> further advance along the road of specialisation and expertise. The teachers now constitute a professional body with established rules and values. The selection of subject matter is determined in large measure by the judgements and practices of the specialist scholars who lead enquiries in the field. Students are initiated into a tradition, their attitudes approaching passivity and resignation, a prelude to disenchantment.[20]

As can be seen, one of the important features of this model is that as the subject develops it becomes more 'academic', both in content and in the attitudes of the teachers. Goodson[21] claims that the direct connection between GCE A-level examinations and the academic tradition is a strong and important feature of our educational system where, intellectually able pupils take these advanced examinations.

This aspect of the academic tradition dominates even over utilitarian features, and one important consideration is that it is a significant step in recruitment and preparation for the professions. The teachers teaching in these areas see themselves as specialists and generally they are less willing to teach across what they see as subject boundaries. Ball[22] in a wide-ranging discussion of the secondary school curriculum, points to differences in the perceived status of subjects. In that discussion he states that an important factor contributing to these differences is the way that some subjects enable pupils to progress to studies of an academic kind in the senior part of the school and then subsequently to go on further to higher education.

Reid[23] points out that universities have high academic status and at the same time they have an autonomy which allows them almost complete control over who they admit into their institutions. Also the majority of school subjects that are straightforwardly related to those in universities are specific disciplines, and they are in turn supported by the universities. As a consequence, those subjects with an integrated nature are often considered less favourably than specialist subjects. Environmental subjects are integrated and have the additional disadvantage, as far as higher education is concerned, that the apparently equivalent subjects that exist in universities that are called 'environmental studies' or 'environmental science', are often actually somewhat specialized courses frequently related to only a rather specialized part of the overall subject area as it is taught in schools. As a result, the universities are less inclined to give strong support to the subject in schools since they do not see it as a relevant preparation for their own courses as compared to the more obvious and direct relationship between many other school subjects and their equivalents in universities.

One of the important factors attempting to promote the academic status of environmental subjects in schools has been the efforts of rural studies teachers. Here, specialists in this area of the curriculum particularly wish to attach their subject to a new examination subject which could be related to higher education.[24] The Hertfordshire scheme was a special bid in this direction, where a case was made for environmental studies as an academic discipline. An A-level syllabus was constructed with examinations set by the University of London Schools Examination Department. As Goodson[25] points out in this context, the claims for a subject to have academic status are best validated through university scholarship, and this provides a considerable part of the basis for academic acceptance.

Environmental studies and environmental science are relatively new subjects in the public examination system and it is still unclear what academic status teachers and pupils consider then to have: whether they consider them as an easy option for the less academic pupils or whether they consider them to be demanding and requiring greater academic ability. Related to this is the acceptability of A-levels in environmental subjects as part of the entrance requirements of universities, and then, related to this, the teachers' perception of the attitude of universities to A-level qualifications in these subjects.

The teachers and teaching

Evidence from the study carried out by Gayford[26] in which particular attention was given to environmental education as an academic subject area, showed that in many of those institutions where GCE examinations were taken in environmental studies or environmental science, less than half (48%) of these institutions (total sampled was 96) had one or more teachers with some type of initial training for teaching the subjects. Often this initial qualification was a PGCE special study or option, part of a BEd qualification or a rural studies speciality. Few (11%) had an initial degree in some form of environmental studies. The attendance at short in-service courses for environmental studies or environmental science teaching was a little more encouraging (59·5%), but rather less satisfactory was the attendance at extended in-service courses, where only about 12% had been involved.

The impression gained from these findings and from further comments made by teachers, is that although recognized courses of training exist, the teaching of environmental subjects is open to interested teachers from other subjects, who do not necessarily have to retrain, in a way that occurs in few other subjects. Comments stating that the subject is taught mainly be a geographer or a biologist were not uncommon. Team-teaching involving teachers from different specialist disciplines occurred in quite a number of cases; but more often the timetable was divided for convenience between two or even more teachers from separate disciplines, and the effect of this appeared to be that a number of viewpoints were given which were never coherently co-ordinated.

The problems caused by the diversity of teachers involved often seems to result in a lack of consensus between teachers about objectives. It appears that environmental subjects attract teachers from different disciplines and varied initial and in-service training. This situation could potentially enrich the educational quality of the teaching. In practice there are difficulties in defining the scope of these environmental subjects because a great deal depends on the background of the teachers involved. Each teacher has the freedom to introduce elements of his own specialist area of knowledge. As a result, the boundaries of the subject become blurred and this is unlikely to enhance the status of the subject as one with specialist content. Also, and importantly from the point of view of Layton's model of curriculum evolution, it is more difficult for teachers from different specialist backgrounds to form a professional group with common interests and to develop environmental education in the way that occurs in many other subject areas.

The motivation for teachers to become involved in environmental education was overwhelmingly through a feeling of concern about the need to care for and protect the environment. 79% of teachers made this claim and 94% considered that environmental education helped to create an understanding of man's responsibility for the environment.

Environmental education and other parts of the formal curriculum

The extent to which interest in environmental education has permeated other areas of the formal curriculum besides GCE O- and A-level courses is also important.[27] This can be assessed in a number of ways, but three that may be considered are (a) the existence of environmental courses lower down the school, i.e. in the first and second year of secondary schools, (b) the existence of CSE and non-examination courses in environmental subjects in fourth and fifth forms, and (c) the existence of traditional examination courses containing a good deal of environmental work.

It appears that where GCE examinations are taken in environmental subjects this certainly does not mean that courses are automatically provided lower down the school. The survey showed that 57% of schools claimed to have courses which contained a substantial amount of environmental material in the lower secondary classes, and of these only 56% of them (i.e. 32% of the whole sample) had more than half of the pupils in these classes involved in these courses. The nature of these courses varied but included separate environmental studies courses, or were a part of social studies, science or special courses for pupils of lower ability.

This leads to the possible conclusion that the general lack of uptake of these subjects further on in secondary school could be partly due to the small number of suitable courses provided lower down the school. In this way, pupils with no background in environmental subjects may be hesitant to choose the subject for examinations.

Few schools (13%) stated that non-examination courses existed for pupils of 14–16 years. However, a larger proportion (56%) reported that they also sat CSE examinations. Both of these results indicate a less than wholehearted commitment to environmental subjects as examination subjects for all abilities up to 16 years.

Another indicator of the degree to which environmental subjects have permeated the curriculum is the extent of strongly based environmental work in other examination subjects in the school. Of the schools surveyed, half (50%) had no courses at all which answered to this description. Of the remainder many of the schools claimed to have several courses with a strong environmental component. These were particularly associated with geography, biology, rural studies and history.

From the results of this part of the study, it appears that the subject is frequently taught by a lone teacher without affecting the rest of the formal curriculum. It is also possible to speculate from this that the spread of environmental subjects owes much to the individual enthusiasm of teachers, which does not spread easily to other colleagues.

It may also be speculated that the existence of environmental studies or environmental science in the formal curriculum could have a further adverse effect in that, by its very existence on the timetable, teachers from other subject areas may feel that responsibility for that part of education is being covered and therefore they can satisfactorily ignore this element in their own teaching. This, together with the fact that environmental studies and environmental science are minority subjects with relatively few pupils taking them, will mean that few children are actually given any environmental education.

Materials

In a another part of the study,[28] it was shown that further problems were related to teaching materials. Texts were written at an unsuitable level, publishers appeared unwilling to cater for a relatively small market, the materials rapidly became outdated and the scope of the subject was very wide. The local nature of environmental materials was also seen as a contributory factor to the teachers' difficulties.

Since there was a general lack of comprehensive texts, teachers often found it necessary to use a wide range of books or prepare extensive materials themselves to supplement commercially produced materials. Even when teachers were prepared to expend the time and effort in creating the necessary materials, pupils were not always willing to make use of what was provided since it was usually less attractively presented than the commercially produced texts that they were used to in other subjects.

It appeared from these studies that successful teaching in these areas requires teachers who are prepared to go to considerable lengths to find local and up-to-date material and who are willing to search in texts in order to find material with a depth of coverage appropriate for the course. Indeed, 60% of teachers claimed to be using materials that were mainly school-devised. These features relating to environmental education materials were considered by those sampled to be a deterrent to many potential teachers of the subjects.

The pupils

Little is known about the attitudes of pupils and their motivation for being interested in environmental subjects. This applies particularly when considering the levels where pupils may exercise a reasonable degree of freedom in choosing the subjects that they wish to study. It seems, however, that the attitudes of pupils and their perception of the academic status of environmental subjects might be an important factor in the uptake of the subject. Teachers were asked to express their opinion about this matter rather than the pupils being directly asked. It appears from these investigations that teachers in very few centres appeared to encourage pupils of relatively low ability to take the subject. The subject was considered to be taxing, with a workload greater than that in many traditional subject areas. However, it is also clear that teachers often do not obtain the pupils that they want since many teachers actually stated that intake is generally limited to the relatively less academic. On the other hand, 62% of the pupils taking the subject at A-level were taking at least two other subjects at the same level. There is no data available about the relative performance of these pupils in the other subjects that they were taking in comparison with others not taking environmental subjects. However, the very fact that many were sitting this number of subjects indicates the likelihood of a reasonable standard of academic ability.

Higher education

Of the students completing A-level studies which included environmental studies or environmental science (in excess of 400 in this survey) 35% went on to higher education establishments, 19% taking environmental courses, and 16% taking other courses. However, of the total that sat examinations at A-level in environmental subjects, 68% would have wished to attend a higher education establishment, 30% of these intending to follow an environmental course and 38% wishing to follow other courses. From this it is apparent that a considerably larger number of pupils would have preferred to follow courses in higher education than those who actually managed to achieve this aim. Furthermore, a larger proportion of those taking environmental subjects at A-level and who wished to follow different subjects in higher education failed to do this. In many cases, it is likely that this was because pupils did not achieve passes which fulfilled the entrance requirements; however, in proportion to cases in other school subjects, there was an expressed opinion by many teachers that A-levels in environmental subjects were positively discriminated against by institutions of higher education. This idea was indeed to some extent supported when the numbers of pupils who had achieved passes at A-level in environmental subjects as well as two or more other subjects, and who went on to higher education establishments to read either environmental subjects or non-environmental subjects, were compared. Here, it was clear that a much smaller proportion of students who had the requisite A-level qualifications and who were seeking a place on a course in higher education faced difficulties, especially when competing for places in environmental courses. However, the situation for non-environmental courses was significantly different.

One of the purposes of a GCE A-level is to serve as a qualifying examination for university entrance. It therefore follows that acceptance of an A-level subject by universities is a major factor in the uptake of a subject at A-level in schools. In 1970, a conference organized to consider the construction of an A-level syllabus in environmental studies[29] was attended by many university staff, and attitudes of universities to such an A-level examination were expressed. Comments at the time indicated that A-levels in environmental studies were not likely to be entirely acceptable for university entrance. It was felt that departments in universities would prefer candidates with physics, chemistry, biology or geography. Uncertainty was expressed over the boundaries of the subject and therefore how environmental subjects would relate to other A-levels or university courses in environmental areas. It was also questioned whether a pass at A-level in environmental subjects was as demanding intellectually as in other subjects and therefore whether it was a satisfactory indication of academic ability.

In a survey, all of the universities in the UK were asked about their formal requirements and their general attitudes and less formal expectations relating to A-level qualifications in environmental subjects.[30] The results were fairly well in line with those obtained by Carson in 1971.[31]

In their formal requirements, of the 80% that replied, quite understandably, all universities regarded A-levels in environmental studies and environmental science as alternatives. They also usually considered geography in its various forms as alternatives to the two environmental subjects. This was not the case with biology, geology or other sciences which could be combined with environmental subjects to fulfil the entry requirements. However, in the less formal and more subjective area concerning preferences for A-level subjects, most expressed definite bias towards more traditional academic subjects as part of the entrance requirement for other courses. A number did cite these environmental subjects as a possible 'third' subject. For their environmental courses there was a preference for environmental subjects at A-level provided that all other aspects of the requirements were satisfied. It was clear that many universities had a preference for physical science or biology at A-level, but several stated that there was a general shortage of candidates with adequate grades in those subjects offering themselves for environmental courses.

Not one of the universities or university colleges stated that they had a specific requirement for A-level qualifications in environmental studies or environmental science for entry to any of their courses; but some stated that for some courses it had equal merit with other acceptable A-level subjects. One of the most frequently cited problems was the concern about overlap of content with other A-level subjects and particular mention was made of geography in this context. Another area of concern was the difficulty that arises in many cases because the students who take these environmental subjects tend to avoid the more traditional sciences in their combinations and frequently offer arts-based subjects in conjunction, which may not be appropriate to the course. Although this may be partly true it is not borne out by the results of a survey of popular combinations carried out in earlier parts of this whole study.[32]

Jackson[33] considered that the three A-level student, be he on the science or arts side, benefits from a broad range of advanced subjects and therefore might well find that a course like the London University A-level syllabus in environmental studies broadens his curriculum very well. This view of the breadth of education provided by subjects at A-level however is not in practice welcomed by universities.

In the part of the survey concerned with the teachers' perceptions of the attitudes of universities, respondents were asked whether they considered that A-level qualifications in environmental subjects were more or less sought after by universities in either courses in environmental subjects or non-environmental areas. The results were fairly well in accord with the findings made by asking the universities in the earlier part of the study.

A number of respondents added comments which showed that many considered that A-levels in environmental subjects should be at least as acceptable as other subjects at A-level, particularly for environmental courses in universities, even if their perception was that they were not at present treated in this way. However, a number of teachers also reported that many interviewers at university appeared to be unaware of environmental studies or environmental science at A-level and pupils often had to inform them of what these subjects were about when they were at interviews. A few reported a conflicting view in which there was considerable interest shown on the part of the interviewers.

It is also of note that while a large number of teachers expressed very positive opinions about the acceptability of the subject at A-level, many added comments which showed that there is a very great deal of uncertainty amongst teachers over this matter. It is also be be supposed that often the teachers have little positive evidence for responding as they did. The result is nevertheless revealing since it indicates the current perceptions of the teachers of the situation.

From the responses of the teachers and also from those of the universities, it appears that both are in some agreement that environmental studies and environmental science are not totally acceptable for university entry. However, it has been shown in an earlier part of this study[34] that usually fairly able pupils are encouraged to take A-levels in these subjects. There is consequently an incongruity in the system which is apparent and this could be one of the reasons for the lack of uptake of the subject. On the one hand, teachers consider that fairly high ability is needed for candidates who wish to take environmental subjects at A-level, and on the other hand universities express caution and would prefer candidates who are able to meet their traditional requirements.

It is probably an important point with regard to the future development of environmental subjects as GCE examination subjects, that whatever the actual policy of institutions of higher education it is likely that the uncertainty in the minds of the teachers has a significant effect on the growth of these subjects.

Conclusions

The situation relating to the development of environmental education within the secondary school curriculum may at first appear somewhat depressing, particularly when environmental subjects are considered as discrete disciplines with separate public examinations. The total number of candidates for these examinations has remained small and the reasons for this are complex, but they are probably to some extent related to the status given to the subjects by teachers, pupils and parents. On the other hand, progress in the promotion of environmental education by incorporating it into other established disciplines has been more successful, notable examples being *Geography for the Young School Leaver*[35] and the *Schools Council Integrated Science Project*.[36] These may form the basis of a model for further development.

It appears from experience over the years that education related to the environment has a great deal to offer in schools. Young people appear to be well motivated towards environmental matters and many schools have incorporated environmental material successfully into their curricula. The environmental dimension of education has the advantage of relevance when applied to the pupils' concerns. Also, it is not especially demanding of resources, thus keeping it within the scope of what schools will feel is possible on limited budgets. Environmental education also provides a useful counterbalance to the significant and active programmes related to the world of work undertaken in many schools. In general it has a broadening effect and in this way can counterbalance some of the important but more directly utilitarian functions of education.

At the same time, it is essential that the environmental aspects of education do not become especially associated with the less able pupils, for it is likely that children of all abilities can benefit from environmental education.

Already, it seems that a good deal of environmental education is going on; but often this takes place in a somewhat uncoordinated way. Few schools appear to have a policy for environmental education and few have anybody on the staff with responsibility for this aspect of education across the curriculum.

Environmental education can take its place in many areas of education and indeed it may in some cases be a dimension of either the whole curriculum or individual subjects within the curriculum. However, it is necessary to consider that the route into the formal curriculum through academic respectability may require too many compromises for some of its advocates, who particularly value the strongly affective elements of this component of education. In a number of cases there is likely to be a more appropriate place for environmental education within the informal school curriculum or even outside school altogether among voluntary organizations in their educational programmes.

References

1. WHEELER, K. (1975) The genesis of environmental education. In Martin G. and Wheeler K. (eds.) *Insights into Emvironmental Education* (Oliver and Boyd, Edinburgh), pp. 7–9.
2. *The Conservation and Development Programme for the UK: A Response to the World Conservation Strategy* (Kogan Page, London, 1983).
3. RICHMOND, J. M. and MORGAN, R. F. (1977) *A National Survey of Environmental Knowledge and Attitudes of Fifth Year Pupils in England* (The ERIC Science, Mathematics and Environmental Clearinghouse, Columbus, Ohio).
4. CARSON, S. McB. (1978) *Environmental Education: Principles and Practice* (Arnold, London).
5. DEPARTMENT OF EDUCATION AND SCIENCE (1979) *Curriculum 11–16*. Supplementary Working Paper by HM Inspectorate (HMSO, London).
6. DEPARTMENT OF EDUCATION AND SCIENCE (1981) *Environmental Education: A Review* (HMSO, London).
7. GOODSON, I. F. (1983) *School Subjects and Curriculum Change: Case Studies in Curriculum History* (Croom Helm, London and Canberra).
8. GAYFORD, C. G. (1983) Environmental studies and environmental science at GCE 'O' and 'A' level. *Review of Environmental Education Developments*, 11, 2, pp. 16–18; GAYFORD, C. G. (1984) The materials used and the pupils involved in environmental science at GCE 'O' and 'A' level. *Review of Environmental Education Developments*, 12, 10, pp. 7–8; GAYFORD, C. G., (1984b) Environmental studies and environmental science: A study of the academic acceptability of these subjects at GCE Advanced level. *Environmental Education*, 20.
9. GOODSON,, I. F. (1983) (see note 7).
10. GOODSON, I. F. (1983) Environmental education for all: Strategies for change. *Review of Environmental Education Developments*, 11, 3, pp. 6–7.

11. GOODSON, I. F. (1975) Urban Studies. In Martin, G. and Wheeler, K. (eds.) *Insights into Environmental Education* (Oliver and Boyd, Edinburgh), p. 71.
12. GOODSON, I. F. (1983) (see note 10).
13. KELLY, A. V. (1977) *The Curriculum: Theory and Practice* (Harper & Row, London), p. 42.
14. SCHOOLS COUNCIL (1969) *Rural Studies in Secondary Schools.* Working Paper No. 24 (Evans/Methuen Educational, London).
15. CARSON, S. McB. (1963) The changing climate. *NRSA Journal*, p. 14; TOPHAM, P. (1968) Rural studies courses in the comprehensive school. *NRSA Journal*, p. 45.
16. REID, W. A. (1972) *The University and the Sixth Form Curriculum* (Macmillan, London), pp. 49, 61.
17. LAYTON, D. (1972) Science as general education. *Trends in General Education*, 25, January, p. 11.
18. LAYTON, D. *ibid.*
19. LAYTON, D. *ibid.*
20. LAYTON, D. *ibid.*
21. GOODSON, I. F. (1983) (see note 7).
22. BALL, S. J. (1981) *Beachside Comprehensive School* (Cambridge University Press, Cambridge), p. 16.
23. REID, W. A. (1972) (see note 16).
24. CARSON, S. McB. (1971) *Environmental Studies: The Construction of an A-level Syllabus* (NFER Slough), pp. 6–7; MYLECHREEST, M. (1975) Whither rural studies? *School Science Review*, 57, 199, pp. 276–284.
25. GOODSON, I. F. (1983) (see note 7).
26. GAYFORD, C. G. (1983) (see note 8).
27. GAYFORD, C. G. *ibid.*
28. GAYFORD, C. G. (1984 a) (see note 8).
29. CARSON, S. McB. (1971) (see note 24).
30. GAYFORD, C. G. (1984 a) (see note 8).
31. CARSON, S. McB. (1971) (see note 24).
32. GAYFORD, C. G. (1984 b) (see note 8).
33. JACKSON, P. (1978) The upper school. In Carson, S. McB. (ed.) *Environmental Education: Principles and Practice* (Arnold, London), p. 234.
34. GAYFORD, C. G. (1984 a) (see note 8).
35. SCHOOLS COUNCIL (1970) *Geography for the Young School Leaver* (Nelson and Son, Surrey).
36. SCHOOLS COUNCIL (1974) *Schools Council Integrated Science Project* (Longman/Penguin, Harlow, Essex).

Subjects for Study: Aspects of a Social History of Curriculum

IVOR GOODSON

In this paper I want to scrutinize the part which social histories can play in studying the curriculum. The introduction briefly examines the use (misuse or non-use) of historical evidence in some theoretical work. Then, by drawing on parts of historical studies I have personally undertaken, I seek to show, albeit in a limited way, how such work allows hypotheses to be examined and reformulated. In this way, a 'sequence to theory' emerges from historical work which not only extends the range of our studies but which, by posing questions about our current theories, can aid in generating new theories and agendas.

Partly because *Knowledge and Control* was an influential starting point for my own historical studies, I want to examine briefly the use that has been made of history by some of the sociologists who contributed to this volume. I am aware that this is to dangerously generalize from the particular; I am also aware of the excellent past and recent sociological work that has employed historical perspectives.

Nonetheless, I think the example will establish some general points, especially as Young and Bernstein have, since *Knowledge and Control*, come to argue for historical work. Young has said that 'one crucial way of reformulating and transcending the limits within which we work, is to see ... how such limits are not given or fixed, but produced through the conflicting actions and interests of man in history'.[1] Likewise Bernstein has argued that 'if we are to take shifts in the content of education seriously, then we require histories of these contents, and their relationships to institutions and symbolic arrangements external to the school'.[2]

However, much of the work of these sociologists to date can be characterized in two ways: (*a*) either history is not used, or alternatively (*b*) history is misused, or to use Silver's elegant phrase 'raided'.[3] Much of the work actually ignores historical background and no evolutionary historical process is provided. Studies develop, so to speak, horizontally, working out from theories of social structure and the social order. When historical evidence is presented it is provided as a snapshot from the past to prove a contemporary point.

The use of Layton's work by Young (1977) is a good example of how history is used in this manner. Layton was describing a particular movement led by Richard Dawes towards 'The Science of Common Things' and its fate in a particular period during the 19th century. Layton was clear that there were striking similarities between many of the issues which engaged science educators in the mid 19th century and those which occupy their latter-day counterparts, but in the first paragraph of his conclusion he warned:

Within the last century and a quarter the social environment of science education has been radically transformed. At the time when Davies and Moseley fought their cause science was a national enterprise of limited scale, operating at the level which Derek Price has termed 'little science'. State and science had not begun to interact in any significant way and the limits of the principles of voluntaryism and *laissez-faire* applied to the growth of scientific activity, were only just becoming clear. Today, in contrast, 'big science' is not only heavily dependent on state patronage, but has become inextricably interwoven into the economic, political and ethical problems of the age. Concomitantly, there has arisen a national system of secular education in which the importance of scientific studies is recognized at all levels.[4]

Young, however, uses Layton's work to question Professor Jevons's *contemporary* view that in science 'we are up against something in the cognitive structure of science itself. A historical snapshot is used to question a view about science today; moreover, the implication is that our conceptions of school science can be understood from evidence of this particular period of conflict. Young is aware that 'it is not possible to draw any direct parallels with science education today' but nonetheless implies, if not parallels, direct continuities: 'what is emphasized is the historical emergence and political character of the most basic assumptions of what is *now* taken to be school science'.[5] In fact, without direct parallels and with no evidence produced of continuities, it is difficult to move to *any* understanding of the basic assumptions of contemporary school science from the specific historical evidence presented from Layton's work.

Clearly the danger of 'raiding' history is that such moves can span centuries of change at all levels of content and context. A more systematic *evolutionary* understanding of how the curriculum is negotiated is therefore needed. One is concerned to ensure that histories make evolutionary connections partly to secure against 'raiding', but more constructively to facilitate the use of such histories in developing theoretical frameworks. A continuity thesis cannot be assumed (as in the Young example) but has to be established over time. It is surely at the centre of the sociological as well as historical enterprise to examine curriculum transformation and reproduction at work over time: such complex undertakings simply cannot be elucidated by 'snapshots' of unique events which may be entirely aberrant and without general significance. The *recurrence* of events, however, can help in discerning explanatory frameworks in which structure and interaction interrelate. One is reminded of the humility of Lowe's comment in his seminal article on the divided curriculum:

> While it is well known that the major educational enquiries of the mid-nineteenth century culminated in an analysis of the educational needs of society by the Taunton Commissioners which in some ways prefigured [this] twentieth century tripartism, it is not widely realised that the evolution of ideas on a structured and hierarchical system of secondary education was both gradual and continuous from that time.[6]

Historical studies, therefore, should seek to describe the 'gradual and continuous' nature of curriculum change and do so in ways which examine negotiation and action. On this view, to seek to provide from the macro level theories of curriculum without related empirical studies of how the curriculum has been negotiated at micro level is a dangerous sequence through which to proceed. This article will argue that pursuing an understanding of the complexity of curriculum action and negotiation over time is a meaningful sequence through which to test, and formulate, theory.

Subjects for study

Having made a polemical plea for the potential of curriculum history in furthering our understanding of schooling, I want to provide some instances of historical work which begin to explore that potential. By citing some of the work in which is brought together in *School Subjects and Curriculum Change*.[7] I hope (*a*) to characterize the kinds of insights and hypotheses which are generated through undertaking curriculum histories, and (*b*) to illustrate the capacity of such histories to aid the examination of sociological theories.

My original interest in undertaking curriculum history grew out of my teaching experience. Certainly after a spell at Countesthorpe (a secondary school recently described as an 'unemulated educational maverick') I was susceptible to the arguments presented by R. A. Nisbet in *Social Change and History*. Nisbet argues that we are often deluded into thinking fundamental social change is taking place because we do not take account of a vital distinction

> Between readjustment or individual deviance within a social structure (whose effects, although possibly cumulative are never sufficient to alter the structure or the basic postulates of a society or institution) and the more fundamental though enigmatic, change of structure, type, pattern or paradigm.[8]

To pursue Nisbet's crucial distinction into the field of curriculum demands, I think, that we undertake historical work. This is true whether we seek to understand how change is contained as 'readjustment or individual deviance' as at Countesthorpe or to analyse more fundamental changes of structure over time.

In the curriculum histories undertaken, I focused on subject groups and subgroups in action. The particular historical context was the emergence of the environment as an influential idea and area of concern, and of environmental education as a viable curriculum possibility. The location of the 'environmental' climate of opinion within a broader structural milieu has been dealt with in a number of studies, but here my particular concern was to understand how subject groups and subgroups responded to the change in 'climate' (one subject advocate spoke of the 'changing climate' and argued that his subject group would have to 'adapt or perish'): beyond this there was a need to investigate the manner in which subject groups scrutinized the new climate for opportunities of promoting their interests, and to examine why it was that one subgroup decided to promote a new subject at A-level in 'Environmental Studies' and other subgroups and subject groups responded so strongly against this initiative as to threaten its viability.

Eventually, a strategy for this historical investigation was designed and divided into three sections which aimed to focus on the conflict over environmental studies in the 1960s and 1970s. Beyond this paramount concern, where possible, the sections were designed so as to test hypotheses and to examine theories which related to the content studied. The first and second sections focused on the origins and evolution of the three subjects involved in the emergence of environmental studies: geography, biology and rural studies. Here the concern was to understand the process of becoming a school subject and patterns of internal change. The third section dealt with 'external relations' between subjects and with the conflict over environmental studies, in particular the moves to promote an A-level syllabus for the subject. Hence the sections evolve chronologically: the subjects are scrutinized under construction and as they pursue status and resources; the groups, traditions and alliances within subjects are analysed; these subjects and groups are then analysed in the culminating conflict over environmental studies. For the purposes of this paper I do not, however, want to provide a summary account of *School Subjects and Curriculum Change* but rather to concentrate on the way in which, in constructing that account, hypotheses were tested and reformulated and theories examined.

Testing hypotheses

In the three sections of the book noted above three main hypotheses are examined within the context of the history of three subjects involved in the conflict over environmental studies:

(*a*) 'That subjects are not monolithic entities but shifting amalgamations of sub-groups and traditions. These groups within the subject influence and change boundaries and priorities'. (See Bucher and Strauss 1961,[9] Musgrove 1968,[10] Young [ed.] 1971, Kuhn 1970,[11] 1972, Waring 1978,[12] 1979);

(*b*) 'That in the process of establishing a school subject (and associated university discipline) base subject groups tend to move from promoting pedagogic and utilitarian traditions towards the academic tradition'. (See Layton 1973,[4] Ben-David and Collins 1966[48]);

(*c*) 'That in the conflict over environmental studies much of the curriculum debate can be interpreted in terms of conflict between subjects over status, resources and territory'. (Again Ben-David and Collins, Musgrove, Young.)

(*a*) Obviously, this pattern would appear most strongly in subjects representing 'fields' rather than 'forms' of knowledge. The history of geography, for instance, shows that in the early stages the subject was made up of a variety of idiosyncratic local versions devised or taught by specialists from other disciplines. During the period in curriculum history that is the concern of this book, the battle over environmental education in the late 1960s and early 1970s, the subgroups within geography can be seen 'pursuing different objectives in different manners'.[9] So much so that in 1970 Professor Fisher wrote that 'The light-hearted prophecy I made in 1959 that we might soon expect to see the full 57 varieties of geography has been almost literally fulfilled, and my personal collection of different categories of geography that have seriously been put forward in professional literature now stands at well over half that number'.[13] At about the same time, the President of the Geographical Association was warning that 'new' geography created a problem because 'it leads towards subject fragmentation', so that ultimately 'the question must arise as to how much longer the subject can effectively be held together'.[14] The potential danger of new versions of geography was touched on by Walford who argued that 'unity within the subject' was 'a basic requirement for its continued existence'.[15]

The tendency to fragmentation in geography through the proliferation of subgroups and sub-versions is a recurrent feature of the subject's history, and was echoed by the Norwood Report's fear about the 'expansiveness of geography'. At this earlier stage, 1943, they saw geography as 'The study of man and his environment from selected points of view'—a definition at that time leading to fears that through its expansiveness geography was becoming 'a "world citizenship" subject, with the citizens detached from their physical environment'.[16] As a result 'by then, geography had become grievously out of balance; the geographical synthesis had been abandoned'. The problem was fairly rapidly addressed, and a decade later Garnett claimed that most departments were headed by specialists so that 'The initial marked differences and contrasts in subject personality had been blurred or obliterated'.[17]

The means by which the fragmentary subgroups were monitored, controlled and periodically unified will be dealt with later. However, in the period of the battle over 'environmental education', two, or more accurately three, major subgroups within the subject were actively concerned: the regional geographers, the field geographers and, the fastest-growing subgroup, the 'new' geographers. The first two groups, representing strong traditions within the subject had large support among school geography teachers. The latter group was largely derived from new developments in the subject within the universities. The first two subgroups were considerably more sympathetic to environmental initiatives than the new geographers. This was because the environmental lobby offered aid and sustenance to the field and regional geographers. Hence, we find eminent regional geographers like Professor Bryan promoting conferences in environmental studies because this expressed more clearly than new geography 'his own life's work and ambitions as a geographer'.[18] P. R. Thomas explained the affection for environmental approaches entirely in terms of the struggle for survival of the regional subgroup[19] and a college lecturer in geography judged that the new crisis among geography subgroups 'caused traditional [i.e. regional and field] geographers to flee into environmental studies for a time'.[20]

This flirtation proved a short-run phenomenon because of the overwhelming desire for fully-fledged academic status among all geographers; because new geography carried within it the seeds of this final acceptance; and because the activities of the Geographical Association and the University Schools of Geography together directed and managed the change towards a new 'geographical synthesis' where once again the subgroups were 'delicately held together'.

The pattern discerned among geography subgroups in the period of environmental education's emergence is partly echoed when considering biology. Again the subject began with a variety of idiosyncratic versions and groupings devised and taught by specialists from botany and zoology. By the

1960s biology had also developed a major subgroup whose concern with ecology and field biology bordered on the new environmental approaches. For a time this subgroup gained considerable momentum from initiatives like the Keele Conference, which saw this version of biology as promoting environmental awareness.

Alongside field biology a subgroup promoting biology as a 'hard science' based in laboratories gained increasing adherents. The rise of molecular biology, symbolized by the Nobel prizewinning work of Crick and Watson in the late 1950s, gave renewed impetus to the work of this group. In the new universities opening up in the 1960s and in many schools following the Nuffield project, this group managed to dominate the versions of biology that were accepted. Hence the 'hard-science' version was embodied in the new laboratories that were then being built and in the departments that were set up.

So dominant did the 'hard-science' group become in biology, that for a time the ecology and field biology subgroup developed defensive connections with environmental studies. As with geography, a number of professors associated with the subgroup appeared at events or in publications sponsored by the National Association for Environmental Education. However, although only a subgroup on the defensive within biology, the field biologists were actively pursuing opportunities elsewhere and secured a dominant position (along with the field geographers) in the field studies movement which grew rapidly as the 'environmental lobby' gained momentum. The field biology subgroup was thereby able to develop important new 'territory' inside the growth area of field studies, which partially compensated for losing the battle for mainstream biology to the hard-science subgroup. By securing this leading role in field studies any permanent alliance with the rural studies groups promoting environmental studies was rendered both unnecessary and undesirable.

In both geography and biology the subgroups allied to distinctive versions of the subject often gathered very different degrees of support according to whether school or university groups were being considered. Sometimes this reflected a time-lag effect as the new versions of the subject only slowly worked their way into the schools with new graduates taking up teaching posts in them. This was, for instance, the case in the battle between the regional geography and new geography groups: a long time after new geography was well-established in universities, regional geography retained the allegiance of the vast majority of school teachers.

In rural studies the varying support according to whether one concentrates attention on school or university groups, was never an issue since the subject was not taught in universities and, beyond certain individuals, there was no academic reference group. The subgroups within rural studies therefore concentrated on particular versions of the subject within schools. In the period when environmental studies was launched, the two main groups were those who wanted to quickly attach rural studies to a new examination subject with some connections in the tertiary sector and those who wanted to retain traditional rural studies as a subject of outstanding appeal to the more 'practical' pupil. The battle which ensued over the name of the subject association and the new subject was essentially a battle between these two subgroups and ended in resounding victory for the first group led by Sean Carson, when the name of the subject and its association was changed from rural studies to environmental studies.

(b) The second hypothesis examined within the book relates to three major 'traditions' discerned in school subjects: the academic, the utilitarian and the pedagogic. As this has been dealt with in considerable detail in *Becoming a School Subject*, I will provide only the briefest of commentaries. It was thought that an evolutionary profile of the school subjects under study would show a progressive movement away from stressing utilitarian and pedagogic versions of the subjects towards increasing promotion of more academic versions. We have already seen when discussing the nature of school subjects that subgroups representing new geography, 'hard-science' biology and examinable environmental studies had come to be leading promoters of their subjects by the early 1970s. The process and rationale behind this outcome require fairly detailed understanding representing as they do the culmination of a contest between a range of well-supported alternative definitions within each of the subjects.

The model of subject establishment towards a culminating 'academic' discipline was found to be closely applicable to both geography and biology. Once successfully promoted as an academic discipline, the selection of the subject content is in both instances clearly considerably influenced 'by the judgement and practices of the specialist scholars in the field'. Subjects defined in this way, require a base of 'specialist scholars' working in universities to continue the definition and legitimation of disciplinary content.

The strategy for achieving this final stage received early recognition in geography. MacKinder's 1903 four-point plan provides an explicit statement of a subject aspiring to academic acceptance:

> Firstly, we should encourage University schools of geography, where geographers can be made . . .
>
> Secondly, we must persuade at any rate some secondary schools to place the geographical teaching of the whole school in the hands of one geographically trained master . . .
>
> Thirdly, we must thrash out by discussion and experiment what is the best progressive method for common acceptation and upon that method we must base our scheme of examination.
>
> Lastly, the examination papers must be set by practical geography teachers.[21]

The key to the strategy was the first point, the establishment of 'University schools of geographers where geographers can be made'. To complete the control of the subject's identity, geography teaching and examination construction was to be placed in the hands only of teachers 'made in the universities'. The mediation between university and school was in geography placed in the hands of the Geographical Association. The Association, founded in 1893, played a central role in the promotion of geography, which in its early days was confined to idiosyncratic school-based versions and had obtained a tentative place in only a few universities.

The close linkage between the growth in schools and the establishment of the subject elicited regular comment in the pages of *Geography*. The

President of the Geographical Association paid homage to 'fruits of inspired teaching', which have led to the 'intense and remarkable upsurge in the demand to read our subject in the universities'. The result has been 'the recognition of our subject's status among university disciplines ... together with the costly provision made available for its study'.[22] The latter point shows the direct link between academic status and resources in our educational system: the triumph of the 'academic' tradition over the utilitarian and pedagogic traditions which played such a prominent part in geography's early days needs to be partly understood in these terms.

The establishment of 'discipline' status inside the universities which had been so systematically pursued since Mackinder's 1903 proclamation provided a range of material improvements in the subject's place within schools. In 1954 Honeybone could claim that 'at long last, geography is forcing its complete acceptance as a major discipline in universities, and ... geographers are welcomed in to commerce, industry and the professions, because they are well educated men and women'.[23] From now on geography could claim its place in educating the most able children, and thereby become established as a well-funded department inside schools, staffed with trained specialists on graded posts. By 1967 Marchant noted that geography was 'at last attaining to intellectual respectability in the academic streams of our secondary schools'.[24] But he noted that the battle was not quite over and gave two instances where the subject was still undesirably taught as a 'less able' option. With the launching of new geography the subject finally attained total acceptance as an academic discipline in universities and as a fully-fledged A-level subject in all schools, with the resources and 'costly provisions' which such status attracts.

In biology, the evolution of the subject is distinguishable from the case of geography because from the beginning there was an associated and well-established university base in the form of botany and zoology. For this reason, and also because from the outset the subject benefited from the side-effects of the influential science lobby, the task of subject promotion never totally resembled geography's 'beginning from scratch'. Biology's task was more to present a case for inclusion within the, by then well-established (and consequently well-resourced), science area of the curriculum. This task was often pursued within the overall arena of the Science Masters' Association, who from the 1930s onwards played an active role in promoting biology. In 1936 an influential biology subcommittee was formed to promote biology syllabuses, and many articles in the Association's *School Science Review* argued the case for biology's recognition as an examination subject for the able student. The problem was best voiced by the Ministry of Education in 1960: 'The place which is occupied by advanced biological studies in schools ... is unfortunately that of vocational training rather than of an instrument of education'. The need to be seen as an 'instrument of education' meant that the promoters of the subject had to move away from the utilitarian towards more academic versions—only then could an A-level subject command sufficient pupil numbers to warrant 'departmental' status and resources in schools. Hence we find the common theme being stressed: biology must be treated 'as a comprehensive discipline in its own right'.[25]

In the final stages in the promotion of biology as an 'academic discipline' in the 1960s, the two main initiatives stressed the subject as a 'hard science' needing 'laboratories and equipment'. In the rapidly expanding universities it was this version of the subject which was widely introduced, thereby establishing the academic discipline base; likewise, part of the Nuffield Biology Project for Schools centred on a crusade in terms of equipment and laboratory staff. With the new generation of biology graduates trained in this hard science at universities, the establishment of the subject as a fully fledged academic O- and A-level subject was finally assured.

Unlike biology and geography, rural studies remained for generations a low-status enclave, stressing highly utilitarian or pedagogic values. This provides confirmation for Ben-David and Collins's contention that the move to a change in intellectual and occupational identity came at the time when the subject was faced with survival problems in a reorganizing educational system stressing academic examinations. The pervasive influence of this tradition can clearly be seen in the following quotation:

> The lack of a clear definition of an area of study as a discipline has often been a difficulty for local authorities in deciding what facilities to provide . . . It has been one of the reasons for the fact that no A-level course in rural studies exists at present.[26]

The Schools Council Working Party in 1968 confirmed this with the broad hint that there was the 'need for a scholarly discipline'.[27]

With no tertiary base and hence no specialist scholars involved, except random specialists from other disciplines, the Hertfordshire strategy was to develop an A-level syllabus from groups working in the secondary schools. This offered the promise of tailoring 'a course to the needs of the kids' and not 'have to meet the requirements of other people's courses'. But the crucial reason in terms of the subject teachers' material self-interest was often frankly admitted:

> I think we had got to prove that environmental studies was something that the most able of students could achieve and do something with . . . if you started off there all the expertise and finance that you put into it will benefit the rest— your teaching ratio goes up etc. and everyone else benefits.[28]

Likewise, another leading advocate admitted that they had seen that:

> the only way to make progress was to get in on the examination racket . . . the exam, was essential, otherwise you couldn't be equal with any other subject. Another thing was that comprehensive education was coming in. Once that came in, no teacher who didn't teach in the fifth or sixth form was going to count for twopence. So you had to have an A-level for teachers to aim at.

The survival rationale was always a strong factor:

> 'I just thought if you're outside this you've had it in schools: it was already happening in some schools where a [rural studies] teacher was leaving, they didn't fill the place, because they gave it to someone in the examination set up'. And beyond survival the reasons for an academic A-level were simply 'because if you didn't you wouldn't get any money, any status, any intelligent kids'.[29]

The Hertfordshire A-level in environmental studies which was ultimately devised is a recognition of the factors defining the aspirations and efforts of these rural studies teachers. What has subsequently been denied is not that environmental studies represents a valid area of curriculum, but that it can thereby claim to be an academic discipline. Such claims, it would appear, are best validated through university scholarship, and without a university base status passage to acceptance as an academic subject has been denied.

MacKinder's strategy of using the school subject base to help bring about the creation of university departments was correctly conceived. As Carson noted at the Offley conference, new contenders for academic status are often placed in an impossible situation since they are asked 'What evidence have you that universities would accept this sort of A-level?' On making enquiries to universities, the reply was 'show us the successful candidates and we will tell you'. A chicken-and-egg situation![30]

(c) A third hypothesis follows on as one moves from consideration of the patterns of internal evolution in school subjects to investigate the role that the pursuit of academic status plays in the relationship between subjects. In continuity with the second hypothesis we would expect established subjects to defend their own academic status at the same time as denying such status to any new subject contenders, particularly in the battle over new A-level examinations.

In the struggle to launch environmental studies as an A-level subject, the geographers reacted strongly, and the biologists much more mildly, following the lines of the hypothesis. MacKinder, the founding father of geography's road to academic establishment, would have understood this. In explaining the geologists' opposition to geography he saw their fear of the new subject making 'inroads in their classes' as the reason for their response and noted that 'even scientific folk are human, and such ideas must be taken into account'.[31] In continuity with this, the geographers strongly opposed social studies, an integrated package that pre-dated environmental studies by several decades. The geographers, it was claimed, 'saw the new proposals as a threat to the integrity and status of their own subject'.[32]

The growth of environmental studies was treated in similar manner by the geographers. The discussion of the Executive Committee of the Geographical Association shows precious little concern with the intellectual or epistemological arguments for environmental studies. The discussion focused on 'the threat to geography involved in the growth of environmental studies'. Indeed, when the possibility of engaging in a dialogue with environmental studies teachers was suggested, 'some members felt that to do so would be tantamount to admitting the validity of environmental studies'.[33] The most overt plea for defence rather than dialogue came in the Presidential Address to the Geographical Association in 1973. Mr A. D. Nicholls laid great emphasis on the 'practical realities' for 'practising teachers'. 'With constant pressure on teaching time, headmasters are ever searching for new space into which additional prestige subjects can be fitted, and the total loss of teaching time to environmental subjects may be considerable'. Beyond these practical fears about the material interests of geography teachers, environmental studies evoked a particularly emotional response among geographers because of its proximity to geography's continuing identity crisis. Nicholls provides an unusually frank admission of the need for territorial defence being placed above any intellectual imperatives:

> Ten years ago almost to the day and from this platform, Professor Kirk said 'modern geography was created by scholars, trained in other disciplines, asking themselves geographical questions and moving inwards in a community of problems; it could die by a reversal of the process whereby trained geographers moved outwards in a fragmentation of interests seeking solutions to non-geographical problems'. Might not this be prophetic for us today? Could it not all too soon prove disastrous if the trained teachers of geography moved outwards as teachers of environmental studies seeking solutions to non-geographical problems?[34]

The fears which geographers expressed so strongly and emotionally about the emergence of environmental studies were not shared to the same degree by biologists. As we have seen, only the field biology subgroup was threatened and they managed to expand into the growing territory of field studies. However, in the negotiations at the Schools Council the Science Subcommittee, which included a number of biologists, joined forces with the geographers in their opposition to the environmental studies A-level. In both subcommittees 'concern was expressed at the heavy overlap between this syllabus and syllabuses in both geography and biology'. The pursuance of this allegation involved a clever strategy. First the committees arued that the A-level must delete 'irrelevant topics'—mainly content not related to geography or biology. Then the committees stated that 'if irrelevant topics were envisaged as removed, the effect would be to reveal how close the resulting syllabus would be to existing syllabuses',[35] in geography and to a lesser extent biology.

A judgement from Carson that the Schools Council subcommittees 'jealously guarded the preserves of their subject' was confirmed by the comments from the Geography Subcommittee when the decision on the A-level was finally announced. They were plainly fairly satisfied with their territorial defence and 'noted with approval that candidates could not take this examination together with geography'. A final point was added that there was 'as yet no indication that universities would be prepared to accept a pass in this subject as an entry qualification for degree courses'.[36] The restriction on environmental studies being offered with geography, together with an initial restriction to a five-year 'experimental' period and to a limited number of schools placed enormous practical obstacles in the way of any widespread adoption of the subject. By ensuring that these obstacles faced the new subject in the early years when the momentum for change was strong, the opponents of the new subject effectively extinguished its chances of establishment in the secondary school curriculum.

Examining theory: an example

Since we began by instancing the non-use or mis-use of history by sociologists who contributed to *Knowledge and Control* it would be instructive to examine their theories with respect to school subjects. This way we can examine an earlier contention that 'to seek to provide from the macro level theories of curriculum without empirical investigation or understanding of how the curriculum has been negotiated at micro level over time is a poor sequence through which to proceed to theory'.[37]

The first point to recognize in Young *et al.* is the assumption in a number of the papers that subjects are monolithic. This is hardly a promising starting point from which to develop the theme that the curriculum is subject to patterns of control by dominant interest groups. The papers in the book reflect Bernstein's contention that 'how a society selects, classifies, distributes, transmits and evaluates the educational knowledge it considers to be public, reflects both the distribution of power and the principles of social control'.[38] Young likewise suggests that 'consideration of the assumptions underlying the selection and organization of knowledge by those in positions of power may be a fruitful perspective for raising sociological questions about curricula'.[39] The emphasis leads to general statements of the following kind:

> Academic curricula in this country involve assumptions that some kinds and areas of knowledge are much more 'worthwhile' than others: that as soon as possible all knowledge should become specialised and with minimum explicit emphasis on the relations between the subjects specialised in and between specialist teachers involved. It may be useful therefore, to view curricular changes as involving changing definitions of knowledge along one or more of the dimensions towards a less or more stratified, specialised and open organization of knowledge.
>
> Further, that as we assume some patterns of social relations associated with any curriculum, these changes will be resisted in so far as they are perceived to undermine the values, relative power and privileges of the dominant groups involved.[40]

The process whereby the unspecified 'dominant groups' exercise control over other presumably subordinate groups is not scrutinized, although certain hints are offered. We learn that a school's autonomy in curriculum matters 'is in practice extremely limited by the control of the sixth form (and therefore lower form) curricula by the universities, both through their entrance requirements and their domination of all but one of the school examination boards'.[41] In a footnote Young assures that 'no direct control is implied here, but rather a process by which teachers legitimate their curricula through their shared assumptions about "what we all know the universities want"'.[42] This concentration on the teachers' socialization as the major agency of control is picked up elsewhere. We learn that:

> The contemporary British educational system is dominated by academic curricula with a rigid stratification of knowledge. It follows that if teachers and children are socialised within an institutionalised structure which legitimates such assumptions, then for teachers high status (and rewards) will be associated with areas of the curriculum that are (1) formally assessed (2) taught to the 'ablest' children (3) taught to homogeneous ability groups of children who show themselves most successful with such curricula.[43]

Young's explanation of patterns of curriculum control therefore hinges on his belief that universities 'control sixth form curricula' through 'their entrance requirements and their domination of all but one of the school examination boards'. Direct control is not apparently meant; rather, indirect control through the shared assumptions into which teachers are socialized.

Curriculum histories present evidence of a more complex process at work. The role of dominant groups shows perhaps most clearly in the victory of the academic tradition in the early years of the 20th century. This victory was embodied in the influential 1904 Regulations and, most significantly, the 1917 School Certificate. *Once established*, however, these curricular patterns (and their associated financial and resource implications) were retained and defended in a much more complex way and by a wider range of agencies. It is therefore correct to assume that, initially, the rules for high-status knowledge reflected the values of dominant interest groups at that time. But it is quite another issue to assume that this is *inevitably* still the case or that it is dominant interest groups themselves who *actively* defend high-status curricula. It is perhaps useful to distinguish between domination and structure and mechanism and mediation.

By focusing on subjects in evolution and the conflict over A-level examinable knowledge the studies in *School Subjects and Curriculum Change* clearly indicate the central role played by school subject groups and subgroups. The most powerful of these agencies are those groups promoting the academic tradition—successfully in geography and biology, but unsuccessfully in environmental studies. These groups *demanded* the creation of an academic discipline based in the universities. The 'academic tradition' subject groups act in this way because of the legacy of curricula, financial and resource structures inherited from the early 20th century (when dominant interests *were* actively defended). Because of this legacy, able pupils and academic examinations are linked and consequently resources, graded posts, and career prospects are maximized for those who can claim academic status for their subject.

The evidence indicates not so much domination by dominant forces as solicitous surrender by subordinate groups. Far from teacher socialization in dominant institutions being the major factor creating the patterns discerned it was much more considerations of teachers' material self-interest in their working lives. Since the misconception is purveyed by sociologists who exhort us 'to understand the teacher's real world' they should really know better. High-status knowledge gains its school subject adherents and aspirants less through the control of the curricula, which socialize, than through well-established connection with patterns of resource allocation and the associated work and career prospects these ensure. The study of curriculum histories argues that we must replace crude notions of domination with patterns of control in which subordinate groups can be seen actively at work. A tentative explanatory framework at this level is provided in the next section.

School subjects and curriculum change: an explanatory framework

1. The structure of material interests: status, resources and careers

The historical investigation of the curriculum conflict over 'environmental studies' suggests the pursuit of material interests as a major explanatory factor in understanding curriculum change. This is not to provide an overarching theory but to suggest that this aspect has been substantially neglected in previous accounts.

1.1. The similar aspirational patterns discerned in the subject histories discussed direct our attention to the structuring of material interests—how resources and career chances are distributed and status attributed.

1.2. Essentially the structure emerged in the period 1904–1917. The 1904 Regulations defined the subjects suitable for the secondary, grammar schools. These were largely academic subjects and they were subsequently enshrined in the School Certificate examinations launched in 1917. From then on these examination subjects inherited the priority treatment on finance and resources directed at the grammar schools.[44]

1.3. The structure has effectively survived the ensuing changes in the educational system. Byrne, for instance, stated 'that more resources are given to able students and hence to academic subjects' (the two are still synonymous), since 'it has been assumed that they necessarily need more staff, more highly paid staff and more money for equipment and books'.[45]

1.4. The material interests of teachers—their pay, promotion and conditions—are broadly interlinked with the fate of their specialist subject communities. The 'academic' subject is placed at the top of the hierarchy of subjects because resource allocation takes place on the basis of assumptions that such subjects are best suited for the able students who, it is further assumed, should receive favourable treatment.

1.5. Hence, in secondary schools the self-interest of subject teachers is closely connected with the status of the subject in terms of its examinable knowledge. Academic subjects provide the teacher with a career structure characterized by better promotion prospects and pay than less academic subjects. Seen from this viewpoint the conflict over the status of examinable knowledge is therefore essentially a battle over the material resources and career prospects available to each subject community.

2. Subjects as 'coalitions'

2.1. The process model developed by Bucher and Strauss for the study of professions provides valuable guidelines for those studying school subjects. Within a profession, they argue, are varied identities, values and interests. Hence professions are to be seen as 'loose amalgamation of segments pursuing different objectives in different manners and more or less delicately held together under a common name at particular periods in history'.[46] The most frequent conflicts arise over the gaining of institutional footholds, over recruitment and over external relations with clients and other institutions. At times when conflicts such as these become intense, professional associations may be created or, if already in existence, become more strongly institutionalized.

2.2. The Bucher and Strauss model of professions suggests that perhaps the 'subject community' should not be viewed as a homogeneous group whose members share similar values and definition of role, common interests and identity. Rather, the subject community should be seen as comprising a range of conflicting groups, segments or factions (referred to as subject subgroups). The importance of these groups might vary considerably over time. As with professions, school subject associations (for example the Geographical Association) often develop at particular points in time when there is an intensification of conflict over school curriculum and resources and over recruitment and training.

3. Subject coalitions in evolution: internal curriculum change

3.1. In the subjects studied, a pattern of evolution can be discerned in the process of becoming a subject. Initially a subject is a very loose amalgamation of subgroups and even of idiosyncratic versions, often focused on pedagogic and utilitarian concerns.

3.2. A subgroup emerges arguing for the subject to become an academic discipline so as to be able to claim resources and status.

3.3. At the point of conflict between earlier subgroups and the proselytizing 'academic' subgroup, a subject association is often formed. The association increasingly acts to unify subgroups into a *dominant coalition* promoting academic status.

3.4. The dominant coalition calls for discipline status and for university departments to be set up to train its disciplinary specialists (see MacKinder's manifesto). Some subjects (for example rural/environmental studies) are blocked at this point (university admissions policies play a role here).

3.5. For the successful subjects a final stage is the creation of a university discipline base. The subject is now defined increasingly by university scholars and it is to the structure of their material interests and resulting aspirational patterns that we must now look to explain curriculum change (for example new geography, molecular biology) and for the source of tensions for the school subject.

4. Changing climates and external relations: defining a new subject

4.1. The emergence of the environmental climate of opinion offered new opportunities for subject groups and subgroups in the promotion of their interests. (I have not dealt with the structural origins of this new climate as other work has attempted this. For example, 'the climate of opinion which made environmental studies a credible label for curriculum innovation in the '60s and '70s is best understood in terms of the historical circumstances of post-war capitalism'.[47])

4.2. In this respect Ben-David and Collins's hypotheses were substantially proven. They argued that for a new subject or discipline:

(a) The ideas necessary for creation . . . are usually available over a relatively prolonged period of time and in several places.

(*b*) Only a few of these potential beginnings lead to further growth.

(*c*) Such growth occurs where and when persons become interested in the new idea, not only as intellectual content but also as a potential means of establishing a new intellectual identity and particularly a new occupational role. They conclude 'the conditions under which such interest arises can be identified and used as a basis for eventually building a predictive theory'.[48]

4.3. Applying this to subject groups and subgroups a number of factors would be relevant:

(i) Subject group/subgroup position in hierarchies of subjects (current power and status).

(ii) Their position regarding resource allocation in schools (current resources).

(iii) Patterns of career and age position of practitioners (current career patterns).

4.4. Subjects with low status and resources and poor career patterns, like rural studies, therefore embraced the opportunity to establish a new intellectual identity and occupational role. Established high-status subjects conversely ignored the opportunity but contested the new contenders' right to claim similar academic status (and thereby establish parity of status and resources and a share therein). Carson has provided the rationale for the rural studies subgroup's move to promote an environmental studies A-level: 'because if you didn't you wouldn't get any money, any status, any intelligent kids'.

Conclusion

In this article I have been concerned to show that curriculum histories can be a valuable complement to, and at times an active agency in the development of theoretical frameworks. The essential value of such histories (as of more general historise) is that they are immersed in the complexity of the social process. They develop from the desire to understand particular events, not from a desire to prove particular theories. The curriculum historian will often travel with her/his ideological and theoretical baggage packed away. This implicitness, whilst avoiding the primacy of theoretical verification, should not, however, limit theoretical aspiration. The curriculum historian *should* be concerned to aid the generation of theories about actions and events in specific historical conditions. In this way the historian can play an important role in the theoretical enterprise and in the making of agendas for further studies.

Of course the specificity of curriculum histories often acts against their capacity for generalization. The model of subject change developed herein clearly has many limitations. What about pastoral systems? falling rolls? whole curriculum planning? Does this apply to subjects like classics, economics or, dare we mention it, sociology? What about subjects where industrial and external forces are more clearly involved? What are the factors behind 'the changing climates for action' that have been discerned?

But beyond the problem of the specificity of curriculum histories lies the problem of the *nature* of curriculum histories. Clearly (following hypothesis 1) history is no monolithic subject or method. In history there are schisms which resemble those in other subjects—notably the disagreement between the 'general law' school of historians and what we might call the 'uniqueness' school.

At one level, the argument for curriculum histories merely reiterates the need to study how the curriculum has evolved and to understand historical background and origins so as to provide a context for contemporary inquiry. By this view historical studies can *extend* the range of our accounts. I would, however, want to go further than this and argue that histories are important because of their potential to *transform* our accounts: to pose fundamental questions and to point towards new agendas for study. For instance, reformulating notions of 'domination', changing the priority given to prior socialization in accounts of subject change or stressing the importance of professional subject groups in the evolution of the curriculum.

The specificity and nature of history leave us with a dual challenge in conducting future curriculum histories. Firstly, where possibly, we must pursue the gradual and continuous nature of curriculum change (certainly in systems as decentralized as ours) so as to illuminate *contemporary situations*; this argues against too rigidly 'periodized' histories. Secondly, we must aspire not only to *extend* our range of data but to contribute to the examination and reformulation of hypotheses and theories, thereby offering the potential for *transforming* our accounts.

These aspirations have, of course, to be set against the limitations of the studies reported here. I hope, however, that some progress has been indicated as well. Curriculum histories point to the evolutionary nature of subjects as coalitions 'more or less delicately held together under a common name at particular periods'. The nature of these coalitions responds both to the structuring of material interests and to the 'changing climates' for action. Because of the manner in which resources (and associated career prospects) are distributed and status attributed, 'academic' subjects groups normally develop as 'dominant coalitions'. The conflict over the status of examinable knowledge therefore becomes the crucial conflict arena where the subject coalitions (and their representative associations) contest the right to material resources and career prospects.

This article suggests that it is a dangerous enterprise to develop theories of curriculum whilst under-using or misusing historical studies. Particularly significant is the constant harking back to the early 19th century for analogies with which to support contemporary theory. Structural and interactional features are not continuous and to assume continuity is to at best oversimplify and at worst wilfully mislead.

A particular problem in those studies which generalize from the early 19th century to the contemporary situation has been identified in the work reported here. Namely that by raiding history in this way sociologists have been returning to a *pre-professional era* with respect to curriculum groups.

The evidence presented here points to the power and importance of professional subject groups; they cannot be dismissed as powerless agencies in the face of structural change.

The use of socialization as a kind of black box theory of causation seems a common but inappropriate device; postulating causation without presenting evidence is poor theorizing, particularly when professionalization has been substantially ignored. The evidence presented here suggests that it is not so much prior socialization as the structuring of material interest which provides the mediating mechanism between structural and interactional levels.

The dominance of 'academicism' can be shown over the last century or more. But historical studies pose questions about *in whose interests* this dominance prevails: professional groups, culturally dominant groups or industrial or financial capital. Academicism may be the past cultural consequence of previous domination rather than a guarantee of future domination. Through immersion in the complexity of social action the curriculum historian (as also the ethnographer of course) places her/himself among the participants in social action as they respond to and scrutinize the social structure. In such a position, by analogy, she/he is well placed to scrutinize macro-theory 'from the complexity below'. In particular, curriculum histories may help in emancipating us from the grosser correspondence and dominance theories and give new meanings to concepts of relative autonomy where people can be recognizably viewed in action.

Acknowledgement

An earlier version of this paper was presented at the conference on 'School Subjects: Histories and Ethnographies' held at St. Hilda's College, Oxford in September 1982.

References and notes

1. YOUNG, M. Curriculum change: limits and possibilities. In M. Young and G. Whitty (eds.) *Society, State and Schooling* (Falmer Press, Brighton, 1977), pp. 248–249.
2. BERNSTEIN, B. Sociology and the sociology of education: a brief account. In J. Rex(ed.) *Approaches to Sociology* (Routledge and Kegan Paul, London, 1974), p. 156.
3. SILVER, H. Nothing but the past, or nothing but the present? *Times Higher Educational Supplement* (1 July 1977), 17.
4. LAYTON, D. *Science for the People* (George Allen and Unwin, London, 1973), p. 166.
5. YOUNG, M. op. cit., p. 245.
6. LOWE, R. The divided curriculum: Sadler, Morant and the English secondary school. *Journal of Curriculum Studies*, 8 (1976), pp. 139–140.
7. See GOODSON, I. F. Curriculum Conflict 1895–1975 (unpublished D.Phil. 1980); Becoming a school subject: patterns of evolution and explanation. *British Journal of Sociology of Education* (July 1981); *School Subjects and Curriculum Change: Case Studies in the Social History of Curriculum* (Croom Helm, London, 1932); Defining and defending the subject. In M. Hammersley and A. Hargreaves, *Sociology of Curriculum Practice* (Falmer Press, Brighton, 1983).
8. NISBET, R. A. *Social Change and History, Aspects of the Western Theory of Development* (1969), quoted in WEBSTER, J. R., Curriculum change and crisis. *British Journal of Educational Studies*, 3 (October 1971), pp. 204–205.

9. BUCHER, R. and STRAUSS, A. Professions in process. In M. Hammersiey and P. Woods (eds.) *The Process of Schooling* (Routledge and Kegan Paul, London, 1976), p. 19.

10. MUSGROVE, F. The contribution of sociology to the study of curriculum. In J. F. Kerr (ed.) *Changing the Curriculum* (University of London Press, London, 1968).

11. KUHN, T. S. *The Structure of Scientific Revolution* (2nd edn, University of Chicago Press, Chicago, 1970).

12. WARING, M. *Social Pressures and Curriculum Innovation* (Methuen, London, 1978).

13. FISHER, C. A. Whither regional geography? *Geography*, 55, 4 (November 1970), pp. 373–374.

14. GARNETT, A. Teaching geography: some reflections. *Geography*, 54 (November 1969), pp. 388–389.

15. WALFORD, R. Models, simulations and games. In R. Walford (ed.) *New Directions in Geography Teaching* (Longmans, London, 1973), p. 97.

16. *Norwood Report* (HMSO, London, 1943), pp. 101–102.

17. Op. cit., GARNETT, p. 368.

18. MILLWARD, R. Obituary: Patrick Walter Bryan. *Geography*, 54, 1 (January 1969), p. 93.

19. THOMAS, P. R. Education and new geography. *Geography*, 55, 3 (1970), pp. 274–275.

20. Interview: Scraptoft (14 December 1976).

21. MACKINDER, M. J. *Report of the Discussion on Geographical Education* (1903).

22. GARNETT, op. cit.

23. HONEYBONE, R. C. Balance in geography and education. *Geography*, 34, 186 (1954).

24. MARCHANT, E. C. Some responsibilities of the teacher of geography. *Geography*, 3 (1968).

25. Ministry of Education *Science in Secondary Schools*. Pamphlet No. 38 (HMSO, London, 1960).

26. CARSON, S. Unpublished M.Ed, thesis (Manchester, 1967), p. 135.

27. *Schools Council Working Paper No. 24*, p. 19.

28. Interview: P. Topham.

29. Interview: S. Carson.

30. CARSON, S. (ed.) *Environmental Studies, the Construction of an 'A' level syllabus* (NFER, Slough, 1971), p. 6.

31. MACKINDER, H. J. Teaching of geography and history as a combined subject. *The Geographical Teacher*, 7 (1913).

32. CHANNON, C. Social studies in the secondary school. *Educational Review* (17 January 1964).

33. Geographical Association. *Notes of Meeting of Chairmen of Section/Standing Committee* (25 September 1970).

34. NICHOLLS, A. D. Environmental studies in schools. *Geography*, 58, 3 (July 1973), p. 201.

35. Letter Schools Council, Hertfordshire File (21 February 1973), SS/L/G/191.

36. Ibid.

37. GOODSON, I. F. *Towards a Social History of Subjects* (in mimeo).

38. BERNSTEIN, B. On the classification and framing of educational knowledge. In M. Young (ed.) *Knowledge and Control* (Collier Macmillan, London, 1971).

39. YOUNG, M. An approach to the study of curricula as socially organised knowledge. In M. Young (ed.), ibid., p. 3.

40. Ibid., p. 34.

41. Ibid., p. 22.

42. Ibid.

43. Ibid., p. 36.

44. See SMITH, M. The evaluation of curriculum priorities in secondary schools 1903–4. *British Journal Sociology of Education*, 1, 2 (June 1980), pp. 153–172.

45. BYRNE, E. M. *Planning and Educational Inequality* (NFER, Slough, 1974), p. 29.

46. BUCHER and STRAUSS, op. cit.

47. GOMM, R. *Environment and Environmental Studies* (monograph).

48. BEN-DAVID, J. and COLLINS, R. Social factors in the origins of a new science: the case of psychology. *American Sociological Review*, 31, 4 (August 1966), p. 452.

Greening the future for education: changing curriculum content and school organization

ANNETTE GREENALL GOUGH

Children and Controversial Issues: Strategies for the Early and Middle Years of Schooling. B. CARRINGTON and C. TROYNA (eds), Falmer Press, London (1988), £19·95, USA $44·00 (cloth), ISBN 1-85000-416-1; £9·95 USA $22·00 (paper) ISBN 1-8500-417-X.

Greenprints for Changing Schools. S. GRIEG, G. PIKE and S. SELBY, The World Wide Fund For Nature and Kogan Page, London (1989), £7·75 (paper), ISBN 1-85091-950-X.

Educating for Peace: Issues, Principles, and Practice in the Classroom. D. HICKS (ed.), Routledge, London (1988), USA $17·95 (paper) ISBN 0-415-01329-1.

Education: In Search of a Future. H. LAUDER and P. BROWN (eds), Falmer Press, London (1988), £19·95, USA $44·00 (cloth) ISBN 1-85000-406-4; £9·95, USA $22·00 (paper) ISBN 1-85000-407-2.

Educating for Global Responsibility: Teacher-Designed Curricula for Peace Education, K-12. B. A. REARDON (ed.), Teachers College Press, New York (1988), USA $12·95 (paper) ISBN 0-8077-2879-9.

The common concern of the five books considered here is perhaps best summarized by the title of one of them: *Education: In Search of a Future*. The future for education that they are seeking can be characterized as 'green', in the broad interpretation of that term. Although the environment usually is seen as the most characteristic (and most often caricatured) green issue, green politics is also concerned with issues of industrial development, peace, racism, social justice, feminism and health – and with the environmental, economic and educational implications of these issues. The 'greening' of society is part of a 'global mind change' (Harman 1988) that many writers and commentators have identified as signalling a paradigm shift towards a more holistic world-view (see, for example, Birch 1990, Capra 1983, Ferguson 1982, Gough 1989, Michael and Anderson 1986).

Each of the books reviewed here explores a number of these issues to varying degrees, particularly with respect to their manifestation in school curricula. Four of them also discuss the purposes of education, relevant content and processes in the curriculum, and changing curriculum paradigms. Issues pertaining to the latter topic have also been the subject of recent articles in this journal, notably, the symposium on 'Curriculum Paradigms and the Post-Modern World' (Doll 1989, Gough 1989, Slaughter 1989). In this symposium Doll (1989: 252) argues for a 'a re-visioning of curriculum' which challenges 'the modernist assumptions on which our present curriculum is founded' and urges us to 'look beyond these assumptions to new ones on which the post-modern paradigm is being built'. However, Gough (1989: 233–234) is critical of certain examples among the plethora of 'green' studies (such as peace studies and environmental education) which 'preach a holistic perspective but still present themselves as separate entities and, worse, ... have preserved the teaching practices that go with a fragmented world view'. Several of the authors of the books reviewed, in particular *Educating for Peace*, and *Green Prints for Changing Schools*, add to the discussion of these issues.

Another issue raised by several of these books concerns the cultural content of the curriculum. This issue was also raised recently by Bowers (1990: 72), who questions curriculum priorities in the light of the ecological crisis and the nature of the culture that is contained in the curriculum:

> formal education involves transmitting culture to the next generation, the question of whether the culture that is to become the basis of thought and behaviour contributes to a further deterioration of critical life-sustaining natural systems should be basic to any discussion of curriculum policy and practice.

Peace education

The contributors to *Education for Peace* go beyond the topic of peace as such and discuss a number of green issues. As Hicks (1988: 8) explains in his introduction, education for peace involves developing the knowledge, attitudes and skills needed in order to explore concepts of peace to enquire into the obstacles of peace (such as violence and war, inequality, injustice, environmental damage and alienation), and to resolve conflicts and explore a range of different alternative futures, particularly ways of building a more just and sustainable world society. Thus, peace is seen as an all-encompassing term which covers numerous green issues.

Hicks (1988: 12) distinguishes between *peace studies* (as a separate subject on the school timetable) and *peace education* (as a subject-matter that may be taught in different places at different times in the timetable). He describes the knowledge, attitudes and skills that equally form the essential core of education for peace. The skills include critical thinking, co-operation and empathy, assertiveness, conflict resolution and political literacy. The teacher is seen as a facilitator rather than an authority, creating a person-centred learning climate which encourages participatory and experiential learning and involves democracy in action. It is therefore not surprising that the authors portray education for peace as being incompatible with many traditional assumptions about learning. Another of the contributors to *Educating for Peace*, Whitaker, draws on A. S. Neill, Carl Rogers, John Holt, and Paolo Freire to argue that peace education is part of a newly emerging 'transformative paradigm' for education. In this new paradigm the key characteristics of that which is seen to be desirable educationally refer to the nature of learning rather than to the content of instruction.

As well as grounding itself in a description of the field articulating and confronting the critics of peace education, *Education for Peace* is a very practical book. Each of the ten case studies (on the topics of conflict, peace, war, nuclear issues, justice and development, power, gender, race, environment and futures) which form the major part of the book (176 pages) provide background notes on each topic, relate the topic to education, describe at least two classroom activities, and include a list of references. All provide useful information on their respective topics and many of the activities are novel as well as being well matched to their topics. Several of the authors make use of role playing in their activities (Burnley, Williamson-Fien, Swee-Hin); others include activities involving developing visions of the future (Fell, Cooke, Slaughter), developing communication skills (Fell, Duczek) and coping with feelings (Yarwood and Weaver, Cooke). I particularly liked Yarwood and Weaver's activity on 'Fighting in School' (in Hicks 1988: 96–97) which, through the simple use of statements on fighting that the students must order through small-group discussion, encourages discussion about fears and expectations in relation to fighting and about how alternatives to fighting might offer more creative and lasting solutions. However, I was disappointed with the race relations activities suggested by Henfrey in that they required teachers to do a great deal of prior reading before being able to use them: for example, 'Teachers will need to acquaint themselves with the major themes in the historical and contemporary experiences of black women from the list of resources given below' (in Hicks 1988: 192). This does not seem to be a particularly practical way to go about encouraging teachers to embark on teaching a new topic. I was also disappointed with the activities suggested by Huckle. One of them required prior teaching on 'the economic and political system of the USSR in outline' and on the two main water transfer projects in the USSR (in Hicks 1988: 211) before students would be able to do the activity; the other seemed a superficial way of considering issues to do with environmental responsibility (through ordering already numbered sketches from a Chinese storybook). These two activities were selected from the *What We Consume* series (Huckle 1988–90) which contains many examples of more imaginative and practical activities than those which accompany Huckle's chapter here.

The main theme of *Education for Peace* is that an education appropriate for the 21st century needs to pay very careful attention to both relevant content and processes in the curriculum. The paradigm which encompasses the necessary conceptual shifts is seen as being quite different from the presently dominant world-view. Each of the authors addresses this theme. A frequently repeated message is the importance of looking at the whole school and asking how much its ethos and organization seek to promote justice and equality of opportunity (see, for example, Fell, in Hicks 1988: 82) as well as focusing on how teachers interact with pupils and how pupils interact with each other to address the attitudes, language and behaviour patterns that are exhibited and which are dysfunctional to a peaceful classroom (for example, Duczek, in Hicks 1988: 174). Richardson (in Hicks 1988: 236) summarizes this theme well: 'The move towards greater participation, democracy, and openness in the individual classroom, for example, has to be strengthened and sustained by analogous changes in school organization and culture'.

The authors of *Education for Peace* are not advocating a new subject for inclusion on the timetable, rather they are suggesting an approach to education. As articulated by Hicks (1988: 245) this approach is 'reconstructionist' (in arguing that education has a role to play in the transformation of society) and person-centred (in arguing that the prerequisite for transforming society is the development of a centred and assertive self-reliance in the individual). They also address the issue of fragmentation, which is seen as a problem of the old world-view, and distinguish it from the holistic character of the emerging world-view (Hicks 1988: 253).

Educating for Global Responsibility takes a different approach from *Education for Peace* but it is also intended to be very practical. It focuses on the same green issue, peace (and again interprets the term broadly), and contains a collection of curricula to demonstrate a range of possibilities for integrating peace education into all areas of the curriculum at all grade levels. According to the editor (Reardon 1988: xix), 'There are as yet no clear and precise limits to nor standards for what is to be included in peace education'. However, the subject-matter offered covers a similar range of content to that selected by Hicks for *Education for Peace*, so there are obviously some common understandings among people interested in peace education. When compared with *Education for Peace*, the notable omissions from Reardon's index are the topics of gender, power and futures, but peace is still interpreted as covering the topics of peace, conflict, war, environment, justice and development.

As this book is primarily concerned with 'teacher-designed curricula for peace education, K-12', which form the greatest part of the book (168 pages), it is not surprising that the book is relatively weak on contextualizing and connecting peace education with wider curriculum issues (10 pages). However, the omission of any consideration of curriculum processes and of the whole-school organizational implications of peace education is a major weakness. Instead, Reardon promotes only content-based approaches, such as the introduction of specific new subject-matter and infusion of peace content into existing subject-matter (Reardon 1988: xxii). But, as Fell notes (in Hicks 1988: 82),

> this does raise the spectre of how we can attempt to build a positive concept of peace within an education system that is hierarchical and full of traps for the unwary, such as awarding a prize for the best essay on co-operation! It is important also to look at the whole school and ask how much its ethos and organization seek to promote justice and equality of opportunity.

In *Educating for Global Responsibility*, the core concepts that govern curriculum content selection and design (including specific learning objectives and the choice of issues to be studied) are stewardship, citizenship and relationship. The fundamental purposes of peace education are seen as the development of the capacities of care, concern and commitment. The 35 selections are organized into four sections corresponding to grades K–3, 4–6, 7–9 and 10–12. Within each section there are several selections which range from single-lesson activities to whole-course outlines. The skills to be encouraged through these activities – such as critical thinking, empowerment, conflict resolution, co-operation and empathy – are very similar to those listed in *Educating for Peace* and Reardon is to be commended for collecting materials for primary as well as secondary grades. Some of the selections are descriptions of what teachers and students actually did, and do not provide much guidance for other teachers on what else could be done to achieve the stated objectives – unless one calls up the curriculum from the *Teachers College Peace Education Resource Bank* (for example, selection 1 on 'Dream-Makers, Peacemakers, and Heroes' and selection 28 on 'Global Studies'). Other selections provide step-by-step instructions for student activities without recourse to much external material (for example, selection 8 on 'Ideas for Teaching Peace in the Elementary Schools' and selection 15 'How Do We Spend Our Money?').

As Carrington and Troyna (1988: 1) note, in *Children and Controversial Issues*, the legitimization of the teaching of controversial issues in primary schooling in the UK (and probably elsewhere) seems a long way off. It is thus a matter for regret that the possible effectiveness of some of what has been included in Reardon's book may be diluted through infusion into schools where the whole school is not party to the curriculum processes which are an integral – and perhaps essential (according to Hicks *et al.*) – part of peace education.

As a final comment on these two books, I find it peculiar that the contributors to *Education for Peace* cite references and resources from a range of countries – in recognition of the widespread interest in, and commitment to, peace education around the world – whereas Reardon's book, with the solitary exception of a Costa Rican reference co-authored by Reardon, confines its sources to the USA – even the one reference to a United Nations publication is published in New York. Given that several of the curriculum selections include the word 'global' in their titles, one is left to wonder how global perspectives can be developed from purely US resources.

Children and controversial issues

By their very nature most green issues are controversial and many teachers are reluctant to deal with them in their classrooms, dismissing them 'as "biased", even subversive, and as an anathema to "good" education' (Carrington and Troyna 1988: 1). In response to this reaction, *Children and Controversial Issues* is designed to provide a theoretical and empirical framework for the teaching of controversial issues to children in primary and middle schools. In addition, the book contains some case-study material on several green issues (racism, peace and conflict, gender) as well as other issues which may be controversial in certain contexts (sexuality, the world of work).

The foci for many of the chapters in this book are England's 1986 Education Act, the 1988 Education Reform Act, the (then) prospect of the National Curriculum, and the implications of these policy initiatives for the handling of controversial issues in primary schools. These foci may limit the potential audience for the book. And, because the implied reader of several of the chapters clearly is British, other readers may have some difficulties understanding the local nuances and acronyms (there is no list of abbreviations and, while foreigners can usually cope with 'HMI' [Her

Majesty's Inspector] and 'DES' [Department of Education and Science] without explanation, it may be mystifying when acronyms such as 'YTS' – Youth Training Scheme – are used without explication). However, readers outside the UK should have no difficulty translating the issues being raised to their own circumstances.

Like Hicks and his contributors, the authors in this volume subscribe to the belief that more attention needs to be given to the 'hidden curriculum' which contextualizes and underpins the teaching of controversial issues. All the authors agree that there is no consensus about what constitutes a controversial issue in the curriculum and that schools should engage critically with various perceptions of reality. They are also 'united in their conviction that politically contentious issues should be addressed' in primary classrooms, and 'they are concerned to demonstrate that primary school teachers' reluctance to consider sociopolitical controversies with their pupils is predicated on largely untenable grounds' (Carrington and Troyna 1988: 5).

The justification for teaching controversial issues in primary grades is provided by chapters by Short, Burgess, Pollard and Jeffs. While Short and Burgess give essential background information from their own perspectives, Pollard and Jeffs argue for major changes to classroom practices. Short provides a well-documented counter-argument to the enduring influence of Piaget's theory of sequential developmental and idealist conceptions of childhood innocence, and demonstrates that teachers of younger children often underestimate their students' degrees of political awareness and understanding. Burgess extends a similar theme by providing an appraisal of primary and middle school curriculum practice through focusing on teachers' professional ideologies and their influence on the characteristics of the 'normal' curriculum. She concludes that the future for the teaching of controversial issues in this sector is not good, resulting from the high status given to the normal curriculum, teachers' perceptions of the innocence of children – 'incapable of developing concepts concerning issues such as racial unrest, political education or sexual behaviour' (Burgess in Carrington and Troyna 1988: 88) – and 'the ideology of achievement' embraced by the Education Reform Act and associated policies.

Rather than adopting a 'not getting involved' approach to dealing with controversial issues in the classroom, Pollard argues for *reflective teaching* as the appropriate conceptual framework for teaching controversial issues. For Pollard (in Carrington and Troyna 1988: 64, 66), reflective teaching 'implies an active concern with aims and consequences, as well as with means and technical efficiency' and 'combines enquiry and implementation skills with attitudes of open-mindedness, responsibility and wholeheartedness'. He sees it as being applied in a cyclical or spiralling process, as in the action research model described by Stenhouse (1975, 1983), Elliott (1981) and others, and based on teacher judgement which is 'informed partly by self-reflection and partly by insights from educational disciplines'. There are also similarities between Pollard's concept of reflective teaching and Schön's (1983, 1987) writings on the reflective practitioner although Pollard does not refer to them. Pollard sees his model of reflective teaching as presenting one way of conceptualizing a new type of professional teaching activity. He then provides a case study of a student teacher working successfully in a reflective mode with a class of 9-year-olds. For Pollard (in Carrington and Troyna 1988: 69), 'discussion of controversial issues in schools can be valuable preparation, per se, for future participation in the democratic process' and he believes that teachers need to approach such teaching in socially aware and reflective ways, as well as being technically competent as 'good teachers'. In this he is promoting a similar message to Kemmis, Cole and Suggett (1983) in their writings on the socially critical school.

In a chapter on preparing young people for participatory democracy, Jeffs (in Carrington and Troyna 1988: 50) argues for sweeping changes to the curriculum process: 'If our present democratic system is not only going to survive but more importantly develop then, from the very onset of schooling, modes of behaviour and attitudes must be formed that are fundamentally democratic'. He sees such reforms as also liberating the classroom teacher from custodial and disciplinary roles, encouraging the reflective and intellectual practitioner, and promoting creativity rather than passivity at the school level. Without phrasing it in terms of calling for a new paradigm for education, Jeffs' arguments are very much in empathy with those of the contributors to *Educating for Peace*. He sees the teaching of democratic and participatory practices as being in the realm of the controversial through their challenging of established educational administration and practices and, possibly, subverting the institutional values of the school; but if a school does organize itself in a way that affirms the values of a democratic society – through devolving power and involving students in decision-making processes – then the school's educational role is not undermined, and the situation is therefore no longer controversial. Jeffs concludes that from the very first days of schooling teachers should 'perceive their role as being initiators of young people into a democratic way of life' rather than being 'warders'.

Together with Jeffs, Singh (in Carrington and Troyna 1988: 7) believes that 'schools can provide a systematic and structured opportunity for young people to explore fundamental questions relating to social justice and equality', which is something other socializing institutions are unable to do. Singh questions the basic premise of any 'neutral chair' approach to the teaching of controversial issues advocated in the well-known UK Humanities Curriculum Project and concludes that the approach is 'a highly problematic one and quite unacceptable in the teaching of controversial issues relating to forms of discrimination' (in Carrington and Troyna 1988: 92). Singh recognizes that teachers cannot take an entirely value-free position, so in its place he argues for a balanced approach 'where teachers put both sides of the argument with equal enthusiasm before their pupils, who in the end will be encouraged to make up their own minds' (in Carrington and Troyna 1988: 100–101)[1].

In the remaining chapters of *Children and Controversial Issues*, the authors focus on particular aspects of the controversial content or situation. McNicholl discusses the teaching of controversial issues in the controversial society of Northern Ireland where the role of education and the school system is a focus for debate. Here, the school system mirrors the overall social and religious divisions in society and two separate but almost identical systems exist. She describes the initiatives which have been undertaken to reduce sectarian divisions and promote mutual understanding at the secondary level – not an easy task when even minimal challenging of existing social and political systems is seen as controversial. She then argues (in Carrington and Troyna 1988: 117) for 'phased integrated education based on the community school model, where the pupils, teachers, community and the participants are totally involved in the learning process' and which places emphasis on the teaching of controversial issues in relation to Northern Ireland in the curriculum. Whitely focuses on bias in books for young children, especially in the areas of race, gender and politics, the kinds of ideological statements being made there, and how teachers may tackle them.

Continuing some of the arguments of Jeff and Pollard, three of the remaining authors in *Children and Controversial Issues* (Carrington and Troyna on racism and political education, Gammage on sex education, and Skelton on gender relations) emphasize the crucial importance of the structure of social relationships within the school. These include the relationships between staff and students and between the home and the school, and they suggest that these relationships could be of more crucial importance than the 'hidden curriculum'. Together with Ross (industry education) and Hicks (peace and conflict) these chapters complete the volume. Each of these chapters describes the dilemmas for teachers handling their particular controversial issue with primary school students and provides examples of suitable curricular and organizational strategies.

Given the general direction of the arguments in *Children and Controversial Issues* it is not surprising that David Hicks is the author of the chapter on peace and conflict. Indeed *Children and Controversial Issues* and *Educating for Peace* cite many references in common and obviously come from a similar philosophical position. In his concluding paragraph Hicks (in Carrington and Troyna 1988: 186) perhaps encapsulates the whole message. For him, the raising of important human issues within the curriculum and encouraging of the introduction of active experiential approaches to learning must continue whatever the political climate, 'for they need to become an integral part of children's experiences in both school and community if we are to achieve, in any form, the better worlds that we may dream of'.

Searching for a democratic future

Although a cursory glance at its contents may suggest that *Education: In Search of a Future* could be somewhat awkwardly placed in this essay, it actually has much in common with the other titles (and, indeed, it adds to the feeling that *Educating for Global Responsibility* is the odd one out, despite its title and overt focus on a 'green' issue). The authors of *Education: In Search of a Future* are concerned with 'the need for the development in democracy in education', and most have written from a democratic-socialist perspective. Their interest in democracy in education, and the need for social change to achieve this, has a great deal in common with the green themes that are advanced in the books edited by Hicks and Carrington and

Troyna, and *Green Prints for Changing Schools*. Fielding (in Lauder and Brown 1988: 72) summarizes this commonality well: 'What lies at the heart of this is a belief about what it means to be and to become more fully human'. All the authors are agreed that the present dominant circumstances in education do not allow participatory democracy or the empowerment of students and teachers, and these attributes are seen as an essential component of the calls for a new paradigm, and the search for a different future – a greener future – for education.

Although published in the same year (1988) as *Children and Controversial Issues, Education: In Search of a Future* actually predates the former, since it is the result of a 1987 seminar on 'Democracy and Education', but its focus is similar: a search for alternatives to policies and implied practices of the then imminent 1988 English Education Reform Act. Like *Children and Controversial Issues* it may now appear to be dated – for example, by references to 'Thatcherite policies in education' (but, from what has happened so far, the Major government's policies in education seem to deserve the label 'Thatcherite'). Indeed, from these books, and from recent articles in this journal (see, for example, Hartnett and Naish 1990, Halpin 1990), readers may be left with the impression that British academics are preoccupied to the point of obsession with the Education Reform Act and the National Curriculum, but this should not detract from the underlying messages of these contributions.

Several of the authors in *Education: In Search of a Future* provide substantial criticisms of the Thatcher era initiatives in education which, according to Lawn (in Lauder and Brown 1988: xiii), involve the adoption of outdated and inefficient management models that have already 'proved to be counter-productive to the needs of modern industrial corporations, and will have equally negative and stultifying consequences for our schools'. Lawn argues cogently that teachers should not accept a management model for education which does not even reflect the most advanced thinking in industry. He refers to the concept of the quality circle, which is a central concept of socio-technical theory and has been applied extensively in industry, for example, by the Volvo corporation. Ball also focuses on Sweden where a democratic system of education has been created within the context of a capitalist economic system. He believes that the Swedish education system represents a practical refutation of the imperatives of 'standards' and 'economic efficiency' as these terms are interpreted by the New Right. Moore also focuses on the arguments concerning the correspondence between education and the social relations of production and concludes that the labour market is also in need of reform.

Fielding argues for a new paradigm for education (in the comprehensive school) based on the concepts of liberty, equality and fraternity. An emancipatory imperative is central to his alternative paradigm. Lauder argues for the development of the comprehensive system of education based on a participatory democratic framework, and Lacey argues that (socialist) education should inform students about the problems and issues confronting Britain and the rest of the world in order to empower students with the knowledge and skills to make their own judgements about the future. Such a proposal resonates harmoniously with the proposals of Kemmis, Cole and Suggett (1983: 18) for schools to become 'socially critical' – places where students develop 'working knowledge [which] is constituted through the interaction of the knower and the object of knowledge in a social, cultural and historical context, and in relation to tasks or projects'. Lacey's equivalent proposals (in Lauder and Brown 1988: 94) are for 'collective intelligence', which he defines as 'a measure of our ability to face up to the problems that confront us collectively and to develop collective solutions', and he argues for critical and constructive education which focuses 'on those problems in our society that will confront [students] as individuals and those problems with the now integrated world system that will afflict them collectively'. Unfortunately, and frustratingly for those interested in reading further on this topic, Lacey's chapter lacks references; but I am sure that such a reference list would have included books by Kemmis, Cole and Suggett (1983) and Giroux (1983, 1990). (Every other chapter in this volume has a reference list, so its omission from Lacey's is peculiar and may bespeak a production error.) Lacey (in Lauder and Brown 1988: 97) also relates much of his writing to 'green' issues: he sees the 'real politics of the global environment' as being more important than the National Curriculum and as providing a real opportunity for the socialist educator to engage in critical practice.

Other chapters in *Education: In Search of a Future* look at various forms of practice and how these could be democratized. Kelly evaluates current practices in science education and examines the prospects for a democratic non-sexist science education. Evans carries out a similar exercise for physical education. White (who also contributes to *Educating for Peace*) describes an attempt to introduce democratic decision making into rules about the use of a playground in a London primary school and concludes that the processes involved in such attempts have illuminated the practical problems and possibilities for others wishing to follow a similar route. Blackman examines the youth training activities of the UK Manpower Services Commission and argues that its policies could be more progressive if the state rather than the market had responsibility for youth training.

Changing schools

While changing the dominant paradigm of curriculum towards one that is more ecologically aware, holistic and feminist is given great importance but discussed in little detail in *Education for Peace* (see Hicks 1988: 254), *Greenprints for Changing Schools* puts considerable effort into arguing for a holistic world-view, as distinct from a 'fragmentationalist' one, and for a holistic curriculum as the basis for education. From this base the authors (Greig *et al.* 1989: 32) review 'how teachers and others involved in education have gone about effecting change so that our schools can better set about educating whole people for a whole planet' through the 'hubs' or locations of the more significant forces for change: personal, institutional (i.e., school), local authorities and other external agencies. They also consider the place of the National Curriculum. The authors (Greig *et al.* 1989: 129) see 'an understanding of change processes in education, and particularly of a holistic model of change' as being central to their task. In so doing they have identified a key point that has been overlooked all too frequently in much of the comparable literature, though not necessarily in all of the titles under review. For example, Richardson (in Hicks 1988: 231–244) also considers curriculum change and has similar focal points to the *Greenprints* authors in his discussion; in a similar way, Pollard (in Carrington and Troyna 1988: 54–70) argues for reflective teaching as a strategy for handling change and Lawn (in Lauder and Brown 1988: 207–220) looks at change in schools through the imposition of management strategies and how these relate to the work of teachers.

The case studies which form the major part of *Greenprints for Changing Schools* treat change as an organic process which occurs whether we planned for it or not. The kind of change on which they are focusing is generally the adoption of a 'world studies' perspective in schools. The theory of change to which the authors subscribe is 'organicism' and they see this corresponding to a holistic world-view and to a curricular and instructional philosophy of 'transformation' (Grieg *et al.* 1989: 45), where education is a process of personal and social development as distinct from a top-down transmission of selected knowledge, skills and values. Such a view, particularly the section on 'Towards organic change in schools' (Grieg *et al.* 1989: 51–52), also has many similarities to 'socially critical' education, as described by Kemmis, Cole and Suggett (1983) and many others, although no references are made to this literature. Indeed, a melding of parts of several of the publications would make interesting and reasonably comprehensive reading on the theory of socially critical education, strategies for achieving it, and some case studies of how socially critical education is being achieved. In such a collection I would include Lacey and Fielding's chapters from *Education in Search of a Future*, Pollard and Jeffs from *Children and Controversial Issues*, much of *Greenprints for Changing Schools* together with selections from Kemmis, Cole and Suggett (1983), some background on action research from McTaggart (1991) and some exemplary literature on teachers as researchers, such as that collected in a recent issue of *Theory into Practice* (29 (3) (1990)).

The authors of *Greenprints for Changing Schools* introduce two other synonyms for the range of issues that I have termed 'green' issues, and that Hicks (1988) and Reardon (1988) have called 'peace education': they are working within a framework of 'world studies' or 'global education' (which they call the more holistic relation of world studies). World studies is denned (Grieg *et al.* 1989: 57) as

> an educational approach which combines a global perspective in the curriculum (concern for issues to do with the environment, development, human rights, peace and co-operation) with a humanistic, child-centred methodology (using participatory, affirming and co-operative techniques in the classroom).

The case studies on personal change document, for each of the teacher interviewees, the significant entry points to the processes of professional transformation, which have as their focal point 'a questioning and re-evaluation of personal values, perspectives and beliefs' (Grieg *et al.* 1989: 58). Quotations from transcripts of interviews illustrate teachers developing a sense of empowerment by changing their teaching practices to ones that are in accord with world studies and experiencing the responsiveness of students to more participatory, interactive learning styles. Another chapter examines strategies necessary for sustaining personal and professional change in schools and for achieving widespread acceptance and internalization of those changes across whole school communities. Here the head or principal is identified by many of the interviewees as having a key role in the promotion or obstruction of change, and several of those who promoted change recognized the importance of creating a supportive and democratic climate for teachers as well as students. Similarly, local education authority officers are seen as important in legitimizing and stimulating change within their ambit by, for example, identifying and publicizing (and thus rewarding) active schools and teachers. A key role is in 'alerting the "conspirators" to each other's work, connecting seemingly separate islands of growth and development [as] a crucial element in achieving the critical mass necessary for social transformation' (Grieg *et al*. 1989: 122). External agencies are seen as being able to make a very important contribution to the process of change towards holistic education, because they attempt to influence schooling through their curriculum resource materials and through the provision of people who can work with students and teachers. It is argued that 'without these agencies' support the movement towards holistic education would be much less substantial, given the relatively little encouragement emanating from mainstream sources of curriculum and professional development' (Grieg *et al*. 1989: 143), but it is also argued that these agencies could be much more effective if they had a clearer understanding of change processes in education. External agencies must contend with obstacles such as access to schools, relationships with school personnel and operational difficulties (including insufficient agency staff and inflexible school timetables), but they can still offer a counter to the 'fragmentationalist tendencies of current curriculum reform' through the holistic perspective they can bring to schools.

Like the other three British titles, *Greenprints for Changing Schools* also considers the National Curriculum, but it treats it in a more positive (albeit critical) way, as a practical problem for classroom teachers that has to be addressed. The authors (Grieg *et al*. 1989: 164) see a more holistic curriculum being pressed for under the National Curriculum and, with environmental education being one of the cross-curricular themes legitimized by the National Curriculum, they believe many schools that previously have shown no inclination to take global themes on board will finally have to address them. They also believe that there is scope for the diversity of teaching and learning styles essential to whole planet education to be pursued. They conclude on an optimistic note (Grieg *et al*. 1989: 167):

> It is up to the teachers and educationalists . . . to cultivate what is beneficial in the national curriculum whilst actively seeking to reverse or redirect those manifestations of reaction it undoubtedly contains. If the current counter-challenge can be successfully exploited by the 'conspirators', the final wholesale breakthrough of an holistic paradigm for education may not be too far distant.

Greenprints for Changing Schools is an important book in the context of this essay, but it is also a frustrating one. It is important because its text weaves together many of the threads from the other titles being reviewed

– the 'green' content, the calls for a new paradigm for education, the need for more participatory and democratic classroom practices –and many of the significant references it cites are also found in the other titles. It is frustrating because, while it is the only book reviewed here that is not an edited collection, its format leads the reader to think that it is one, with chapter endnotes being called 'references' and with no overall reference list – just a very selective bibliography at the end and a very weak index that does not even include the authors of the references cited.[2]

Conclusion

The authors in each of the potentially disparate volumes reviewed in this essay are arguing for a different future for education from that which can be anticipated from the path now being trodden by most educators. Some argue for a change in content, but most are equally if not more concerned with changing the educational paradigm within which our conceptions of pedagogy and curriculum are located. Education for the future they all write about should be more holistic, more human, more socially critical, more participatory and more democratic – that is, the content as well as the learning processes and the school organization should be more 'green'. It is certainly a very different vision from the one that currently can be found in most places around the world, but it is equally certainly one that is worth pursuing. However, the nature and extent of this vision does not emerge from reading just one of these books: rather, each contributes to the image of a whole that is much more than the sum of its parts, which, in the context of 'greening' education, is just as it should be.

Notes

1. It should be noted, however, that recent classroom observations (including unpublished research by John Fien, Griffith University, Australia) indicate that such an approach can alienate the teacher from the class.
2. On this point the Falmer Press deserves praise: among the books reviewed here their books – *Children and Controversial Issues* and *Education: In Search of a Future* – have by far the most comprehensive indexes.

References

BIRCH, C. (1990) *On Purpose – A New Way of Thinking for The New Millennium* (Kensington, NSW: University of New South Wales Press).

BOWERS, C. A. (1990) Educational computing and the ecological crisis: some questions about our curriculum priorities. *Journal of Curriculum Studies*, 22 (1): 72–76.

CAPRA, F. (1983) *The Turning Point: Science, Society and the Rising Culture* (London: Fontana).

DOLL, W. E. (1989) Foundations for a post-modern curriculum. *Journal of Curriculum Studies*, 21 (3): 243–253.

ELLIOTT, J. (1981) Action research: a framework for self-evaluation in schools. *Teacher–pupil Interaction and the Quality of Learning Project*. Working Paper 1 (Cambridge: Cambridge Institute of Education).

FERGUSON, M. (1982) *The Aquarian Conspiracy: Personal and Social Transformation in the 1980s* (London: Granada).

GIROUX, H. A. (1983) *Theory and Resistance in Education* (South Hadley, MA: Bergin and Garvey).

GIROUX, H. A. (1990) *Curriculum Discourse as Postmodernist Critical Practice* (Geelong: Deakin University Press).

GOUGH, N. (1989) From epistemology to ecopolitics: renewing a paradigm for curriculum. *Journal of Curriculum Studies*, 21 (3): 225–241.

HALPIN, D. (1990) The Education Reform Act Research Network: an exercise in collaboration and communication. *Journal of Curriculum Studies*, 22 (3): 295–297.

HARMAN, W. W. (1988) *Global Mind Change: The Promise of the Last Years of the Twentieth Century* (Indianapolis, IN: Knowledge Systems).

HARTNETT, A. and NAISH, M. (1990) The sleep of reason breeds monsters: the birth of a statutory curriculum in England and Wales. *Journal of Curriculum Studies*, 22 (1): 1–16.

HUCKLE, J. (1988–90) *What We Consume* (Richmond, UK: WWF United Kingdom and Bedford College of Higher Education in conjunction with The Richmond Publishing Company).

KEMMIS, S., COLE, P. and SUGGETT, D. (1983) *Orientations to Curriculum and Transition: Towards the Socially-critical School* (Melbourne: Victorian Institute of Secondary Education).

McTAGGART, R. (1991) *Action Research: A Short Modern History* (Geelong, Victoria: Deakin University Press).

MICHAEL, D. N. and ANDERSON, W. T. (1986) Norms in conflict and confusion. In H. Didsbury (ed) *Challenges and Opportunities: From Now to 2001* (Washington DC: World Future Society), 114–124.

SCHÖN, D. A. (1983) *The Reflective Practitioner: How Professionals Think in Action* (New York: Basic Books).

SCHÖN, D. A. (1987) *Educating the Reflective Practitioner: Toward A New Design for Teaching and Learning in the Professions* (San Francisco: Jossey-Bass).

SLAUGHTER, R. A. (1989) Cultural reconstruction in the post-modern world. *Journal of Curriculum Studies*, 21 (3): 255–270.

STENHOUSE, L. (1963) *Authority, Education and Emancipation* (London: Heinemann).

STENHOUSE, L. (1975) *An Introduction to Curriculum Research and Development* (London: Heinemann).

Globalization and environmental education: looking beyond sustainable development

BOB JICKLING and ARJEN E. J. WALS

This study contends that environmental education is being significantly altered by globalizing forces, witnessing the effort to convert environmental education into education for sustainable development. This internationally propagated conversion can be challenged from many vantage points. This study identifies anomalies that have arisen as international organizations such as UNESCO have championed this conversion, and discusses issues arising from these anomalies in light of the nature and purposes of education. This study presents a heuristic that has helped one to support a better understanding of the relationships between sustainable development, environmental thought, democracy, and education.

Globalization and neo-liberalism

It has been argued that many of the world's people live in what may be described as a corporatist society with soft pretensions to democracy. Globalization affects them in tangible and intangible ways. The neo-liberalist forces that tend to shape and frame globalization in terms of markets and opportunities for growth result in power slipping away from citizens to corporate elites. In examining this argument, Saul (1995, 2005) suggests that globalization refers to the rise of economic ideologies embodied by the corporate sector, and to the erosion of grassroots democracy. In fact, he argues that the corporatist movement was born in the 19th century as an alternative to democracy. He is not alone. As Crossley and Watson (2003: 103) assert:

> It is the executive directors of these powerful banks [i.e. International Monetary Fund and World Bank] and transnational corporations (TNCs) that can direct, or at the least influence, the policies of individual countries and national economies by integrating them into regional or global economies, and

by making it increasingly impossible for them to regulate and control their own affairs.

In effect, Crossley and Watson are pointing out that multinational corporations are now richer and stronger than nation states, and international trade agreements are more powerful than the will of elected governments.

Globalizing ideologies and the corresponding material effects are also having an impact on education. The powerful wave of neo-liberalism rolling over the planet, with pleas for 'market solutions' to educational problems and universal quality-assurance schemes, are homogenizing the educational landscape.[1] Ross (2000: 12) illustrates this development when describing the impact of the corporate curriculum:

> While your local high school hasn't yet been bought out by McDonalds, many educators already use teaching aids and packets of materials, 'donated' by companies, that are crammed with industry propaganda designed to instil product awareness among young consumers: lessons about the history of the potato chip, sponsored by the Snack Food Association, or literacy programmes that reward students who reach monthly reading goals with Pizza Hut slices.

Goodman and Saltman (2002: 68) characterize BP-Amoco's iMPACT middle-school science curriculum[2] as a diversion from the company's core business. 'Amoco's curriculum produces ideologies of consumerism that bolster its global corporate agenda and it does so under the guise of disinterested scientific knowledge, benevolent technology, and innocent entertainment'.

In response to this and other manifestations of globalization, Gough (2000: 335) seeks ways 'in which diverse local traditions can be sustained and amplified transnationally while resisting the forms of cultural homogenization for which McDonalds and Hilton Hotels are emblematic'. This search becomes even more important when powerful agencies such as the World Bank propagate this corporate agenda by shaping educational policies and influencing international research agendas in neo-liberalist ways (e.g. Crossley and Watson 2003). For institutions such as the World Bank, education appears simply and solely about preparing individuals to join the local labour market to nourish the global marketplace and satisfy corporate needs. As a result, education is less and less seen as a public good, and the state's role in providing citizens with the best possible education is diminished. Put another way, Saul (1995) claims that these neo-liberal authoritarians are fond of order and contemptuous of legitimate doubters.

While the public sector becomes more privatized, the private sector is being reframed as essential for public well-being. The emergence of socially responsible corporations is fuelled by a demand for kinder and gentler companies that are in tune with people, planet, and profit (the so-called 'Triple-P' bottom line). Companies are orchestrating their own education and training schemes for developing their 'human resources' as 'Triple-P' jugglers. They can even be certified as a company that is environmentally sound by applying for International Organization for Standardization (ISO 14001) certification.[3] The environmental education sector, increasingly dependent upon private funding to support its work, at least in some countries, sees the corporate world as a new 'market' with new 'customers' or 'clients'.[4]

As environmental educators and researchers, we follow these trends with suspicion. They seem to engulf education and make it a contributor to, or even a catalyst for, more exploitation of 'human' and 'natural resources', as the 'P' for profit silently has become an undisputed component of the triple bottom-line. Even if education does not become absorbed by these trends, and if it is able to help people to reflect critically on what is happening to the planet and to themselves and provide some space for alternatives, the economic forces of consumerism are so much bigger that these efforts may all be in vain. After all, for the 10% of the Earth's population that uses well over 90% of its resources, the drive to consume is greater than the drive to sustain (Brown 2005). Orr (2003) describes this imbalance as walking north on a southbound train. Although environmental educators do good and important work, they are still passengers on this accelerating train moving in the opposite direction (Orr 2003). Although we take issue with the use of 'north' and 'south' in his analogy, we recognize the phenomenon Orr describes. The coupling of globalization and neo-liberalism affects all education. In this paper, we examine the idea of education for sustainable development that is presented by UNESCO as the successor of the more established environmental education. We are concerned with the impact of this coupling of globalization and neo-liberalism on environmental education in particular. We believe that some of our concerns, and some of our responses to these concerns, transcend the field of environmental education.

Converting environmental education to education for sustainable development

Many believe that the effects of globalization in education are positive. As some contend, 'There is no greater context for educational change than that of globalization, nor no grander way of conceptualizing what educational change is about' (Wells *et al.* 1998: 322). In this spirit, Waks (2003) suggests that the impact of globalization upon curriculum will lead to fundamental change as opposed to the incremental and piecemeal change that characterized the 20th century. For Waks, fundamental changes imply changes not only in subject-matter selection, but also in instructional methods, technology utilization, organization, and administration.

It is not surprising that there are those who applaud educational policy decisions arising from global initiatives to 'improve' education. Although they tend to recognize difficulties and challenges, they basically have faith that good educational change can arise from the creative tensions and uncertainties which accompany the multicultural context and vague language. UNESCO's Education for All movement, millennium goals, and its decade for Education for Sustainable Development, for instance, are seen as opportunities for educational change.

Education for sustainable development was launched by the report of the World Commission on Environment and Development (WCED) (1987), *Our Common Future*. It was propelled forward by the 1992 World

Conference on Environment and Development in Rio de Janeiro, and was the focus of attention again at the World Summit on Sustainable Development at Johannesburg in 2002. Throughout this period, with the assistance of numerous additional conferences, concerted efforts have been made to transform environmental education into education for sustainable development. In December 2002, the United Nations passed *Resolution 57/254*, that declared a Decade of Education for Sustainable Development beginning in 2005. Interestingly, aside from the preamble that recalls the 1992 United Nations Conference on Environment and Development, the resolution makes no reference to 'environment', 'environmental', 'ecology', or 'ecological'. Fundamentally important elements for many educators are not present in this resolution.

The response by the environmental education community to these concerted efforts to convert environmental education into education for sustainable development has been varied (Hesselink *et al.* 2000). Some, who for a long time have claimed that environmental education has to examine issues related to inequity, North–South relationships, and sustainable use, welcome the move as a legitimization of their interpretation of environmental education. Others reject the move, somewhat ironically for the same reason. They suggest that environmental education is a well-established field that already examines the issues education for sustainable development is supposed to examine. 'Why throw away the baby with the bathwater?' they seem to say. Others, including ourselves, display more principled resistance. They question globalizing trends based on a vague and problematic concept such as sustainable development. They also see a downside to the homogenizing tendencies of these global policy movements and take offence at prescriptive constructions such as 'education *for* sustainable development' that reduce the conceptual space for self-determination, autonomy, and alternative ways of thinking. Although these three types of responses can all be found in the environmental education community, by and large education for sustainable development has become widely seen as a new and improved version of environmental education, most visibly at the national policy level of many countries.

We regard as problematic the emergence of education for sustainable development in educational policies and the pressure on the environmental educators around the world to re-frame their work as contributions towards sustainable development. Globalization, we fear, can be viewed as a process that strengthens the instrumental tendencies of environmental education to promote a certain kind of citizenship, particularly one that serves, or at least does not question, a neo-liberalist agenda. At the same time, globalization can also be seen as a process that allows powerful world bodies, such as the World Bank, the World Trade Organization, and UNESCO, to influence educational policy agendas on a global scale with lightning speed.

With these developments in mind, we make the following arguments. First, we identify anomalies that have arisen as world bodies such as UNESCO and the International Union for the Conservation of Nature and Natural Resources (IUCN) have championed the conversion of environmental education to education for sustainable development. Second, we discuss these anomalies in light of an emancipatory interpretation of education. Third, we present a heuristic that emerged while improving our own understanding of relationships between sustainable development, environmental thought, democracy, and education. We hope that the heuristic will be helpful to others when wishing to examine their own frames for making sense of sustainable development, and other global issues.

Anomalies

Sometimes policy follows innovation; at other times innovation follows policy. Many trends in education seem policy-driven, rather than innovation-driven. Trends such as lifelong learning and competence-based education are inspired by national and international policies (and corresponding economic incentives). Environmental education is no different. The conversion from environmental education to education for sustainable development may be seen as a policy-driven transition. However, externally triggered change often results in resistance from within, as illustrated by the outcomes of an international, on-line debate on education for sustainable development referred to as the 'ESDebate' (Hesselink *et al.* 2000). In all, 50 invited experts from 25 countries registered for the debate and its five rounds of questions. Through a series of provoking (although somewhat leading) questions, organizers of the debate were able to elicit many ideas about potential meanings of education for sustainable development, examples of good practices of education for sustainable development, and ideas for implementing sustainable development education.

An anomaly was revealed when participants were asked, 'Should ESD [education for sustainable development] be abolished as a concept?' (Hesselink *et al.* 2000: 49). More than half of the responses were yes. Of course, ideas cannot be abolished, but this reaction suggests some strong misgivings about the appropriateness of making sustainable development the new focal point of environmental education. Some of those with misgivings suggested that making sustainable development the new aim of environmental education, and using education for sustainable development as an instrument to change people's behaviour in a pre-determined direction, would leave less space for reflective self-determination about educational outcomes, autonomous thinking, and exploration of more contextual pathways towards a 'better' world.

While there is a constellation of ideas as to what sustainable development might entail, the lack of consensus about the implications of an exact meaning in variable contexts prevents global prescriptions. Forcing consensus about an ambiguous issue such as sustainable development is undesirable from a democratic perspective and is essentially 'mis-educative'. Democracy depends on differences, dissonance, conflict, and antagonism, so that deliberation is radically indeterminate (Saul 1995, Goodman and Saltman 2002). The conflicts that emerge in the exploration of sustainable development, for instance, reveal the inevitable tensions among the Triple Ps (people, planet, profit) or the three Es (efficiency, environment, equity). From a learning perspective, these tensions are prerequisites rather than barriers to education (Wals 2007).

A spokesperson for United Nations Environment and Development, United Kingdom Committee (UNED-UK) (Education events 2002) who attended the World Summit on Sustainable Development in 2002, observed there was no critical exploration of notions of sustainable development. It was as if engaging in this discussion could potentially ruin the 'whole idea' and slow down its world-wide implementation. The focus of this international gathering, instead, seems to have been on how to *promote* education for sustainable development, and how to set standards, benchmarks, and control mechanisms to confidently assess progress towards its realization. Rather than discussing and exposing underlying ideologies, values, and worldviews, the general consensus at the World Summit on Sustainable Development, and the many meetings that were organized in its slipstream, seemed to be that educators have passed the reflective stage, and that they must roll up their sleeves and start implementing! However, it can also be argued that at best they are implementing a chimera—a fanciful illusion—or worse. It could also be argued that many educators have become agents in a trend towards economic globalization.

These examples indicate that there is cause to be somewhat uncomfortable with the sustainable development agenda. We believe that a similar discomfort might be justified when considering other transitions in education that are inspired by global agendas.

Education and sustainable development: suitable alliances?

The ESDebate (Hesselink *et al.* 2000) showed that environmental educators are divided on how to respond to the emergence of education for sustainable development. Some appear quite comfortable with the term and seek to infuse it with meaning, or use it to examine issues underrepresented by traditional environmental education. Others are clearly worried by the continued sustainable development focus, and express concerns about the ideological and globalizing nature of this agenda and stress the need to nurture alternatives. From another perspective, Sauvé (2005), in her analysis of trends in environmental education, has concluded that education for sustainable development is just one of 15 trends or currents in environmental education. Yet others, while recognizing limitations to this terminology, seek to accommodate pragmatically the global political agenda. As a tentative step in this direction, Smyth (1999) spoke about education 'consistent' with *Agenda 21* (United Nations Conference on Environment and Development [UNCED] 1992)—the action plan arising from the World Conference on Environment and Development in 1992.

How educators and curriculum theorists respond to these varied perspectives about education for sustainable development will depend on how they think about 'education' and the role education plays, or needs to play, in society. It will also depend on their image of 'educated persons', and their interactions within respective societies—in particular, the perceived role people are to assume in decision-making processes.

Conceptions of education

Whether or not 'sustainable development' is an appropriate aim for education depends on how education is conceptualized. Reflecting on educational theories in the 20th century, scholarly analyses have articulated tensions between dominant tendencies and emerging perspectives that were increasingly prevalent as the century came to a close (Shepard 2000). These tensions, and ultimately contesting views about the nature of education, can be characterized in several ways. One such characterization is to think of education as being essentially *transmissive*—that is, the transmission of facts, skills, and values to students. Here, content and learning outcomes are predetermined and prescribed by a small group of experts. Learning is, by and large, a closed process, a unidirectional transmission of information from the teacher to the student. Education is about social reproduction and social efficiency. Working within this transmissive perspective, much contemporary rhetoric now rests on the assumption that education is an instrument for getting one's 'message' into impressionable young minds—for implanting a particular agenda. In this case, education leads to an authoritatively created and prescribed destination. Engineered by governments, special-interest groups, or industry, education inculcates the preferred message, agenda, ideology, or consumer preference.

In contrast to this transmissive view of education, some theorists[5] argue that education is increasingly reflected in emergent and more transactional, or *transformative*, perspectives sympathetic to cognitive and socio-constructivist learning theories. Here, knowledge and understanding are co-constructed within a social context—new learning is shaped by prior knowledge and diverging cultural perspectives. Such a socio-constructivist, transformative learning mode of education is more open and provides some space for autonomy and self-determination on the part of the learner. Knowledge is not fixed, cut up in pieces and handed over, but rather (co)created by transacting with prior tacit knowledge, the curriculum, and other learners. In this sense, a function of environmental education is to enable students to become critically aware of how they perceive the world with a view to fostering citizen engagement with social and environmental issues and participation in decision-making processes.

Shaull (1970: 15) articulates this sentiment as follows:

> Education either functions as an instrument which is used to facilitate the integration of the younger generation into the logic of the present system and bring about conformity to it, *or* it becomes 'the practice of freedom', the means by which men and women deal critically and creatively with reality and discover how to participate in the transformation of their world.

Taking a somewhat less binary view than Shaull's on the functions of education, we maintain that education, including environmental education, is not just about social reproduction, but also, and perhaps foremost, about creating the ability to critique and transcend social norms, patterns of behaviour, and lifestyles without authoritatively prescribing alternative norms, behaviours, and lifestyles.

Conceptions of the 'educated' citizen

Much traditional debate has turned on whether education is about social reproduction or about enabling social transformation, and this debate is reflected in the way educators imagine the educated citizen interacting within society. If social reproduction is the inherent expectation, then citizens should work efficiently within existing frameworks. Taking this view of the 'educated' citizen, we expect to see individuals well prepared to accept their role within society and the workforce. They are obedient, deferential, and compliant as they take their place within hierarchical and authoritative social structures and power relationships. From this vantage point, individuals are content to participate in democratic processes at electoral intervals while daily choices are made by decision-makers and their supporting bureaucracies.

If enabling social transformation is the inherent expectation, then we would expect to find 'educated' citizens who are active participants in on-going decision-making processes within their communities. They would be democratic practitioners in the sense that democracy is more than selecting a government, but rather:

> a mode of associated living, of conjoint communicated experience. The extension in space of the number of individuals who participate in an interest so that each has to refer his [or her] own action to that of others, and to consider the actions of others to give point and direction to his [or her] own, is equivalent to the breaking down of those barriers of class, race, and national territory which kept [people] from perceiving the full import of their activity. (Dewey 1966: 87)

Dewey suggested that more numerous and more varied points of contact denote a greater diversity of stimuli to which an individual can respond. We, too, put a premium on variation in persons' actions.

In 1916, Dewey spoke of the role of democracy in finding balance between individualization and a sense of community and belonging. According to his view, democracy in education is crucial in realizing a sense of self, a sense of other, and a sense of community; it creates space for self-determination, as individuals or members of groups, and greater degrees of autonomous thinking in a social context. As such, education cannot be reconciled with notions of deterministic, instrumental, and exclusive thinking as embodied by the international policy statements on education for sustainable development. Instead, we can imagine the 'educated' citizen enacting democratic practices in a caring community that shares features with other communities but is also unique (Shepard 2000).

Using the two composite conceptions of education, and the two corresponding views of an educated citizen, we have constructed a heuristic as pictured in figure 1. For continuing heuristic purposes, we suggest that the dynamic framed by ideas represented along these two intersecting axes creates interpretive possibilities within each of the four quadrants delineated. These interpretive exercises can serve to frame and reframe perspectives on education for sustainable development.

Education is a complex and messy business and a two-axes heuristic is not sufficient to capture the shape and scope of the entire enterprise. Nonetheless, the heuristic we present here, as an analytical tool, has provided us

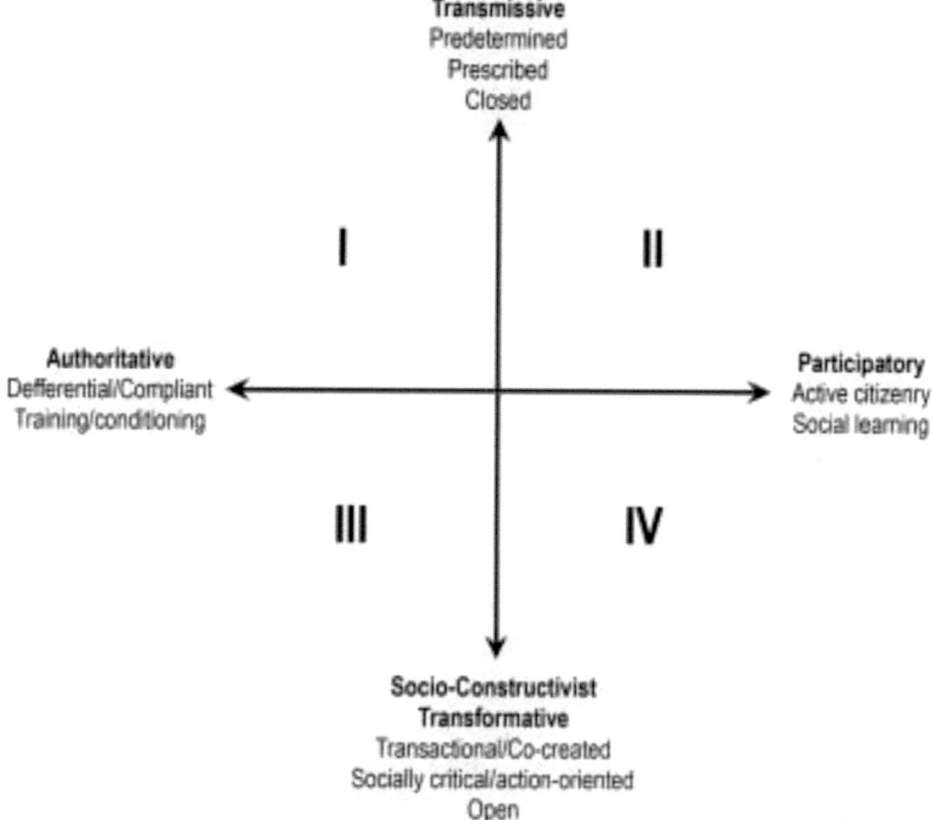

Figure 1. Positioning of ideas about 'education' alongside the social role of the 'educated person'.

with access to important questions and has increased our reflexivity. That we have called it a heuristic, rather than a framework, is important. As a heuristic, its intent is generative—to engage people with educational tensions related to the sustainable development agenda and to challenge them to frame and reframe their own perspectives and questions. Although we work from the position that sustainable development can be seen as only one of many stepping stones in environmental thinking, we believe that this does not negate the generative potential of the heuristic. In the section that follows, we discuss some educational implications associated with these dynamics.

An emergent heuristic

Out of the preceding discussion, three realms of possibility have emerged that can serve to focus discussion around relationships between sustainable development and education (see figure 2). We describe Quadrant I as 'Big Brother Sustainable Development', reminiscent of Orwell's (1989) meta-phor for extreme state control, where even language used by citizens was

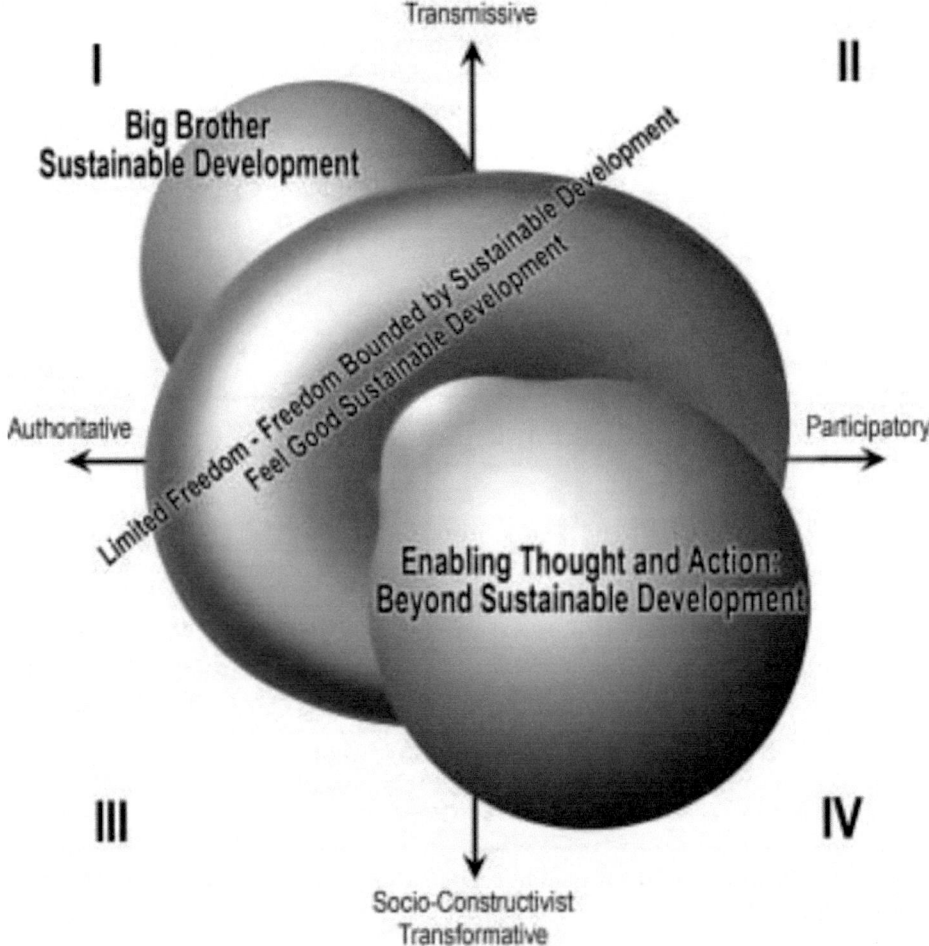

Figure 2. **Positioning sustainable development in education within two force fields.**

controlled by the 'thought police'. Quadrants II and III differ in some characteristics, but, in the end, share important qualities. In Quadrant II, participatory approaches to learning are taken up, yet the approach delineated by this conceptual space also tilts towards transmissive goals. In Quadrant III, socio-constructivist or transformative goals are moderated by authoritative approaches to teaching. It can be argued that while participatory learning and socio-constructivist goals promise possibilities for conceptually transcending education for sustainable development, the transmissive and authoritative tendencies still constrain possibilities. And, as such, both Quadrants II and III suggest a kind of 'feel-good sustainable development', in the sense that citizens are given a limited, or false, sense of control over their future and their ability to shape the future while in fact authorities of all kinds remain in control. We will discuss these two Quadrants together under the heading 'Limited freedom—freedom bounded by sustainable development'. Finally, we will discuss Quadrant IV possibilities under the heading 'Enabling thought and action: beyond sustainable development'.

Relationships such as these do not occur in such flat, one-dimensional, depictions. With this in mind, we have redrawn the heuristic in a three-dimensional fashion to convey more multi-dimensional and dynamic relationships. However, both representations are simply tools for reflecting on the nature, goals, and processes associated with education for sustainable development and, for that matter, environmental education.

Big Brother sustainable development

When located within Quadrant I of figures 1 and 2, bounded by a transmissive approach to education and an authoritarian and hierarchical view of social interactions, this area can be called 'Big Brother sustainable development'. This characterization is reminiscent of the Orwellian metaphor for the ever-present, and ever-powerful, state—in which directives are to be followed and deviants are to be eliminated. Here 'authorities' have determined the correct course of action and the purpose of education is to implement this course. Often, standardization techniques are used to create a consistent and unambiguous message and benchmarks that can be used measure progress towards the pre-determined goals and objectives. Think, for instance, of national and even global standards for sustainable development, or elements thereof, such as the International Organization for Standardization (ISO 14001), environmental standards for business and industry (ISO 2007), International Federation of Organic Agriculture Movements (IFOAM) standards for organic agriculture,[6] and the Forest Stewardship Council (FSC) labelling for hardwood.[7]) Not unlike Orwell's 'thought police', bodies are created to check and control these, often externally created, standards.

In Quadrant I, education is a tool or an instrument among several other instruments (e.g. economic and legal ones) that can help realize the sustainable development agenda. Under the influence of governments, special-interest groups, or industry, 'education' inculcates the preferred message, agenda, ideology, or consumer preference. This 'destination view' of education is both instrumental and deterministic in that some segment of society decides what is best and it uses education as a tool to disseminate its conception of 'best'. The problem faced by advocates of education for sustainable development, who find themselves working in conceptual space delineated in varying degrees by transmissive goals and authoritative approaches to learning, is that such determinism is not consistent with how many environmental educators interpret education. While education can also be thought of as a process that can enable social transformation (as would, for example, be described in varying degrees in Quadrant IV space), critics find it anathema that education should serve such pre-determined ends. Sustainable development is a normative term and placing it as a desired outcome is sometimes seen as reducing education to a struggle between contesting ideologies, or to indoctrination (Jickling 1992, 2001, Hattingh 2002).

Some educators have tried to examine issues of indoctrination through definitional stipulations and qualifications, sometimes requiring awkward circumlocutions (Jickling and Spork 1998). Hopkins (1998: 172), for instance, advocates education that 'should be able to cope with determining and implanting these broad guiding principles [of sustainability] at the heart of ESD'. From the perspective of those working to varying degrees in conceptual space delineated by transformative goals and participatory approaches to learning (Quadrant IV), it can be argued that using education to 'implant' guiding principles is essentially 'un-educative' at best and 'mis-educative' at worse. After all, from a transformative perspective, education is more about teaching students how to think than what to think. Yet, there are educators, especially under the wings of powerful global organizations such as UNESCO and IUCN, who pragmatically adopt and promote education for sustainable development because it conforms to the aspirations of many non-educators who set policy agendas and control funding opportunities.

For example, there are some educators who take inter-governmental approval of *Agenda 21* as an example of global leadership with education as a legitimate tool for implementing this agenda. This seems to be the message reflected by the spokesperson for UNED-UK (Education events 2002) who attended various stakeholder events at the World Summit in Johannesburg in 2002 and reported the main interest and focus of these events was in finding ways to take the education for sustainable development agenda forward.

Another example is found in the Fundamental Programme Principles of the newly developing University of the Arctic (University of the Arctic 2000: 7) that promote 'an understanding of sustainable development...'. This language, too, can be read as authoritarian, prescriptive, and, as such, deterministic. While 'sustainable development' is a social construct that warrants study, it may not be so important that it should be elevated to the status of privileged doctrine. Some in the circumpolar world who may find Næss's work on 'deep ecology' more compelling (e.g. Næss 1989).

From a transformative perspective, *a priori* elevation of a particular pathway or destination to the status of privileged doctrine is objectionable. Alternative perspectives, however controversial, should not be at an *a priori* disadvantage; the dissonance created by exposing learners to a wide range of perspectives is what triggers reflection and meaningful learning. From a more transmissive perspective, however, this might not be seen as problematic—that is, as long as the pathways and destinations are seen as non-controversial, widely supported, agreed upon, and legitimate.

Furthermore, Dobson (1996) provides challenges to perceptions of stability around the meaning of sustainable development. According to his research, there were, at the time, more than 300 available definitions for sustainable development and sustainability. It appears that after nearly a decade of work to bring meaning to these terms, there was less coherence and understanding, and perhaps even more divergence, than recognized by global organizations, non-governmental organizations, and national governments promoting education for sustainable development. With this in mind, it seems that arriving at some common understanding that can be transmitted globally, irrespective of whether this is even desirable, is more remote than ever and perhaps an illusion.

Aside from the determinism inherent in this destination view of education, it can also be argued that this view also rests on the assumption that governments and governmental organizations can and do provide environmental leadership. In an era of scepticism, and indeed cynicism, few take seriously the idea that governments generate vision (Cleary and Stokes 2006). As one critic suggests, when so many environmentally-aware politicians favour 'sustainable development', one suspects they do not understand what it means (Cairncross 1991).

We do, however, have a caution. Some may be quick to denounce the approach to education described by Quadrant I of our heuristic that tilts in degrees towards transmissive goals and authoritative approaches to learning. It should be noted that many educators emerged from an educational system much as has been described. This has been a foundational experience for many persons now committed to transformative educational practices. People apparently are able to overcome the boundaries of the systems they are part of, even when such systems do not encourage them to do so.

Limited freedom—freedom bounded by sustainable development

The areas delineated by Quadrants II and III of Figures 1 and 2 provide more freedom for creating new understandings and citizen participation—in Quadrant II through participatory approaches to learning, and in Quadrant III through socio-constructivist, or transformative, aims. Yet this is a bounded freedom—freedom still framed by the language of sustainable development. For some, like aspirant academics at the University of the Arctic (2000) who value academic freedom, this might seem like a 'feel-good freedom', yet it is a false freedom.

Educators may choose to work in territory demarcated by these two quadrants for pragmatic reasons. Government support and funding often are more readily available for projects aligned with the sustainable development agenda. Such pragmatism does not necessarily preclude the emergence of transformative learning activities. Yet, as we have suggested, the knowledge and value bases of sustainable development and sustainability are variable, unstable, and questionable—characteristics that tend to be ignored, denied, or at least downplayed in (inter)national policy-arenas and documents promoting education for sustainable development. It is perhaps the denial or downplaying of these characteristics that is most problematic because it easily leads to a false globalizing consensus that denies diverging perspective and negates contextual differences. However, when these characteristics are recognized, a sustainable-development orientation may become a stepping stone to transformative learning about existentially relevant issues.

When handled with care and reflexivity, recognition of the unstable and questionable value bases can enhance the concept's educational potential. Dreyfus *et al.* (1999) suggest that 'sustainable development talk' potentially brings together different groups in a society searching for a common language to discuss environmental issues. Where different ways of looking at the world meet, dissonance can be created and learning is likely to take place—learning at the edge. This dialogue also allows the socio-scientific dispute character of emerging knowledge and values to surface. Participation in such a dispute is an excellent opportunity to learn about a highly relevant, controversial, emotionally charged, and debatable topic at the crossroads of science, technology, and society (Dreyfus *et al.* 1999). This appears a sound argument provided that all voices *could* have equal access to these conversations; otherwise, in the absence of such dissonance, the outcomes may well retain Orwellian overtones.

The more ominous, and Orwellian, outcomes can be anticipated when the interests of groups with radically different ideas about what should be sustained, are masked by illusions of shared understandings, values, and visions of the future. Put another way, sustainable development talk can lead people in the direction of Orwell's (1989) famously satirical notion of 'double-think' whereby ordinary citizens can increasingly hold in their minds contradictory meanings for the same term and accept them both (p. 223). The power of universal discourse in reducing meaning to a minimum is such that, as in *Nineteen Eighty-Four*, antagonistic concepts can be conjoined in a single phrase ('war = peace', 'sustainability = economic growth') or concept (i.e. 'sustainable development'). Big Brother's 'Newspeak' was designed not to extend but to *diminish* the range of thought, and this purpose was indirectly assisted by cutting the choice of words down to a minimum' (p. 313). In Newspeak, concepts capable of opposing, contradicting, or transcending the status quo were liquidated. As a result of this devaluation of language, the people in *Nineteen Eighty-Four* found themselves in a state of linguistic dysfunction that was exactly what Big Brother wanted (Jickling 2001). Seen this way, 'sustainable development' and 'sustainability' tend to blur the very distinctions required to evaluate an issue thoughtfully. Comparing the sustaining of ecological processes with the sustaining of consumerism reveals inconsistencies and incompatibilities of values, yet many people, conditioned to think that sustainable development is inherently good, will promote both at the same time.

Two examples, one from Northern Canada and one from Southern Africa, illustrate the concerns raised herein. In many parts of the world, conflicts between those who wish to develop resources and those who wish to protect ecosystems are common. In Northern Canada, there is a particular large wilderness area as yet without roads and home to healthy populations of carnivores, spawning salmon, and myriad other species that define the ecology of this intact watershed. Those wishing to protect the area oppose the building of a road to enable development of a mine. A powerful argument in their campaign is based on the need to sustain ecological processes and the ecological integrity of the watershed. Another interest group called 'Concerned Atlin Residents for Economic Sustainability' also mobilized to defend mining from ecological activists (Simpson 1999). A similarly powerful argument, based on an appeal for sustainable economic activity and jobs, was launched. Interestingly and ironically, what unites the environmental community arguing for regional ecology and the mining community for economic development is the word 'sustainability'. Their differences are absorbed by use of this single term and the concept is rendered a cliché. With public approval, both environmental advocates and mining promoters can use the term sustainability to support radically different ideas.

Similarly, Price (2002) provides a critique of 'sustainability' as used by the Electricity Supply Commission of South Africa (ESKOM). For this Commission sustainability is a process that 'provides the strategic framework for projections of supply-side and demand-side options that will need to be implemented to meet future energy demands' (ESKOM 2000, cited in Price 2002: 79). In Price's assessment, sustainability for ESKOM has more to do with the business of energy production than environmental sustainability. In this instance, it seems that the interests of the company, and not the Earth, are being sustained. She suggests that, when sustainability is used to serve the interests of business rather than the Earth, ideological contradictions are presented which legitimate and ameliorate the premises of domination. For her, it is important to both show the complexity of environmental issues and to break open for debate some of the hegemonic discourses currently being used in support of sustainable development.

Interestingly, Dobson (1996) notes that most of the work done in developing the more than 300 definitions for sustainable development and sustainability has been undertaken in the private sector. When we step back to look at the overall picture, we might well ask how effective this economic sector has been in neutralizing those struggling to promote ecologically-oriented versions of sustainability and sustainable development. We might also ask, 'Why do governments, multiple stakeholder groups, and industry representatives continue to favour sustainable development in light of such confusion over meaning?' To answer this question, we are drawn to work by Chomsky, particularly as presented in the film *Manufacturing Consent* (Achbar and Wintonick 1992).

In *Manufacturing Consent*, Chomsky spoke about diversions, or activities that people pay a great deal of attention to—professional sports, for example. Diversions, he suggests, are useful to governments and other power elites because they can distract people away from paying attention to other important issues and then doing something about them. A critical citizenry is difficult to control. However, group cohesion can be fostered through inculcation of what he calls irrational jingoism—narrow-minded self-promotion, or chauvinism. One example of this might be found in the proceedings from an international conference held in Thessaloniki in 1997 and hosted by UNESCO and the Government of Greece. In spite of a variety of extant critiques of education for sustainable development, an observer is hard pressed to find citations of such critiques in the voluminous and somewhat self-promoting document (Scoullos 1998).

It is possible that sustainable development is so well tolerated because it, too, is a diversionary concept. We can see anecdotal support for this thesis in the young and disillusioned who find these terms hollow and incapable of imposing sanctions on government or industry. And, sustainable development is, indeed, contentious when expressed as an aim of education. As the examples in the previous section show, philosophical or political evaluations—mediations between contesting values—cannot logically be subsumed by sustainable development. Yet, with continued promotion of sustainable development, jingoism and doublespeak effectively converge and capacity to think is diminished.

In some responses by environmental education scholars to education for sustainable development, at least two examples of work point beyond the limited, or constrained, freedom described by the Quadrants II and III of figures 1 and 2. These scholars have clearly been influenced by debates about this concept. Yet, while working near the edge, they have chosen, at least some of the time, to work within the bounds of sustainable development. The first scholar is the late statesman of Scottish environmental education, John Smyth. Smyth (1999: 79), who sought to avoid some of the problems associated with education for sustainable development, proposed a compromise through use of the term 'education consistent with Agenda 21'. Later, however, he framed many of his recommendations in the language of 'sustainable development education' and 'sustainability' (Smyth 2002). He did, however, suggest that educators need an initiative to meet, discuss priorities, and 'provide a non-adjectival label for what is being done, acceptable to all; and at best a vision of education to capture imagination and commitment' (Smyth 2002: 13). While we recognize that in a pluralistic society 'acceptable for all' is neither possible nor desirable, we feel that, in the context of his comments, Smyth was seeking more inclusive language that would allow room for 'respectful dissensus' or 'dissent' between environmental educators and advocates of education for sustainable development. Finding such terminology seems an important imperative, and a possible antidote to the hegemonic influences of the sustainable-development discourse. This idea foreshadows a theme picked up again in the next section.

The second scholar, Scott (2002), in a short paper titled 'Education and sustainable development: challenges, responsibilities, and frames of mind', responds to some of the issues raised here and provides some hints for resolving them. Scott's arguments call for a choice *within* sustainable development education (our emphasis); schools and teachers must be free to mediate government policy, free from *a priori* certainties about outcomes and instructions about what and how to teach. He contends that doing nothing is not an option, and that schools ignoring sustainable development will prove an inadequate response. In the context of these comments, he posits four responsibilities for educators (Scott 2002: 2):

(1) To help learners understand why the idea of sustainable development ought to be of interest to them;

(2) To help learners gain plural perspectives on issues from a range of cultural stances;

(3) To provide opportunities for an active consideration of issues through appropriate pedagogies which, for example, might begin from learners' and teachers' different interests, helping pupils understand what they are learning and its significance; and

(4) To encourage pupils to continue to think about what to do, individually and socially, and to keep their own and other people's options open.

To do less, he suggests, would seem neglectful, while doing much more risks indoctrination.

In examining these comments, we note that Scott shows concern for previous critiques about the doctrinaire tendencies of education for sustainable development. Point 3 suggests constructivist predilections, and point 4 is a harbinger of a transformative education and educated citizens who are actively engaged in social issues. Yet, point 1, insisting that sustainable development ought to be interesting, does not seem essential to the other three points. Points 2–4 feel good, but the freedoms inherent in them still appear bounded—still *within* sustainable development education. It seems that, if one is consistent in valuing academic freedom, and the kind of intellectual liberty that Scott (2002) holds as a cornerstone value, it is odd to pre-suppose that 'sustainable development' ought to be of such privileged importance, especially when there are alternative organizing frameworks for examining environmental and social issues.

Enabling thought and action: beyond sustainable development

The final emergent force-field in our heuristic, as represented by the boundaries of Quadrant IV in figures 1 and 2, concerns itself with emerging, or enabling, environmental thought. Inspired by socio-constructivist and transformative views about education and actively engaged citizens, it points to possibilities *beyond* sustainable development. Here, sustainable development is seen as just one stepping-stone in the continuing emergence of environmental thought. These emergent concepts will be useful to discuss, critique, and employ as devices to stimulate effective and creative dissonance across disciplinary boundaries. They will not be seen as pre-eminent ideas, or organizational frameworks, but simply as more or less useful conceptual tools. As we think about essentials for inclusiveness and collaboration with this realm of enabling environmental thought, we can move forward from John Smyth's thoughts.

Smyth (1999, 2002) has consistently sought to find language that can reach across conceptual divides and generate respectful dissensus and dissent. This is precisely the intention that guided much of the thinking about, conceptualizing of, and writing of Canada's national environmental education plan, *A Framework for Environmental Learning and Sustainability in Canada* (Government of Canada 2002). Nobody is claiming that this is a perfect document, and some might complain that too much emphasis was given to the term sustainability. However, it is the result of efforts to find a conceptually acceptable compromise with more inclusive language. Environmental learning is joined with sustainability by 'and', acknowledging it as a significant social construct of our times but not as an objective or end-statement about education or even of environmental education. Educators should not feel overwhelmed by pressure from this document to frame all of their work in terms of sustainability and sustainable development, yet those who seek to infuse these terms with meaning, and use them as learning aids, are invited to do so. And, this does not seem a neglectful document in spite of having avoided sustainable development.

Another example of avoiding, or ignoring, sustainable development is found in Bowers' (2002) paper 'Toward an eco-justice pedagogy'. What adds particular interest to this document is that it was first presented at a seminar, 'On the possibility of education for sustainable development', at the European Conference on Educational Research in 2000. Bowers writes about 'eco-justice' and explores 'right relationships'—questions of philosophy, ethics, and justice. Like others working in this heuristic location, he appears to feel no need to place these issues within the sustainable development agenda. He is interested in cultural perspectives, active engagement with contemporary issues, and action, without feeling the need to convince anyone that they ought to do so within a sustainable development framework.

When Carson (1962) wrote *Silent Spring*, no one had heard of deep ecology. When Næss (1973) coined the term 'deep-ecology', nobody had heard of the term sustainable development. When sustainable development became popular (WCED 1987), ecofeminism, with origins in earlier feminist writing, was beginning to make inroads into fields such as environmental ethics and education. And now, drawing on those who have gone before him, Bowers gave increasing attention to eco-justice. In other words, environmental thought and environmental ethics is dynamic. It is evolving, and people have no idea where they might go next, and they do not know what new language and metaphors will—or ought to—shape policies of the future. With this in mind, educators wishing to operate in Quadrant IV of our heuristic would find it counterproductive to build a sustainable development fence around environmental thinking.

Once again, we propose a little caution. Many will feel confident about the educational prospects arising from working in this quadrant. Yet, confidence can breed dogmatism, and a dogmatic insistence on the correctness of the Quadrant IV approach would be wholly inconsistent with the quadrant's own premise. By the same token, we recognize that in the description of this quadrant we privilege 'environmental thought' over other kinds of thought, which some might argue is also limiting. We are happy with this for now, but recognize that we should always be prepared to explain, defend, and re-evaluate our positions.

Investigating the debates

We find that the sustainable development agenda within environmental education is problematic. We view education for sustainable development as a product and a carrier of globalizing forces. This globalizing agenda has instrumental and deterministic tendencies that favour transmissive arrangements for teaching and learning over more transformative ones. In the process, traditional (e.g. environmental education) and alternative (e.g. eco-justice) ways of engaging people in existential questions about the way human beings and other species live on this Earth run the risk of being marginalized or excluded. The same holds true for individuals and communities wishing to deal with such questions in a self-determined, relatively autonomous, and contextually grounded way.

With this in mind, we have created a heuristic for educators wanting to investigate the education-for-sustainable-development debate. We are most concerned about tendencies towards obedience—acquiescence in the face of hegemonic discourses. As Saul (1995: 194) notes, 'Equilibrium [character-ized by resistance to ideology], in the Western Experience, is dependent not just on criticism, but on non-conformism in the public place'. So, this heuristic is a critical tool that can be used to critique current discourses, eval-uate new initiatives, and find one's own place within present debates, but also to support non-conformism. We encourage readers to adapt, develop, or re-invent this heuristic to suit their needs, and to aid in the evaluation of education initiatives concerning poverty, health, social justice, development, and other global agendas.

Acknowledgement

The authors thank the four anonymous reviewers of this paper for their very helpful suggestions.

Notes

1. See Apple (2000), Goodman and Saltman (2002), and Kenway and Bullen (2001).
2. See Danker (2004).
3. For information, see International Organization for Standardization (ISO) (2007).
4. E.g. Hirayama (2003) and Weisbrod (1998).
5. E.g. O'Sullivan (1999, 2002) and Shepard (2000).
6. For information, see IFOAM (2007).
7. For information, see FSC (2007).

References

Achbar, M. and Wintonick, P. (dirs) (1992) *Manufacturing Consent: Noam Chomsky and the Media* [video recording] (Montréal, QC: Necessary Illusions and National Film Board).

Apple, M. W. (2000) *Official Knowledge: Democratic Education in a Conservative Age* (New York: Routledge).

Bowers, C. A. (2002) Toward an eco-justice pedagogy. *Environmental Education Research*, 8(1), 21–34.

Brown, L. R. (2005) *Outgrowing the Earth: The Food Security Challenge in the Age of Falling Water Tables and Rising Temperatures* (London: James & James/Earthscan).

Cairncross, F. (1991) *Costing the Earth: The Challenge for Governments, The Opportunities for Business* (London: Business Books in association with Economist Books).

Carson, R. (1962) *Silent Spring* (Boston: Houghton Mifflin).

Cleary, M. R. and Stokes, S. C. (2006) *Democracy and the Culture of Skepticism: Political Trust in Argentina and Mexico* (New York: Russell Sage Foundation).

Crossley, M. and Watson, K. (2003) *Comparative and International Research in Education: Globalisation, Context and Difference* (London: RoutledgeFalmer).

Danker, S. (2004) A+ for energy. *The BP Magazine*, 2, 40–44. Available online at: http://www.bp.com/sectiongenericarticle.do?categoryId=9011036&contentId=7021215, accessed 21 May 2007.

Dewey, J. (1966 [1916]) *Democracy and Education: An Introduction to the Philosophy of Education* (New York: Free Press).

Dobson, A. (1996) Environmental sustainabilities: an analysis and a typology. *Environmental Politics*, 5(3), 401–428.

Dreyfus, A., Wals, A. E. J. and van Weelie, D. (1999) Biodiversity as a postmodern theme for environmental education. *Canadian Journal of Environmental Education*, 4, 155–176.

Education events (2002) Education events in Johannesburg that were attended by Education Coordinator of UNED-UK. Sustain[ed]: Education for Our Common Future. *Stakeholder Forum*, 4, 4–7.

Electricity Supply Commission of South Africa (ESKOM) (2000) Environmental report, 2000(Johannesburg, South Africa: ESKOM). Available online at: http://www.eskom.co.za/enviroreport(1)/index.htm, accessed 14 April 2007.

Forest Stewardship Council (FSC) (2007) FSC homepage. Available online at: http://www.fsc.org/en/, accessed 21 May 2007.

Goodman, R. T. and Saltman, K. J. (2002) *Strange Love, or How We Learn to Stop Worrying and Love the Market* (Lanham, MD: Rowan & Littlefield).

Gough, N. (2000) Locating curriculum studies in the global village. *Journal of Curriculum Studies*, 32(2), 329–342.

Government of Canada (2002) *A Framework for Environmental Learning and Sustainability in Canada* (Ottawa, ON: Government of Canada). Available online at: http://www.ec.gc.ca/education/framework/pdf/EC-ELS-E.pdf, accessed 2 April 2007.

Hattingh, J. (2002) On the imperative of sustainable development: a philosophical and ethical appraisal. In J. Hattingh, H. Lotz-Sisitka and R. O'Donoghue (eds), *Environmental Education, Ethics and Action in Southern Africa* (Pretoria: Human Sciences Research Council Publishers), 5–16.

Hesselink, F., van Kempen, P. P. and Wals, A. (2000) *ESDebate: International On-line Debate on Education for Sustainable Development* (Gland, Switzerland: International Union for the Conservation of Nature). Available online at: http://www.iucn.org/themes/cec/extra/esdebate/intro.html, accessed 18 September 2007.

Hirayama, K. (2003) Corporate environmental education in Japan: Current situation and problems. *The International Review for Environmental Strategies*, 4(1), 85–95.

Hopkins, C. (1998) Environment and society: education and public awareness for sustainability. In M. Scoullos (ed.), *Environment and Society: Education and Public Awareness for Sustainability*, Proceedings of the Thessaloniki International conference organized by UNESCO and the Government of Greece, 8–12 December 1997 (Athens: University of Athens, Mediterranean Information Office For Environment Culture and Sustainable Development, and Ministry for the Environment, Ministry of Education), 169–172.

International Federation of Organic Agriculture Movements (IFOAM) (2007) IFOAM homepage. Available online at: http://www.ifoam.org/, accessed 21 May 2007.

International Organization for Standardization (ISO) (2007) ISO homepage. Available online at: http://www.iso.org/iso/home.htm, accessed 21 May 2007.

Jickling, B. (1992) Why I don't want my children to be educated for sustainable development. *Journal of Environmental Education*, 23(4), 5–8.

Jickling, B. (2001) Environmental thought, the language of sustainability, and digital watches. *Environmental Education Research*, 7(2), 167–180.

Jickling, B. and Spork, H. (1998) Education for the environment: a critique. *Environmental Education Research*, 4(3), 309–327.

Kenway, J. and Bullen, E. (2001) *Consuming Children: Education—Entertainment—Advertising* (Maidenhead, UK: Open University Press).

Næss, A. (1973) The shallow and the deep, long-range ecology movements. *Inquiry*, 16, 95–100.

Næss, A. (1989) *Ecology, Community and Lifestyle: Outline of an Ecosophy,* trans and rev D. Rothenberg (Cambridge: Cambridge University Press).

Orr, D. W. (2003) Walking north on a southbound train. *Conservation Biology,* 17(2), 348–351.

Orwell, G. (1989 [1949]) *Nineteen Eighty-Four* (London: Penguin Books).

O'Sullivan, E. (1999) *Transformative Learning: Educational Vision for the 21st Century* (London: Zed Books).

O'Sullivan, E. (2002) What kind of education should you experience at a university. *Canadian Journal of Environmental Education*, 7(2), 54–72.

Price, L. (2002) Industry and sustainability: a re-view through critical discourse analysis. In J. Hattingh, H. Lotz-Sisitka and R. O'Donoghue (eds), *Environmental Education,*

Ethics and Action in Southern Africa (Pretoria: Human Sciences Research Council Publishers), 74–84.

Ross, A. (2000) The mental labor problem. *Social Text,* 18(2), 1–31.

Saul, J. R. (1995) *The Unconscious Civilization* (Concord, ON: Anansi).

Saul, J. R. (2005) *The Collapse of Globalism: And the Reinvention of the World* (Toronto, ON: Viking Canada).

Sauvé, L. (2005) Currents in environmental education: mapping a complex and evolving pedagogical field. *Canadian Journal of Environmental Education,* 10(1), 11–37.

Scott, W. (2002) Education and sustainable development: challenges, responsibilities, and frames of mind. *Trumpeter,* 18(1). Available online at: http://trumpeter.athabascau.ca/index.php/trumpet/article/view/123/134, accessed 26 March 2007.

Scoullos, M. J. (ed.) (1998) *Environment and Society: Education and Public Awareness for Sustainability,* Proceedings of the Thessaloniki International Conference Organized by UNESCO and the Government of Greece, 8–12 December 1997 (Athens: University of Athens, Mediterranean Information Office For Environment Culture and Sustainable Development, and Ministry for the Environment, Ministry of Education).

Shaull, R. (1970) Foreword. In P. Freire, *Pedagogy of the Oppressed,* trans. M. B. Ramos (New York: Continuum), 9–15.

Shepard, L. A. (2000) The role of assessment in a learning culture. *Educational Researcher,* 29(7), 4–14.

Simpson, S. (1999) I challenge you to an open discussion. *The Whitehorse Star,* 19 February, 11.

Smyth, J. (1999) Is there a future for education consistent with Agenda 21? *Canadian Journal of Environmental Education,* 4, 69–82.

Smyth, J. C. (2002) *Are Educators Ready for the Next Earth Summit?* The Millennium Papers: Issue 6 (London: Stakeholder Forum For Our Common Future). Available online at: http://www.stakeholderforum.org/publications/millennium/millpaper6.pdf, accessed 14 April 2007.

United Nations (2002) *Resolution 57/254. United Nations Decade of Education for Sustainable Development* (New York: United Nations). Available online at: http://portal.unesco.org/education/en/file_download.php/299680c3c67a50454833fb27fa42d95bUNresolutionen.pdf, accessed 14 April 2007.

United Nations Conference on Environment and Development (UNCED) (1992) *Agenda 21 Earth Summit; The United Nations Programme of Action from Rio* (New York: UN Department of Public Information).

University of the Arctic (2000) *An Integrated Plan for the Implementation of Bachelor of Circumpolar Studies, Arctic Learning Environment, and Circumpolar Mobility Program* (Rovaniemi, Finland: University of Lapland). Available online at: http://www.uarctic.org/prggen02_j0f58.pdf, accessed 14 September 2007.

Waks, L. J. (2003) How globalization can cause fundamental curriculum change: an American perspective. *Journal of Educational Change,* 4(4), 383–418.

Wals, A. E. J. (ed.) (2007) *Social Learning Towards a Sustainable World* (Wageningen, The Netherlands: Wageningen Academic Publishers).

Weisbrod, B. A. (ed.) (1998) *To Profit or Not to Profit? The Commercial Transformation of the Nonprofit Sector* (New York: Cambridge University Press).

Wells, A. S., Carnochan, S., Slayton, J., Allen, R. L. and Vasudeva, A. (1998) Globalization and educational change. In A. Hargreaves, A. Lieberman, M. Fullan and D. Hopkins (eds), *International Handbook of Educational Change, Part One* (Dordrecht, The Netherlands: Kluwer), 322–348.

World Commission on Environment and Development (WCED) (1987) *Our Common Future* (Oxford: Oxford University Press).

Environmental Studies Courses in Colleges of Education

W. E. MARSDEN

In his article "Environmental Studies: a New Synthesis?", Rolls suggested that if Environmental Studies were to become a feature of work in Colleges of Education, "there should be some consensus as to what these studies cover and their main purposes."[1] The intention here is first to present a survey of what is in fact practised or proposed in Environmental Studies Main courses in Colleges. Curriculum courses were not considered. The survey shows that rapid progress has been made since Martin's survey of 1968,[2] almost double the number of Colleges now offering Main courses. Nineteen syllabuses were examined, culled from Handbooks of Institutes of Education or received directly from Colleges, to whom thanks are due. The courses were either in operation in the 1969–70 session or, in the case of some of the Colleges in the Liverpool A.T.O., were about to start in September 1970.[3] The Thornbridge Hall course, described in detail elsewhere,[4] was not included, on the grounds that it is not strictly an elective Main course. Three of the syllabuses received[5] were excluded from the more detailed analysis because their major stress was on topics more central to Social than Environmental Studies, topics which included the individual in a variety of social rôles, various social groupings, and a number of socio-political and socio-economic issues. There is no implied deprecation of the value of this type of course in their omission. Of the sixteen remaining syllabuses, three were entitled "Environmental Science"[6] and one "Regional Studies"[7]. These were included in the survey because their content and approach were in the same mainstream as that of the larger body of syllabuses labelled "Environmental Studies."

The syllabuses were considered in terms of content, aims and associated approaches. The limitations of the survey are obvious. A syllabus is not equivalent to a course of study, though one presumes that it sets guidelines. In interpreting another's syllabus, one is quite likely to misjudge what is actually to be translated into course work. The degree of detail garnishing each syllabus varied, particularly in relation to aims, though most were fairly comprehensive over content. It seemed desirable to take account of the differing emphases given to the main components within each syllabus. To do this, and to provide something more than a general impression, a simple matrix was constructed, the various syllabus components forming the horizontal axis and the list of Colleges the vertical. A loading of one was given to a component mentioned, and of two to one strongly emphasized. Some degree of subjectivity, both in classifying components and in placing what was present in the syllabuses in the appropriate category, was inevitable, but at least the summary is not wholly impressionistic.

Table 1

Components	Sub-components	Nos. of Colleges (*max.* 16)	Weighting (*max.* 32)
Environmental	(a) Astronomy	1	1
Sciences	(b) Geology	16	18
	(c) Geomorphology	16	18
	(d) Petrology	13	14
	(e) Hydrology	1	1
	(f) Meteorology/Climatology	11	12
	(g) Ecology	14	21
	(h) Rural Studies	3	5

The sub-components were also grouped into Earth Sciences (b—f) and Life Sciences (g—h). By assessing the strength of these larger groups as a whole, and not by adding up the separate components, a weighted figure of twenty-three for the Earth Sciences and twenty-one for the Life Sciences resulted. In the latter there was a striking emphasis on Ecology and a neglect of Rural Studies (defined here as including such areas of study as horticulture and animal husbandry). The Environmental Sciences component as a whole is clearly securely established as a central feature of Environmental Studies courses.

This rather unorthodox arrangement of sub-components is an attempt (*a*) to reflect the variations in outlook of different syllabuses, a number of which presented their content in areal as well as, or rather than, in systematic terms; and (*b*) to isolate the essentially geographical characteristics of the syllabuses, particularly those areas in which the Phenomenal and Behavioural Environments, as defined by Kirk,[8] interlock. Hence areas of study often classified as Physical Geography have been placed in the Environmental Sciences section, though the overlap is obvious.

Aspect 1 relates to traditional studies of land use, farms and villages in a rural setting, and not only town studies, but also a consideration of industries and communications in an urban area. The figures reveal a relative neglect of urban areas, particularly as the environmental sciences and Aspect 3 have predominantly rural connotations.[9]

Aspect 2 is concerned with a concentric areal approach, and shows an intense concentration on the local area, though a significant number of Colleges emphasize the need to contrast local with other British environments, presumably those susceptible to field visits. Environments in a world sense are largely neglected. There are additionally, however, a small number of Colleges who look at the world systematically in terms of major international problems (see below).

Aspect 3 can fairly be categorized as the "Hoskins approach", deservedly influential, but complicating here in its inextricable fusion of the geographical and historical elements in the landscape. Overlapping with settlement studies, it covers also the study of place names, field patterns, and various

Table 2

Components	Sub-components	Nos. of Colleges	Weighting
Geographical	Aspect 1 Contemporary Systematic Human Geography Settlement/ Occupations		
	(a) Rural	15	25
	(b) Urban	12	19
	Aspect 2 Contrasting Regional Environments		
	(a) Local	15	27
	(b) British	10	15
	(c) World	6	9
	Aspect 3 Cultural Landscape Evolution	13	20

Table 3

Components	Sub-components	Nos. of Colleges	Weighting
Historical	(a) Archaeology (i) Prehistoric remains	7	11
	(ii) Industrial	6	6
	(b) "Historical Monuments", including, public buildings, castles and homes.	8	11

details specified later in the historical section. It is included as a geographical component largely because of its strong environmental bias. One also has the impression that the approach is valued more by historical geographers than by historians. The syllabuses demonstrate that it is appreciated by environmental educationists.

Although clearly elements of the Cultural Landscape, these are included in this category because they can be considered in part as "historical entities" for their own sake, without necessarily referring them to the environmental setting. In some cases, an interest in period architecture links this component with the aesthetic (No. 5). A striking feature was the absence of overt reference to the historical component in seven syllabuses, though one of these included a background history course.

Contemporary communities were examined in both rural and urban settings. The study of past societies was predominantly local or British, though in one case a look at the ancient civilizations of the Middle East and the Mediterranean was indicated. The figures suggest that sociological matters are either regarded as peripheral to Environmental Studies courses, or perhaps reflect a feeling that they require the expertise of trained social scientists. Martin's survey of 1968 showed that at that time few social

Table 4

Components	Sub-components	Nos. of Colleges	Weighting
Sociological	(a) Community Studies	6	9
	(b) Past Societies	4	5

Table 5

Components	Sub-components	Nos. of Colleges	Weighting
Aesthetic		5	7

Table 6

Components	Sub-components	Nos. of Colleges	Weighting
Recreational		7	10

Table 7

Components	Sub-components	Nos. of Colleges	Weighting
Applied	(a) Conservation and Planning	14	22
	(b) International Problems	8	13

scientists were contributing to Environmental Studies courses.[10] The lack of attention to the social area in the syllabuses examined here does not support his inference, however, that geographers and historians feel adequate, or inclined, to fill this gap.

Particularly stressed was the aesthetic appreciation of the environment and the artist's response to it, whether as painter, poet, photographer or architect. This area involved not only a study of the work of Wordsworth and Walpole, Turner and Lowry, but also, in one case, a training in the use of various media as means of investigating die visual aspects of the landscape.

The use of the environment for leisure activities in general, and for outdoor pursuits (including fell-walking, rambling, camping, orienteering, rock climbing, boating and ski-ing) in particular, was strongly emphasized in three cases, in one of which the outdoor pursuits component appeared as the core element of the Environmental Studies course.

Conservation and planning topics included congestion, pollution, dereliction, land management, the confrontation and interaction between town and country, and the functions of such organizations as the Nature Conservancy, the Countryside Commission, the National Trust and the Civic Trust. International problems such as the population explosion,

Table 8

Components	Sub-components	Nos. of Colleges	Weighting
Background	(a) Scientific/Mathematical	4	6
Subjects	(b) Historical	4	6

Table 9

Components	Sub-components	Nos. of Colleges	Weighting
Professional		9	17

under-development, and conservation on a world scale were more widely considered than the specific environmental settings in which they occur. The case studies approach thus seemed more popular in systematic than in areal terms, which would appear the more logical arrangement in a Social rather than an Environmental Studies course. On the evidence of the syllabuses, the conservation issue emerged as a third major element of Environmental Studies courses, with strong links with two other core components in Geography and Ecology.

These were basic courses in disciplines providing background skills and concepts useful in the study of more overtly environmental subjects. They included Physics, Chemistry, Physiology, Genetics, Mathematics, Social, Economic and Political History, and Demography.

The professional content was possibly regarded as too obvious to be worthy of mention, for it was absent from the syllabuses of almost half the Colleges. Here, as in some syllabuses, it will be considered under aims.

Aims and Approaches

So variable was the weight attached to the statement of aims in the syllabuses, ranging from complete neglect to a fuller coverage than was found under content, that no assessment of relative importance of different aims could be made, except in a very general sense. A classification of aims, however, has been attempted. It should be emphasized that the aims should be seen in the context of Main courses in Environmental Studies in Colleges, and not of Environmental Studies courses in general, though separating the two is often difficult and sometimes unnecessary.

1. Intellectual

The sequence presented below is an attempt to synthesize the range of intellectual aims gleaned from the various syllabuses, starting with reasonably specific objectives and finishing with global aims.

(a) *The Development of Skills*

   (*i*) In the field: such techniques as compass and map reading, field sketching, photography, slope measurement, identification of plants, social surveying, and first aid and camping skills, among others;

   (*ii*) in the library, museum and record office: the meaningful use of old maps, directories, and a variety of documents and books;

   (*iii*) in the classroom and laboratory: preparatory and follow-up techniques, including equipment construction, map making, map interpretation, the provision and interpretation of audio-visual materials, and the preservation and classification of specimens;

   (*iv*) professional skills: involving the students in preparation of appropriate materials for use in school.

(b) *The Promotion of Scientific Methods of Enquiry*

Emphasis was placed on the use of inductive methods, through reconnaissance, observation, measurement and recording, largely entailing on the spot investigation. Some syllabuses also specified the employment of indirect methods, including the employment of maps and photographs for the study of external areas.

(c) *The Formation of Concepts*

One syllabus contained the exhortation "to get the students reflecting on general principles." Training in techniques of analysis, synthesis, comparison and correlation were variously indicated as means to this end. Two Colleges were courageously specific in identifying the central concepts in which they were interested. Under the heading "General Environmental Concepts" one listed: the nature and assessment of evidence; scale (in time and place); diversity of form and structure (as in rocks, plants and human artefacts); succession; population pressures; the biosphere and ecosystems; and inter-relationships among living organisms. Another (an Environmental Science syllabus), under the heading "Unifying Principles", specified: the circulation of materials; energy relationships; interaction (between living organisms and between these and their environment) and evolution, and diagrammatically attempted to relate these to particular topics and areas of study.

(d) *Understanding the Environment*

The various intellectual processes described above are intended to set in motion a cognitive sequence which culminates in the student to a greater or lesser degree possessing "a more informed and thoughtful consciousness of the part played by the environment in shaping human experience", or achieving "an awareness and appreciation of the environment as a whole". These quotations summarize a widely accepted aim of Environmental Studies courses.

(e) *Acculturation*

In the wider context, the environment is seen as part of the cultural heritage, which it is a purpose of an educational system to pass on. A handful of Colleges expressed the hope of instilling "a personal reaction" or of stimulating "a wide variety of aesthetic experience" in relation to the environment. This search for a personal response is perhaps a reflection of the view that Environmental Studies provides a valuable contribution to a "liberal education" and acts as a necessary counterweight to any over-stress laid on scientific methods of investigation.

2. Subject Integration

This is clearly linked with intellectual ends, and seemed widely regarded as a necessary prologue to the development of environmental understanding. "The breaking down of the subject barriers" was frequently quoted as an aim, even though it would seem logically a means rather than an end. Variations of opinion were largely expressed in terms of intensity of utterance, ranging from the deft: "an appreciation of the permeability of traditional subject boundaries"; to the doctrinaire: that Environmental Studies is "educationally more satisfying, more interesting and more challenging than the more traditional forms of study."

3. Motivation

The notion that exposure to the environment, and the opportunity to investigate it at first hand, acts as a stimulant to purposeful educational activity was aired less in the syllabuses than might have been expected, though it is no doubt implied in the widespread requirement from students of a Special Study, based on first-hand investigation of a limited area. The Special Study can be regarded as a motivational as well as an evaluative device. In a covering letter, the director of one Environmental Studies course, in which stress was laid on the heightened relevance of an arrangement involving participation in environmental problems through association with, for example, the activities of the Countryside Commission, emphasized the motivational value. "Students are very keen and we are constantly gratified by the time they devote to their studies and by the quality of material they present."

4. Socialization

Though rarely mentioned in the syllabuses, it is worth noting in this context that one College expected students "to take some part in local activities of which one must be related to provision for the leisure of children and adolescents, and one must be a local pursuit of some adult interest." Another wished to develop in students "a sense of values and proper attitudes towards the environment", a feeling probably shared, though not stated, by other Colleges.

5. Vocation

Most Colleges pointed to the advantages of integrated courses in the training of teachers for primary school work. A summary of the prevailing climate of opinion is contained in the statement that "in content and method the course will prepare the student for an effective teaching career during which he will be able to contribute to developments in Inter-disciplinary Enquiry, exploit the opportunities of the school environment and engender a spirit of scientific enquiry in his pupils . . ." This opinion is reflected in Rolls's concluding comment, that Environmental Studies makes sense ". . . as a part of the general education of students whose futures lie more with the teaching of children than with the teaching of a subject."[11]

Discussion

The second part of this article is put forward as a contribution to the debate concerning the nature of Environmental Studies and its worth in curriculum reorganization, particularly at College level. As the subject (as distinct from the philosophy underlying it[12]) is in a youthful stage in its evolution, it would be untimely if any impression of a developing orthodoxy, or still worse of the hardened "establishment opinion" that has tended to ossify the content and approach in many traditional subject areas, emerged.

The contents of the syllabuses looked at earlier provoke discussion over the balance of components and the relationship between Environmental Studies and other areas of study, notably its special relationship with Geography. A wider, and often critical, climate of opinion prompts a defence of the place of Environmental Studies as an academic course in Colleges of Education.

Links between Environmental Studies and other Subjects

The survey of Main course syllabuses revealed three nuclear components of Environmental Studies courses, namely the Environmental Sciences, Geography, and an Applied element, closely bound up with the conservation issue. A unifying theme might be found in the ecological approach, relevant in varying degree to all three elements.[13,14,15,16] Clearly some central concepts are drawn from this source. The peripheral components of the syllabuses were Sociological, Aesthetic and Recreational, and it might be suggested that these are worthy of wider consideration, at least as optional elements in courses.

The position of History is somewhat equivocal. In many discussions it is recorded as a third, presumably equal, member of the Geography-Biology-History triumvirate, but this could be in part an arrangement of convenience. As the local area is placed both in time and space, there is clearly a considerable overlap between "environmental" History and Geography. It may be that the classification used previously under-estimated the amount of historical content present in syllabuses. On the other hand, History by its nature is much less concerned with environment than Geography. Has not therefore History a more central inter-disciplinary niche in "Social Studies" or "Contemporary Studies"? This is not to underestimate the wealth of historical interest "in the field",[17,18] nor to undervalue the contribution of historical perspectives and techniques of study to an understanding of environments; but it would argue against the too facile adoption of a one-third, one-third, one-third arrangement of time between Geography, Biology and History.

In contrast to History, there is a massive overlap of component disciplines in Environmental Studies and Geography, as the construction of a simple Venn Diagram would show. The two subjects rank as "fields" rather than as "forms" of knowledge in Hirst's definition,[19] relying heavily on the "derived" rather than the "indigenous" concepts in Harvey's,[20] though Harvey attempts to identify indigenous concepts in Geography which Hirst presumably would not regard as discrete to that subject. So far, most environmental educationists seem to have been at pains to play down the close connections between Environmental Studies and Geography. Watts, for example, evades rather than explores this issue, though no doubt consciously in that he feels that "environmental learning situations ought to be approached directly and not as conjunctions of 'subject' materials." None the less, his comment that "All Environmental Studies occur in a setting of Geography . . ."[21] seems the same as saying "All Geographical studies occur in an environmental setting . . .", and is semantically confusing.

Rolls remarks that "Geography is not concerned primarily with the identification, evolution and interaction of the living part of our environment",[22] the "living part" presumably referring to plants, animals and man. This would in fact make a reasonably serviceable definition for those who look upon Geography as a study of "human ecology". Thomas, in proposing that Geography, as "an integrated discipline at the heart of Environmental Studies, is better placed than any of the other constituent nuclei to co-ordinate and unify the larger body of studies of which it is the core", does not explain his preceding assertion that "any equation of Geography and Environmental Studies will destroy the identity of the first and restrict the development of the second."[23]

There is certainly a significant distinction, at least in degree, between standard practice in Environmental Studies and Geography. The "environmental" approach involves the allocation of a much greater proportion of time to a consideration of the local area, whereas in Geography external environments have the stronger emphasis. Watts in effect transposes this difference into one of kind when he quite unequivocally equates Environmental Studies with Local Studies, in affirming that dramatic stereotypes cannot be imported from outside—"somehow the colour has to be found in the neighbourhood itself."[24] This viewpoint stresses that direct contact with the locality confers the crucial advantage of providing regular opportunities for first-hand field study, whereas Geography is more closely welded to the classroom.

Without questioning the benefits of the local studies approach, they must be balanced against the pitfalls. Bluntly, the excision of external environments leaves a yawning gap in educational provision for children or students. Many years ago Bertrand Russell looked to Geography to diminish "the tyranny of familiar surroundings over the imagination."[25] More recently, in the Social Studies context, Taba has pleaded for a "cosmopolitan" rather than an "ethnocentric" approach.[26] Bruner has drawn attention to the difficulties of generalizing from the familiar.[27] Nearer home, the draft of the School Council Welsh Committee's Curriculum Development Project on Environmental Studies suggests that "the attitudes and skills developed locally can serve as a yardstick for the examination of non-local environments . . . The study of the locality is not the end product, rather it is a scheme of reference against which wider studies are viewed."[28] This would support the case for a concentric approach in Environmental Studies, a method neatly explored for school use in Masterton's book.[29] Apart from the deeper issue of widening perspectives, there is the timetabling problem that (at least for the same pupil or student) Environmental Studies and Geography must be competitors in the curriculum, in view of their overlap of content and concept. Environmental Studies cannot be regarded as an adequate replacement if it becomes synonymous with Local Studies.

For both Environmental Studies and Geography there is now available a well-chosen and comprehensive battery of teaching materials and techniques which make it possible to uphold the spirit if not the letter of first-hand enquiry. Two leading protagonists of the sample (case) studies approach have referred to it as "field work in the classroom".[30] It is backed up by the use of maps, diagrams, photographs, literary descriptions, gramophone records or tapes, films and film strips, and statistical material. Martin has suggested that "perhaps the reluctance (of environmental educationists) to study environmental areas overseas is partly a fear of conflict with the field of Geography."[31] It would be a pity if this fear were to be taken as a reason for disregarding a field of such richness and relevance. The burden of this argument is in no sense against Environmental Studies as such, but against the tendency of some practitioners to neglect wider perspectives. At the same time, the study of overseas environments should clearly not play so large a role as in Geography courses. In this context, therefore, it is suggested that the difference of approach between Environmental Studies and Geography should be regarded as one of degree and not of kind.

In terms of "boundary definition", however, there is perhaps a more clear-cut distinction between the two, though even this may be greater in principle than in reality in the sense that many practising geographers ignore frontier limitations. Relatively speaking. Geography has narrower terms of reference than Environmental Studies. In the field, for example, the environmentalist is an opportunist who takes what the locality has to offer. Its elements are examined for their own sake; "because they are there". "Localities have their own processes of selection",[32] writes Watts. The geographer is more concerned to impose a framework. Hence the environmentalist may study a building as an entity in itself, paying due attention to its architraves as well as its overall architectural style, its historical evolution as well as its present setting. The geographer is more strictly concerned with its site and functional relationships. Contrary to the popular conception that the environmentalist uses the holistic approach in viewing a landscape, he often in fact seems to pay more attention to its minutiae. Paradoxically, it is on the traditional and much-criticized geographical "Cook's Tour" that this somewhat romantic outlook prevails.

Environmental Studies Courses in Colleges

The wider content coverage and more flexible framework of Environmental Studies, its closer contact with reality, and the fact that the locality can contribute to "a more illuminated and realistic view of the past and as a pattern for further investigations of more purely scientific matter"[33], give the subject a great appeal for use at certain stages of education. Environmental Studies has been proposed as a desirable alternative to more traditional subjects at various educational levels:

(a) Primary School (5–11);
(b) Middle School (9–13);
(c) Junior forms in Secondary School (11–13);
(d) Secondary School as a whole:
　　(i) for children pursuing courses to C.S.E. and G.C.E. level (examinations in Environmental Studies having already been established in some areas);
　　(ii) for "Newsom" children;

(e) College of Education:
　　(i) in Curriculum or "Third Area" courses;
　　(ii) in Main and B.Ed, courses.

(f) University, though here courses labelled "Environmental Studies" tend to have an individualistic flavour and a distinctly "applied" quality.

It would seem sensible to prepare a special case for each of these eventualities rather than a blanket one to cover them all. The various advantages of Environmental Studies for Primary and Middle school curricula are apparent in the writings of Watts, Hopkins,[34] and others. In relation to group (d)(ii), more research seems at present to be going into the development of Humanities/Social Studies frameworks (see Stenhouse,[35] and James[36]) than of Environmental Studies, and this may be right and proper, particularly if the latter neglects its sociological and international contexts. For group (d)(i), strong cases have been made by subject specialists for a continuation but amelioration of present curricular arrangements (see Bull,[37] and Burston[38]). In these, it is suggested that unless clearcut advantages for more revolutionary changes can be advanced for particular groups, energies should first be devoted to advancing curriculum reform within existing subjects. These are issues for debate elsewhere, however.

In Colleges, the establishment of Environmental Studies Main (and a few B.Ed.) courses has been accompanied by apprehension concerning the "academic respectability" of the newcomer at these levels. It is ironic that some geographers are in the forefront of the attack, for they will need no reminding that their own subject had a similarly hesitant reception when attempts were being made to establish it in the University curriculum in the 1870's. It must be said that traditionalists may have had their case strengthened by some overzealous remarks from the other side. These have included:

(a) basing part of the case for Environmental Studies on the transparent stratagem of contrasting enlightened modern methods of teaching Environmental Studies with bad traditional practice in existing subjects, a clear example of confusing the nature with the performance of a discipline; and

(b) categorizing Environmental Studies vaguely as an "approach" rather than a subject. This is linked with the elevation of what is good practice at Primary level to a College course methodology, thus confusing what is largely a predisciplinary approach with an interdisciplinary approach.[39] Bruner's stimulating but hazardous assertion that "there is an appropriate version of any skill or knowledge that may be imparted at whatever age one wishes to begin teaching, however preparatory the version may be"[40] is a dubious proposition if inverted. While there is point in Gill's suggestion that the B.Ed. student should be involved in learning by methods "analogous to those by which children actually learn, including first-hand investigation and thinking in terms of theme-centred enquiry"[41], these would seem more appropriate to the professional part of the course, as his article subsequently confirms. Ausubel presents a chilling account of the effects of "learning by discovery" methods improperly applied in the U.S.A. "Two final by-Products of this point of view were deification of the act of discovery associated with the inductive and incidental learning methods of teaching, and extrapolation to the secondary school and university student of the elementary school child's dependence on recently prior concrete, empirical experience in the comprehension and manipulation of ideas."[42] The syllabuses looked at earlier seem to uphold the view that College courses should have a meatier diet than one of "preparatory versions" or of "naive emulation of the scientific method."[43]

The "academic" case for Environmental Studies might be reinforced if it were more widely regarded as a form of Inter-disciplinary Enquiry, and that term clearly distinguished from the more woolly concept of "integration". At present, the two are often used as synonyms.[44] Interdisciplinary Enquiry recognizes the integrity of the constituent disciplines, whereas "integration" has at times seemed to apply to them the taint of dirty washing. The image of Watts that the academic subjects can be seen as mere "rows of books in libraries"[45] implicitly if not intentionally devalues their contribution. The shibboleth of the subject barrier as a kind of curricular Bastille to be stormed is understandable as an emotive reaction to the mistakes of the past, but the "appreciation of the permeability of the traditional subject boundaries" appeals as a more considered and productive construct. It recognizes that the boundaries are imperfect and for some purposes inappropriate, but seems to connote that they cannot be dismissed as mere "historical accidents", though the place of some in the curriculum may be. The subject disciplines are indeed relatively efficient ways of sorting out the complexities of reality. As Bantock writes: "The world presents to us an undifferentiated mass of data which only the mind can organize into manageable proportions, through a series of models and conventions which make thinking possible, let alone fruitful. Subject areas are made up of these models and conventions . . ."[46]

There would seem to be no logical conflict between this point of view and the presence of Environmental Studies, if the latter is looked upon in cross-disciplinary terms. As we have seen, as in Geography, it encompasses the ways of thought of a number of extraneous disciplines. The "environmental concepts" specified in one of the College syllabuses were in fact derived concepts. These contributing disciplines provide the prerequisite "lenses" for a meaningful perception of the environment.[47] The "integrating" locality, on the other hand, is a concept-illustrating[48] and not a concept-forming medium. Merely taking what the locality has to offer, though helpful at the Primary stage, may well lead to an over-emphasis on trivial detail, with its subsequent amalgamation more in the nature of a collage than a true synthesis. This would come into the category of predisciplinary work. In contrast, designing an excursion as part of a programme of work to test an ecological concept, some aspect of central place theory, or the impact on an environment of decisions taken in the past, would represent a constructive use of the locality in an inter-disciplinary sense, and would seem a more appropriate method for College use. Pring has gone further and argued that in some areas of understanding, particularly where practical decisions are involved, it is essential to use the procedures of more than one distinct discipline. ". . . one might recognize the autonomy of different bodies of knowledge while at the same time recognizing the problems inherent in their synthesis. And one might make provision in the curriculum for this synthesizing element by concentrating, for at least part of the time, on areas of thought and decision-making where such integration is indispensable."[49]

If due regard is paid to the constituent disciplines and the skills and concepts they embody, the argument of principle against Environmental Studies as worthy of academic acceptance would seem to be deflated. The main difference between it and established subjects (apart from Geography) lies in its derivative and synthesizing character. Analogous arrangements, such as West European or Latin American studies, seem to be accepted at University level. In any case, is not academic worth a function of the quality of personnel undertaking a course, rather than its content?

A negative and important argument remaining, however, is that Environmental Studies represents a less efficient way of organizing subject disciplines. Other things being equal, Inter-disciplinary Enquiry means either that (a) a smaller number of skills and concepts per constituent discipline can be developed and examined in depth; or (b) an equivalent number can be looked at in less depth; or perhaps (c) an equivalent number can be looked at in equal depth but less variedly illustrated. Subject specialists suggest that these in their different ways represent serious losses, particularly for intelligent students. Graves has argued the case at Secondary School level, at least for pupils in group (d)(i), along similar lines. He points to the ever-present difficulties of evolving an ordered sequence of topics even in one subject in such a way that "principles and concepts learnt are gradually enriched and developed",[50] which could reach unmanageable complexity where Inter-disciplinary Enquiry was instituted. Attempts at simplification could result in superficiality. There is also the question of the preponderance of intellectual capital invested in and committed to established subjects, a problem touched upon in the Newsom Report. "An historian turned reluctant geographer, or vice versa, is not likely to inspire pupils who take a great deal of rousing at the best of times. Both subjects can, as we shall see later, offer many of the same values. If the head decides that the cobbler had better stick to his last, we shall neither be surprised nor unduly distressed."[51]

These are not, however, clinching arguments, and it would seem that an individual or group commitment could well be an appropriate final arbiter of choice. So long as either alternative is conducted with enthusiasm, efficiency, and integrity, the balance of advantage appears to sway according to which group of children or students the courses are aimed at. In College terms, it might be fair to conclude, supporting Rolls's "commonsense" argument[52], that

(*a*) in Curriculum courses for prospective Primary teachers, the advantages of Environmental Studies are overwhelming;

(*b*) in Main courses, the "academic" case is evenly poised and that for the Primary group the vocational advantage of Environmental Studies is a strong reason for its recommendation;

(*c*) in B.Ed. courses, the balance of advantage may favour the traditional arrangement, though not powerfully enough to inhibit those who wish to introduce inter-disciplinary courses, conditionally upon the provisos stated earlier being observed. This in a way upholds the current "need for relevance" argument which, however, carried to its logical conclusion would presumably create a dichotomy of Curriculum courses and B.Ed. courses, with traditional Main courses obsolete for the Primary group.[53]

Finally, one would hope that environmental educationists in Colleges would act as guides away from the entrenched position. Can any useful purpose be served by the accusations of "empire building" which ricochet from side to side? That "some writers and teachers of Environmental Studies seem to regard the approach as merely a colonial extension of their own subject empire"[54] no doubt carries the same element of truth as "It appears that it is the teacher trained in economics, sociology, classics or some other subject not in the school curriculum, or who has attended courses in curriculum development, who tends to see in the leadership of an I.D.E. team a convenient vehicle for his advancement."[55] It is fitting here to recall Wiseman's entreaty, made in another context, for "a rapprochement which will give to education the best techniques from both camps appropriate to the particular purpose, leaving the backwoodsmen of both wings to their unproductive sniping."[56]

Notes

1. I. F. Rolls, "Environmental Studies: a New Synthesis?" in *Education for Teaching*, No. 78 (1969), p. 21.
2. G. Martin, "Environmental Studies in Colleges of Education", *Occasional Paper No. 1. Society for Environmental Education*, p. 2.

3. The Colleges concerned were: Bingley, Bordesley (Birmingham), Charlotte Mason (Ambleside), Chorley, Chorley (Blackburn Annexe), Eaton Hall (Retford), Edge Hill (Ormskirk), Kesteven, City of Leicester, Lough-borough, I. M. Marsh (Liverpool), Mather (Man-chester), Notre Dame (Liverpool), Ponteland, Ports-mouth, Poulton-le-Fylde, St. Katharine's (Liverpool), St. Paul's (Cheltenham), and Nottingham Regional College of Technology (Education Department). Their geographical distribution is interesting, only two lying south-east of a line from the Wash to the Bristol Channel, and most of the others within or on the fringe of Highland Britain.

4. R. R. Cumming et al., "Some Factors in the Planning of an Environmental Studies Course to Degree Level in a College of Education" in *S.E.E. Bulletin No. 3*, Vol. 2 No. 1. (1969) pp. 15–34

5. Kesteven (Social and Environmental Studies), Lough-borough (formerly Environmental Studies, now Social Studies), and Nottingham Regional College of Technology (Environmental and Contemporary Studies).

6. Chorley, Ponteland and Portsmouth.

7. Charlotte Mason.

8. W. Kirk, "Problems of Geography" in *Geography*, Vol. 48, Part 4. (1963), pp. 364–70.

9. For the possibilities of biological field work in urban areas, however, see M. Smith and D. A. E. Spalding, "An Approach to the Development of Field Studies in Urban Areas" in *Journal of Biological Education*, Vol. 2, No. 3 (1968), pp. 223–24.

10. G. Martin, op. tit. p. 3.

11. I. F. Rolls, op. cit. p. 27.

12. D. G. Watts, *Environmental Studies* (Routledge and Kegan Paul, London, 1969). As there are some critical references to points of detail later, it is fair to record here the generally stimulating quality of this pioneering book.

13. E. Caulton, "An Ecological Approach to Biology" in *Journal of Biological Education* Vol. 4, No. 1 (1970). pp. 1–10, and especially p. 6 and p. 9.

14. P. J. Newbould, "Production Ecology and the International Biological Programme" in *Geography*, Vol. 49, Part 2. (1964), pp. 98–104.

15. S. R. Eyre, "Determinism and the Ecological Approach to Geography" in *Geography*, Vol. 49, part 4. (1964), pp. 374–6.

16. D. R. Stoddart, "Geography and the Ecological Approach" in *Geography*, Vol. 50, Part 3. (1965), pp. 242–51.

17. W. A. L. Blyth, "Field Studies in the Teaching of History" in M. Dilke (ed.) *The Purpose and Organization of Field Studies* (Rivingtons, London, 1965), pp. 38–60.

18. T. Corfe (ed.), *History in the Field*, (Blond Educational, Leicester, 1970).

19. P. H. Hirst, "Liberal Education and the Nature of Knowledge" in R. D. Archambault (ed.) *Philosophical Analysis and Education*, (Routledge and Kegan Paul, London, 1965), pp. 128–31.

20. D. Harvey, *Explanation in Geography*, (Arnold, London, 1969), pp. 117–27.

21. D. G. Watts, op. cit., p. 86.

22. I. F. Rolls, op. cit., p. 25.

23. I. Thomas, "Rural Studies and Environmental Studies" in *S.E.E. Bulletin No. 3*, Vol. 2 No. 1. (1969), p. 14.

24. D. G. Watts, op. cit., p. 91.

25. B. Russell, *On Education* p. 207. (Allen and Unwin, London, 12th Impression, 1957).

26. H. Taba, *Curriculum Development: Theory and Practice* p. 273. (Harcourt, Brace and "World, Inc. New York, 1962).

27. J. S. Bruner, *Towards a Theory of Instruction* p. 93. (Harvard University Press, 1966).

28. Schools Council Welsh Committee Draft, "Curriculum Development Project on Environmental Studies." Occasional Paper No. 3, p. 2. Society for Environmental Education.

29. T. H. Masterton, *Environmental Studies: a Concentric Approach*, (Oliver and Boyd, Edinburgh, 1969).

30. M. Long and B. S. Robertson, *Teaching Geography* p. 113. (Heinemann, London, 1966).

31. G. Martin, op. cit., p. 4.

32. D. G. Watts, op. cit., p. 89.

33. Schools Council Welsh Committee, op. cit., p. 2.

34. M. F. S. Hopkins, *Learning through the Environment*, (Longmans, London, 1968).

35. L. Stenhouse, "The Humanities Curriculum Project" in *Journal of Curriculum Studies*, Vol. 1. No. 1, (1968). pp. 26–33.
36. C. James, *Young Lives at Stake* (Collins, London, 1968).
37. G. B. G. Bull, "Interdisciplinary Enquiry: a Geography Teacher's Assessment" in *Geography*, Vol. 53, Part 4. (1968), pp. 385–6.
38. W. H. Burston, "The Study of the Curriculum" in *Bulletin* of the University of London Institute of Education, New Series No. 13. (1967), pp. 27–8.
39. D. Lawton, "The Idea of an Integrated Curriculum" in *Bulletin* of the University of London Institute of Education, New Series No. 19. (1969), p. 8.
40. J. S. Bruner, op. cit., p. 35.
41. N. M. Gill, "Integrated Studies in the B.Ed. Course" in *Universities Quarterly*, Vol. 24, No. 2. (1970), p. 197.
42. D. B. Ausubel, *The Psychology of Meaningful Verbal Learning* (Grune and Stratton, New York, 1963), p. 140.
43. D. B. Ausubel, ibid., p. 141.
44. As in R. A. Pring, "Curriculum Integration" in *Bulletin* of the University of London Institute of Education, New Series No. 20. (1970), pp. 4–8. An interesting discussion of the notion of integration.
45. D. G. Watts, op. cit., p. 18.
46. G. H. Bantock, *Culture, Industrialisation and Education* p. 37. (Routledge and Regan Paul, London, 1968).
47. C. James, "The Open Curriculum" in *Bulletin* of the University of London Institute of Education, New Series, No. 19 (1969), p. 12.
48. D. G. Watts, op. cit., p. 9.
49. R. A. Pring, op. cit., pp. 5–6.
50. N. J. Graves, "Geography, Social Science, and Interdisciplinary Enquiry" in *Geographical Journal*, Vol. 134, Part 3. (1968), pp. 392–3.
51. Ministry of Education, "Half Our Future" (*Newsom Report*) (H.M.S.O., 1963), p. 124.
52. I. F. Rolls, op. cit., p. 27.
53. J. S. Hynds, "The College Curriculum and the Primary Teacher" in *Bulletin* of the University of London Institute of Education, New Series No. 20 (1970), pp. 15–19.
54. D. G. Watts, op. cit., p. 86.
55. G. B. G. Bull, op. cit., P. 383.
56. S. Wiseman, *Examinations and English Education*, p. 134. (Manchester Univ. Press, 1961).

Environment in the curriculum: representation and development in the Scottish physical and social sciences

HAMISH ROSS

Scottish official curricular texts, including guidelines and examination papers, are analysed for representations of 'self and the environment'. The environment is represented as fragmented when it is the curricular focus and is only 'whole' when it is background context; 'human–environment relations' are dualized; and the value of 'environment' lies dominantly in its use by humans (although there is a much less clear possibility that it might have inherent value). These representations lend themselves to the kinds of dominant and abusive relationships with environment that the same official curricular text hopes to counter. The assumed need for publicly shared understandings may drive this representation, through processes in which students understand environment by its 'parts', by generalized models of relationship, as being shallowly causal and progressively 'other', and not as contingent, local, or privately experienced. The desire for such a public world-view may in turn be driven by historical efforts to use education to tackle social inequality. The purpose of undertaking such a detailed analysis is to create space for progressive and incremental curricular development rather than for revolutionary revision.

Introduction

There has been long-standing interest in the nature of a curriculum for environmental education (see, e.g. Covert 1969: 12, Orr 2004: 94–98) and an interest in changing the academic 'official curriculum', where one exists. However, Scott and Gough (2003: 83) are right to question the credibility of 'Damascene salvationism' approaches to environmental education, not least because there are policy tensions built into any educational change (Sterling 1996a), and this is true of official curricular change too. So as well as declaring what an official curriculum should look like under ideal conditions, there is a case for analysing, in some detail, more precisely what is wrong with its current condition. For a variety of reasons, outlined by Goodson (1994), not a great deal of this detailed work has been done (though see, for example, Chenhansa and Schleppegrell 1998).

This paper attempts such an analysis, based on reading official curricular text from Scotland using ideal types. In a sense I reveal what is already known (about, for example, the fragmentation of representations of 'environment' through curricular division). However, I hope that greater precision about how such things are articulated in curricular text will create sufficient extra insight that space for stepwise adjustment becomes possible. I suggest, for example, that in official curricular text, contrary to its own stated claims, the 'environment' is fragmented when it is the curricular focus but is 'whole' when it is background context; that 'human–environment interrelations' are dualized; and that the value of 'environment' is either explicitly instrumental or, implicitly, has a possible inherent value that need not be entertained in practice. I go on to discuss some of the historical and philosophical drivers behind this precise articulation in a particular official curricular text, in the hope that recovering such history (from the 'forgetting' and 'neutralizing' that official text itself institutes) will create still more space for change. I suggest that the representation of 'our place in environment' in the Scottish curriculum might be a product of assumptions about epistemological division, child development, and assessment, and that these derive from historical efforts to reduce social inequality. The Scottish official curriculum is currently under major review (Scottish Executive 2004), so this work is particularly relevant at the present time and in this particular place, but it is likely that the analysis here will apply to other 'western-modernist' cultures.

Methods

I have used ideal types as background sensitizing devices for my reading of curricular text. The reasons for using the ideal types are: (1) they provide analytical purchase in my reading of the curriculum; (2) they try to make explicit my own interests and values (Weber 1949: 93, Giddens 1971: 60); and (3) they open up the possibility of identifying a dominant discourse in the curriculum. I use two ideal types, defined in opposition to each other, as alternative representations of how the curriculum, and perhaps students, might perceive environmental relations. I refer to them as *type 1* and *type 2* (see table 1).

It is important to be clear about the purpose of the types and to avoid being distracted by them. They are not the ontological focus, although some readers will relate them to concepts that have been integral to debates

Table 1. The two ideal types used in this study.

Type 1	Type 2
Environment	*Environment*
Reducible to 'parts' and the sum of these in simple mechanistic interaction	Greater than the sum of parts Richly emergent
Society (incl. self)/environment	*Society (incl. self)/environment*
Dualistically separate Environment is 'other' to society/self	Relationally dependent Co-evolving and co-integral
Value of environment	*Value of environment*
Resource to meet basic human needs (instrumental relation) 'Parts' (see above) are substitutable	Emotional/care-laden relation Environment has inherent value

around education about/for the environment and sustainable development. Similar constructions have been developed by others. Naess (1973) contrasts a 'compact in milieu' (*type 1*) model with a 'relational, total-field image' (*type 2*). There are overlaps with 'reductionist/technocentric' and 'holistic/ecocentric' cultural paradigms (Sterling 1993: 79, O'Riordan 2000: 67), with Clark's (1991) Gestalt's I and II, with various authors' references to 'sustainable development' and 'sustainable growth' modes (Fien and Trainer 1993, Palmer 1998: 92), 'dominant social' and 'new environmental' paradigms (Sterling 1996b: 20), and with contrasting political ideologies (Huckle 1996: 10) and epistemological approaches.

However, it is dangerous to take the types further than their position as an analytical tool. The 'truthfulness' or accuracy of the types, or even our subjective understandings of what they might mean, are not directly at stake in this paper. *As far as possible I am aiming to discuss the curriculum's text rather than my own.* The ideal types are a partially-evolving background to a single-reader, grounded content analysis (Strauss and Corbin 1998) of 900 pages of official text. The categories and patterns identified from the text were not constrained by the types and are described in the analysis section below. I believe these to be strong patterns, but caution that their apparent strength is partly derived from a narrow focus of interest *as represented by* the ideal types. The texts studied were:

- the *Environmental Studies 5–14 Guidelines* (Learning and Teaching Scotland [LTS] 2000)—the part of the official curriculum for 5–14-year-olds in Scotland that includes the natural and social sciences;[1] and
- Syllabuses, examination papers, and examiner marking instructions[2] for Standard Grade qualifications in the years 2001 and 2002 in the subjects of science, geography, modern studies, and contemporary social studies. (Modern studies and contemporary social studies are mixtures of political and social studies.) Other texts largely ignore environment but will be mentioned where they engage directly with it.[3]

Compulsory schooling in Scotland is from ages 5–16 years. Children attend primary schools from age 5–11 and follow an official curriculum that includes the *Environmental Studies 5–14 Guidelines* mentioned above. They then move on to secondary schools from age 12. For the first 2 years of their secondary education, most continue to follow the same official curriculum as they did in primary schools and the *5–14 Guidelines* were an attempt to build official curriculum coherence across the institutional transition between primary and secondary schools. However, in secondary schools these curriculum guidelines are much more likely to be interpreted for delivery through discrete 'subjects', such as geography and science, whereas in the primary school the likelihood remains of interpreting them in thematic 'topics' that might cross such 'subject' boundaries. So, despite the common curriculum, many children still experience a significant difference on transition between primary and secondary school. From age 14 they then take a range of subjects—typically seven or eight in total—for 2 years (ages 14–16) towards examination. Their choices include a compulsory core (English, mathematics, science(s), and, in some schools, a modern language) and optional subjects that are offered in such a way as to maintain breadth of study. For example, most pupils must take at least one of the social subjects studied in this paper (geography, modern studies, contemporary social studies) or history. Standard Grade is the main form of examination at this stage. After this end to compulsory schooling, many will stay on to pursue further qualifications up to age 17 or 18.[4]

The concern in this study is with official curricular text. It is therefore rather more focused on some idea of curricular *intent* (for want of a better word) and more loosely concerned with what the curriculum *achieves*. However, this relation is worth considering. Since subject-specific figures are not routinely published, it is difficult to provide straightforward data on the number of pupils affected by the official curriculum discussed in this paper. I can safely claim that all ordinarily schooled 5–14-year-olds are exposed to the *5–14 Environmental Studies Guidelines* discussed here. This is far from the case with the 14–16 qualifications, however. Approximate figures (Munn *et al.* 2004) suggest that, of any given cohort, more than 20% study modern studies and more than 30% study geography. More than 80% will study one of these or history (which is itself taken by 30%). The number studying contemporary social subjects is relatively insignificant. The numbers taking the science Standard Grade discussed in this paper are also small and probably declining (~ 25%; see MacGregor 2003) with the rest of the cohort taking qualifications in biology, physics, or chemistry as separate subjects. Science Standard Grade was analysed rather than biology because of its more general relation to the wider research of which this forms part (Ross and Munn 2007) and because a greater proportion of the syllabus concerns environmental matters. In addition, the study has all the limitations of relying on one kind of data. There is much room for interpretation between what official curriculum documents intend students should learn and what they learn in practice (Ozga 2000: 10, Pring 2000: 122, Cotton 2006: 78), let alone between such learning and environmental attitudes and behaviour. Nevertheless, the official curriculum is worthy of study, expressing, as it does, what is valued for one reason or another in society (Moore 1982: 53), and it would be hard to argue that it has *no* ultimate influence on how students perceive their relationship with the environment.

Findings

Environmental holism in curriculum text

The ideal-type construction asks the following question: To what extent is 'environment' represented (and perhaps therefore understood and/or studied) as a holistic, richly articulated totality (*type 2*), or alternatively as a sum of independent, mechanically interacting parts (*type 1*)? The word 'environment' is used widely in the general rubric of the *5–14 Environmental Studies Guidelines* ('children can be encouraged to appreciate the interrelationship of all living things and their environment' (LTS 2000: 4)). However, the *Guidelines* are broken down into attainment outcomes that correspond to post-14 school 'subjects' such as geography and chemistry.[5] Since these post-14 subjects drive the divisional structure of the rest of the curriculum (and do so in the current curriculum reform proposals (Scottish Executive 2006: 30)), I now turn to them for more detailed analysis. In them, what representation of a holistic environment is there exists in the textual 'white-space' *between* subject disciplines. The enterprising student might access such a representation while metaphorically, say, walking between geography and science classes, but this has nothing to do with the educational intent of the curricular documents (or its interpretation by specialist teachers, to whom this 'between-subject' whole is perhaps even less accessible than it is to their students).

How is this fragmentation achieved despite the official curriculum's general rubric of understanding an environmental whole? A useful starting point is to recognize that subject disciplines differentiate themselves both by 'how to study' and 'what knowledge content is studied'.

In the case of 'how to study', it may be that multiple investigative approaches to the same environment could be beneficial in developing an appreciation of its totality. For example, geography examination papers use bio-climatic regions (such as 'equatorial rainforest' and 'tundra') as a background context while assessing students' climate graph-interpretation skills (SQA 2002d: 10/11). Science examination papers provide textual climate information while assessing reading-comprehension skills about Antarctic Island habitats (SQA 2001f: 5). Surely students must therefore appreciate that habitat and climate are part of one whole environment. However, the environment here is only a background context to skills-learning, and is not the focus. So any construction of a holistic environment that might lie around or between these two disciplines is only implicitly offered, and then only to students who are studying both disciplines. At post-14, students may not be studying both disciplines. Even for pre-14 students, any holistic representation's role is to make skills-learning seem relevant—its wholeness is not an ontological focus. A holistic representation is therefore both 'there' and 'not there' in the curricular text.

Of course, the environment is the direct focus when we consider what 'knowledge content' is to be studied. Subjects such as geography and science do explicitly focus on 'the environment', but how is it represented? My reading is that it runs against the construction of the environment as a holistic totality in a number of ways. First, the knowledge-bases of different subjects are divided between them to some extent, even if it is broadly acknowledged that 'it's all the same environment', and we need to explore how this affects the representation of environment. Secondly, 'the environment' is fragmented *within* a subject discipline: in science, for example, life is in part approached through its many classifications and characteristics (LTS 2000: 56–57); in geography, 'landscape' is a series of 'features' ('pyramidal peak, hanging valley, arête, corrie' (SQA 2001e: 3)) rather than a functionally interdependent totality.

Of course, there is more to environmental content than knowledge understanding of the environment and the *relation* of parts is also expected. However, there is a subtle process at play here, too. It might be called 'the articulation of shallow causality', in which the student is not expected to offer understanding that extends more than a single causal step. For example, the marking, i.e. grading, instructions for a test question asking how ox-bow lakes are formed state:

> Possible answer may include: Ox-bow lakes are formed in the lower course of the river (1 [mark]) where meandering happens (1 [mark]). There is erosion on the outside of the bend (1 [mark]) and sediment builds up on the inside of the bend (1 [mark]), cutting off the Lake (1 [mark]). (SQA 2001a: 3)

These answers are not intended to be exclusive but they are sufficient. Candidates are not required to take a deeper step into explaining, for example, why the lower course of the river is 'where meandering happens'. Analysing process in the environment in this way—to a shallow depth of causality—contributes to the fragmentation of its representation.

Surely some curricular text explores deeper causality in the environmental whole. What of the study of ecology (which is almost the definition of environmental holism), or of systems approaches to the environment, both of which are evident in the curriculum? My analysis suggested that learning here focuses on specific types of interdependence (for example 'food chains', 'competition', and 'abiotic adaptation'; 'give examples of feeding relationships found in the local environment ... construct simple food chains' (LTS 2000: 56)). I would argue that the sense of totality is lost here, replaced by models that break the whole of environment into generalized interdependencies. Figure 1 (SQA 2001g: 11) is not about a richly articulated whole, and neither are representative visual models of, say, 'how a waterfall develops' (SQA 2002c: 5), of 'an area of glacial deposition' (SQA 2002d: 8), or the hydrological cycle, which equally exclude the contingent and background clutter of a whole environment and its process relationships. Although the curriculum claims to be focusing on students' understanding of environment as 'interdependence', it could perhaps be more accurately described as focusing on the student's skill in understanding abstract models of such interdependence. Of course, such modelling is a part of scientific enquiry, but we can question why it is so dominant in these texts.

All of this—the breaking of environment into parts, into single-step causal links between them, or into generalized models of relationship—could be driven by a particular epistemological model for the curriculum, that is the search for the categorical, universal, and generalizable. However, it also happens to suit the needs of the simple assessment arrangements from which the above examples are drawn. Nameable parts, shallow causal analysis, and generalized relationship-types all turn the idea of environment—which might have been a richly articulated and complex whole allowing of multifaceted interpretation—into something that amounts to a finite set of 'right answers'. I will discuss this possible explanation for the representation further below.

Human–environment relations

The types ask whether the environment is represented as something dualistically separate from society, or something in which society is embedded. For 5–14-year-olds in Scotland, teachers should organize learning that includes:

Look at the food web below.

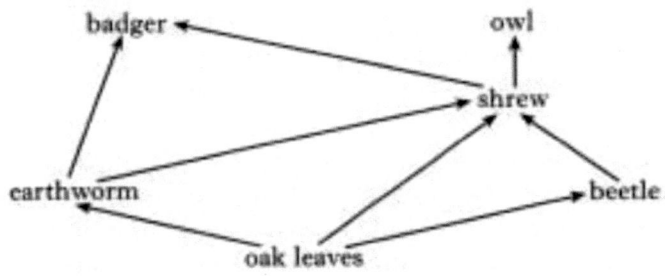

(a) Where do the oak leaves get their energy from?

....

(b) Using the food web, write down a food chain containing **four** organisms.

....   →   ....   →   ....   →   ....

(c) Omnivores are animals which eat both plants and animals.

Name the omnivore in this food web.

....

Figure 1. Question from science Standard Grade examination paper (Scottish Qualifications Authority 2001g: 11) © Scottish Qualifications Authority.

> thinking through the various consequences for living things and for the environment of different choices, decisions and courses of action; the importance of the interrelationships between living things and their environment; participating in the conservation of natural resources and the sustainable use of the Earth's resources. (LTS 2000: 62)

In the detailed attainment targets, however, the environment is to be progressively abstracted away from students' direct experience so that:

> in the early years priority will normally be given to contexts that are immediately within the experience of the pupil. This will usually be the child's local environment, although, on occasion, environments they experience through visits or the media may also contribute to studies ... In the later stages of primary, pupils will increasingly relate their own experiences within familiar environments to those of less familiar ones ... By S1/S2 [i.e. the first 2 years of secondary school] pupils will increasingly study less familiar environments *but will continue to relate them to their own experiences* as they further develop a systematic knowledge and understanding of patterns and processes across the Earth's surface (LTS 2000: 31; emphases added).

The process continues beyond the age of 14. Despite the geography advice that students develop 'awareness of problems facing the environment at all scales from local to global' (SQA 2001i: 40), 'Scotland' is the smallest area within which the key ideas are to be studied, and only one of 17 key ideas is assessed at this scale. All others are studied and assessed at UK, Western European, or global scales. The environment is something that becomes progressively more remote, more 'other' and 'separate' with age, even while students are supposedly intellectualizing it as something in which they are embedded.

Moreover, curricular division, which I have discussed above as driving a fragmented representation of environment *per se*, also separates environment and society. The 'geography' strands within the *5–14 Environmental Studies Guidelines* are divided into 'human', 'physical', and 'human–physical inter-actions' (LTS 2000: 32). Even such 'interactions' strands are dualized: 'describe how technological activity can affect the needs of *people and the environment*' (LTS 2000: 71; emphasis added). Division into post-14 subject curricula also splits humans and the natural environment. Geography includes the environment, modern studies does not. Geography maintains 'The physical environment' and 'The human environment' division, and 'interdependence' is a 'concept' in the advice (SQA 2001i: 15) but it is not assessed. The 'physical environment' at its most interdependent, in relation to humans, is a resource, a constraint, or a subject of conflict. Contemporary social studies separates out the environment in assessment, and one exam paper asks about the separate merits of genetically modified crops for 'the environment', 'the farmer', and 'the consumer' (SQA 2001d: 2–3).

This latter kind of separation seems so ontologically obvious that the above argument contrasting it with a more interdependent representation of human–environment relations could be dismissed as linguistic hair-splitting. However, some kinds of environmentalism (notably the variant of 'deep ecology' discussed by Naess (1973)) adopt a gestalt-based ontology in which subjects are *constituted by* their relation with surrounding wholes rather than as entities in themselves. This is an extreme *ideal-type 2* position that is so 'interdependent' that 'interdependence' loses its meaning. Such an ideal representation must be considered at least potentially legitimate among advocates of 'students understanding interdependence', and so separation of 'people' or 'farmer' and 'environment' in curriculum text matters.

However, does such an analysis of human–environment relations in the curriculum with such an extreme position get us anywhere; a position that seems to dissolve words like 'human' and 'environment' and obviate 'inter-dependence'; a position that Bookchin (e.g. 1981: 154–156) calls the 'environmental kitsch' of 'spectral oneness', and Payne (1999: 376) 'one armorphous blob'? I think it does. For me it has been an analytical tool. In retreating from it—in denying its credibility, in wanting to resolve the world into components again—there seem to be numerous places where boundaries might be drawn. The curriculum studied here persistently draws a boundary around humans.

The next example both emphasizes this and perhaps offers an alternative—a boundary around 'life'—that has the merit, for this argument, of being more obviously arbitrary, even in our conventional ontology. In this case then, *ideal-type 2* decentres us enough, I hope, to see a pattern in the curriculum that might otherwise be viewed less critically. Even to discuss representations of 'human–environment relations' is to adopt a representation, in company with the curriculum itself.

The idea of shallow causal analysis (discussed in the sections above) has an impact on the representation of a relational interdependence between society and environment also. The expected depth of analysis again limits the representation of such interdependence:

> Water speed and pollution are two environmental factors which can affect where animals are found. Give *one* other environmental factor. (SQA 2001g: 16; emphasis in original)

The anticipated 'environmental factors' include 'light, temperature, pH of water, light temperature [sic], weather (*not* hunting, predators)' (SQA 2001b: 6; emphasis in original). Direct interrelation of species (predators', including humans (hunting)) is therefore *not* an 'environmental factor', even though (human) pollution *is* an environmental factor. This is a matter of causal analytical depth—if we take a single analytical step, the environment can be constructed as the immediate life-free milieu surrounding life. If we took another step, that milieu (pollution) is caused by life, we make pollution part of 'environment' while its cause is not, and we create a duality between humans and environment, possibly between life and environment. Similarly, the geography official curricular text refers to 'Natural disasters' (e.g. SQA 2002c: 16), implying a superficial analysis of causal foundations and a separate milieu that surrounds humans.

Where causal analysis does bridge the human–environment duality, it is around environmental 'issues'. The causality remains shallow; for example, the anticipated answers to the question 'What are the main causes of desertification?' (SQA 2002d: 12) are:

> One mark for each valid point. Two where developed. Maximum 1 mark for list. E.g. failure of rainfall (1[mark]) and high evaporation rate (1[mark]). Overgrazing by cattle/goats/sheep (1[mark]). Cultivation where rainfall low (1[mark]). Clearing of woodland areas (1[mark]) for firewood (1[mark]). Or any other valid point. (SQA 2002a: 5)

It is *sufficient*, then, for the candidate to make one causal link (rather than pursue why cultivation is taking places in areas of low rainfall, or why overgrazing is happening) even though deeper analysis might be credited as points that are 'developed'. In addition, the bridge is between the environment and 'human activity', abstracted from personal involvement (the progressive abstraction from 'environment' discussed in sections above also applies to the 'self' in curricular text (Ross and Munn 2007)).

Just as curricular division and shallow causality are, at least in part, responsible for the fragmentation of 'environment' discussed in previous sections, they also help fracture humans from environment here. Moreover, as I have shown, 'environment' becomes increasingly remote from students' direct experience as they progress through the curriculum. This, too, might suit the needs of assessment (which becomes more influential for older students) because the local and personally 'known' environment is less reducible to publicly accepted 'right answers' than the remote but shared environment. However, I also want to mention the possibility that the pattern might also draw on assumptions about child development. I will return to both of these possibilities in my discussion.

The value of the environment

The official curriculum declares that the environment should be valued, but in what ways? The ideal types ask whether it has only resource value to humans or whether there is some emotional or care-laden relationship, or even if it has its own inherent or intrinsic value.

Where the value of the environment is *explicit*, at all levels in the curriculum, I read it as mostly instrumental (its value is rooted in human use of it). In Standard Grade geography assessment, for example, environmental value is often described through resource potential or constraint on human activity:

> Study the course of the River Dee and its valley between [map references] 866010 and 929035. Describe how the *physical* features of the river and its valley have influenced land use. (SQA 2002b: 4; emphasis in original)

In geography the environment is also often presented as having monetary value in relation to agricultural production or tourism, and the income derived from these.

In science, too, human considerations are an important means of valuation:

> One advantage of recycling waste is that it *saves energy*. Give *two* other advantages of recycling waste. (SQA 2001g: 2; emphases in original)

Anticipated responses include 'any two from: *reduces* pollution; *saves* money; *saves* resources (1 mark each)' (SQA 2001b: 2; emphases in original). In contemporary social studies, the environment is sometimes represented with financial value, for example a pie chart showing how 'Scotland spends money on improving the environment' (SQA 2001c: 2). Technological studies, too, bring finance and environment into close relation (e.g. SQA 2001h: 13).

Despite the syllabus advice to 'develop informed attitudes and values relating to the care and conservation of the environment' (LTS 2000: 4), nowhere in the *5–14 Environmental Studies Guidelines* outcomes or Standard Grade assessments is any intrinsic, spiritual, or emotional valuation of the environment explicitly offered. (In some cases, potentially non-anthropocentric values (such as 'pollution') may be interpreted as being bad *per se*.) Asked for advantages and disadvantages of changes in farming since 1950, the anticipated answers include 'wildlife has suffered' and 'chemicals ... can be bad for wildlife' (SQA 2001a: 4), but these are relatively rare. My claim is that, where explicitly addressing the value of environment, the curricular text representation is *dominantly* instrumental.

The implicit in the text also rewards consideration here. First, the representation of environment as the sum of individual 'parts', and progressive study from the local to the remote, discussed in previous sections, has an effect on the value of those parts. 'The' meander bend in my local river becomes 'a' meander bend ('any' meander bend), one of a homogenous set. Implicitly, then, its value is reduced by its substitutability. However, the implicit in the text also allows *the possibility* that environment as a whole might have intrinsic/inherent value:

> for a selected land-use change or industrial process, describe possible effects, good and bad, on the landscape/environment, e.g. tropical forest clearance, improvement of derelict land for leisure use. (LTS 2000: 33)

> explain how living things and the environment can be protected and give examples. (LTS 2000: 56)

'Intrinsic value' is half-avoided, however, in even these *5–14 Environmental Studies Guidelines* attainment targets by not focusing on *why* protection is needed, or what constitutes 'good and bad'. I hope to show below that this is not just a point of absence, but of imbalance.

Conservation is a revealing case. Standard Grade geography's aims include 'to develop a concern for the conserving and developing of the earth's surface for the welfare and happiness of its inhabitants' (SQA 2001i: 10) and contemporary social studies is concerned with the environmental consequences of human use of resources and with conservation. The Standard Grade science advice defines conservation as:

> a way of meeting our basic needs without destroying the environment. Examples of conservation (e.g. replanting areas after timber extraction; maintaining hedges and trees around fields; protecting endangered species). (SQA 2001j: 34)

Again, although it is clear that we should meet needs 'without' destroying the environment, there is nothing here that says *why* 'not destroying the environment' is important. And this, I argue, reveals an inequality in the *explicitness* of rationale. The question of value—the *why* question—is treated differently on the resources side ('meeting our basic needs') than it is on the other side (e.g. 'protecting endangered species'). 'Provide X without destroying Y' appears to relate the value of X and Y but it doesn't really. 'Meeting our basic needs' is so essentially, self-evidently, self-justifying, that 'protecting endangered species' needs some kind of equivalent explicit justification, especially since the oppositional phrasing of the definition suggests that an un-destroyed environment is not one of our 'basic needs'. Instead, its justification is left hanging. The hanging might imply the possibility that the environment has intrinsic value. If so, this is not explicit. At the very least, there are 'other inhabitants' (which the geography syllabus mentions and which might be non-human) that could have basic needs too. Why not declare that 'Conservation is a way of meeting our basic needs and the basic needs of other living things'? This would be a move towards a *type 2* representation of inherent value.

The above point is difficult to make: in official curriculum text, the environment might have intrinsic value, as humans similarly have intrinsic value, but we are not similarly forced to address the possibility. It is this imbalance, between the solidity of our intrinsic value and the flickering absence of other intrinsic value that is revealing. It would be reasonable for the reader to fill the void with the pervasive, explicit representation that the value of environment lies in its use to humans.

Discussion

Much of the preceding sections may seem like an over-analysis of words with various meanings, like 'environment' and 'natural', or of texts over which there was much less hand-wringing in production than there is in this analysis, or of curricular division when any text must be divided in some way (if only into sentences or pages or lessons). Moreover, the representation that I have outlined from the text here would certainly be undermined in school lessons (where geographers would explore the deeper causality of 'natural disaster' or desertification, and scientists would distinguish between 'the environment' and 'environment [-al factors],' both would acknowledge that 'when examiners use this phrase they mean …', and everyone might question the value of protecting endangered species).

However, this does not alter the fact that these texts are as they are, and that they must have *some* influence. If, in writing them, we choose to rely on ambiguity or open-ended interpretation, then we are choosing not to value precision. We are also choosing to divide the curriculum in one way rather than another. At best, we are choosing to accept that the resulting representation of human–environment relations does not matter, or that the dominant interpretation of it is an acceptable one, even if others are possible. This might be a subconscious cultural production, but if the official curriculum was 'all just words' then changing them would not be the high-stakes business it usually is. The pattern here, I believe, is neither random nor unintended.

That pattern in the above sections can be summarized as follows. Natural and social science official curricular text for 5–16-year-olds in Scotland:

- either represents 'environment' as background context falling between curricular divisions and imperatives such as skills-development; or represents it in the foreground as a divided collection of 'parts', with generalized models of interrelationship, and shallow causality;
- represents 'human–environment interrelations' as dualized, again through fragmentation and shallow causality, and the environment as progressively 'other' (as students get older and/or 'more able'); and
- either explicitly represents the value of environment in dominantly instrumental terms; or represents its value implicitly as having a potential inherent value that need not be entertained.

I want to discuss in wider terms the roles of curricular division, assessment, and developmental models in driving this pattern and, in turn, their roots in Scottish curricular development. My aim is to build a hypothesis that the development of socially inclusive education has led to an officially or publicly 'shared' representation of environment, and that this representation is damaging to our relationship with environment.

The development of a publicly-shared representation

In practice the structure of the official curriculum divisions studied here tends to be dominated by the Standard Grade examination subjects, even if the official curriculum for pupils younger than this is, in theory, a very differ-ent matter (Carr *et al.* 2006: 14–15). The subject structure of this influential Standard Grade curriculum (Consultative Committee on the Curriculum 1977) ultimately owes itself to the work of Hirst (e.g. 1969). Hirst asserted: that the ultimate objectives of education concerned the progressive develop-ment of the rational mind (Hirst 1969: 47), for its own sake; that further-more '[t]he concepts on which our knowledge is built form distinctive networks of relationships' (p. 51) (which are referred to as 'distinctive ways of knowing and interpreting evidence' by the Consultative Committee on the Curriculum (1977: 23); and that it is in these distinctive kinds of think-ing that the student should be educated, if not necessarily in individual subject disciplines or through the 'ruthless butchery' of specialization (Hirst 1969: 157). It is nonetheless the case that subject disciplines were in fact used to apply Hirst's 'domains of thinking' (Consultative Committee on the Curriculum 1977: 33) and Carr *et al.* (2006) argue that Hirst has been applied too instrumentally.

There are strong theoretical and historical relationships between the assessment regime and curricular division (Committee to Review Assessment 1977: 30). Since the late 1960s, more inclusive and comprehensive approaches to education led assessment away from norm-referenced approaches (that tended to be about the identification of elites) to criterion-based approaches (that were more suited to inclusive education). For crite-rion-referencing to be secure, well-defined curricular domains are required (Simpson 2006: 17), so the development of criteria-referenced assessment (Standard Grade was an early example of its use), essentially in order to reduce social inequality, reinforced the value of the kind of curricular divi-sions discussed above. Since such subject knowledge was previously available only to elites, its spread to a wider set of social groups was seen as socially inclusive.

It was also convenient in the face of practicality and conservatism (Simpson 2006: 30–32). The reproduction of power relations in curricular division is well theorized (e.g. Bernstein 1971). And at a practical level the Scottish curriculum, whatever its fundamental bases were or will be, has been enacted through 'traditional school subjects and specializations within the teaching profession which form a UK-wide, and indeed, an international constituency' (Simpson 2006: 26). That is, interested actors in the educa-tional field actively maintain the official curriculum.

What all of the above adds up to includes a set of processes through which a publicly-shared representation of the world is devised and promul-gated. It is this 'public' representation that I think is important. To be able to name the parts (such as 'meander bend', 'canopy' or even 'pollution source') of any given 'environmental system' is to demonstrate acquisition of what is publicly shared, as is the ability to understand generalized process relationships (such as 'the glacier system' or 'a food web'). It makes no difference that the question is phrased 'for a local environment you have studied ...', because it is what comes next that is being publicly assessed ('name the *primary producer*'). The alternative would be to ask for the privately-known relationship between a student and his or her environment. The ability to explain single causal steps is also to demonstrate a grasp of finite, public, 'right answers', for the world is non-linearly complex in its causality, and further steps would quickly increase the level of contingency over the direction that the explanation could take. If we accept that there are myriad private journeys that a student might take through the causal world, we can only publicly share a limited range of setting-off points. Finally, formal testing becomes more important in the Scottish system with age, so perhaps students are likely to learn material that is more testable with age. It might be that Scottish students can be more readily assessed on their understanding of the publicly shared Brazilian rainforest (or other global issue) than they can on their privately known or local environment. This might explain the progressive distancing of environment from self and society that is part of the pattern in official curricular text.

There are other possible explanations. The official curriculum might be partly driven by ideas of child development, such that students are progressively expected to be able to handle the world more abstractly. However, this need not exclude parallel efforts to encounter it more personally—the dominance of abstraction in the text for older students merits challenge, even if abstraction *per se* is defensible. Or perhaps the official curriculum studied here is driven by a particular ontological and epistemological position—probably an idealized scientific one, of the search for the real, its categorization, its generalization and universality, even in the social subjects—so valorized as to suggest that the representation of self in environment discussed in this paper is, in fact, just how the world is. However, this is not to deny the promulgation of a publicly-shared world-view; it is just one way of justifying it. And Bonnett (2004: 88–90) argues that the kind of Hirstian influence described above has in fact given way in the face of, *inter alia*, a utilitarian approach to education, albeit with similar results: the promotion of the acquisition of 'abstract "skills" and information-processing strategies' in a fragmented context of pre-specified content suited to measurable outcomes.

We might ask 'so what?' Surely the development of a publicly shared world-view is a good thing, and its reproduction can be seen (as it has been in the history offered above) as a means of social inclusion. It may be rather less open to the charge of reproducing power relations than what went before it. What's wrong with it?

What's wrong with the publicly-shared representation?

Eco-feminists would argue that even Hirst's underlying philosophical divisions lead to a disconnection of our relations with environment, specifically a dualized culture/environment world in which the latter is instrumentalized and ultimately, therefore, subject to hegemonic and abusive power relations (Plumwood 2002: 17). It is the results of this abuse that pushed the environment up our cultural agenda and ironically (and possibly hypocritically) into our official curricular rhetoric. Plumwood (2002: 4) claims that reason, as used in western culture, is a dominating rational*ist* form based on Cartesian dualisms and that it serves to actively deny our corporeal embeddedness in environment, and therefore promote our instrumental use of it as an object. The primacy given by Hirst (1969) to the development of the rational mind in education, while allowing that other dimensions of human development might be acquired *en route*, reflects this priority of reason and is captured by the Consultative Committee on the Curriculum (1977: 21) in setting out the curriculum:

> the furtherance of social, vocational and environmental education is not incompatible with disciplined learning, since it is largely through such learning that these other educational needs are to be met.

In this sense, then, we are educating for a separated and instrumental understanding of the environment that is a sum of mechanical, substitutable parts. Its legacy is in the generalized models of interrelationship (such as 'energy pyramid') or in the progressive abstraction of 'environment' from what is locally 'related to' towards more distant ideals. The absence of located, let alone local, experience in official curriculum text may also in part be due to this priority for distinctive ways of thinking, overriding the more complex 'knowing' of one's own environment. It is for this reason, perhaps, that the environment is only represented holistically as a cluttered, messy, background context, in situations where the development and assessment of skill (of the more pure 'way of thinking') is in the foreground (see also Bonnett 2004: ch. 7).

Although the application of Hirst in official curriculum text cannot be laid at Hirst's door, Plumwood also highlights that the connection between ecological crisis and rationalism is not only the reification of rationalism itself but of the very divisions at the heart of Hirst's (1969: 151) epistemology. Our instrumental valuing of environment, and the denial of our embeddedness in it, can be seen as an expression of our perceived subject-object relation with it. These are supported by an idealized science of objectivity in which there is an absolute separation between the investigating or knowing 'subject', located in the mind, and the investigated (or known) 'object' located in the body and all environment beyond it. The result is a monological (as opposed to dialogical) relationship between student and studied, knower and known, human and environment. Such an ontological and epistemological ideal of science may or may not exist in the daily practice of scientific research (Ziman 2000: 155, Plumwood 2002: 42), but it is strongly embedded in the status, legitimization, cultural understanding, and school teaching of science (Bonnett 2004: 92). Plumwood (2002: 52) holds out hope for 'the humanities' as representing subject–subject enquiry, in which environment could be seen as possessing the kinds of characteristics that we possess ourselves (including, for example, inherent value).

However, this hope is ill-fated in the parts of the Scottish official curriculum studied here, for three reasons. First, this subject–subject approach is more true of the university study of social subjects than their school study where objectification is also central in the social subjects—even 'the self' is an object of study (Ross and Munn 2007). Secondly, some of the distinctive kinds of thinking that Hirst refers to (physical sciences, mathematics) are given higher priority in schooling, culture, and the curriculum than others (human sciences, morals, religion, and philosophy). And even if all this was not the case, Plumwood (2002: 50) makes the point that subject–subject forms of enquiry tend to be restricted to, or at least dominated by, dialogical relation with *other rational subjects*, principally other human ones. So subject–subject, dialogical understandings of our relations with environment fall between the stools of the sciences and the humanities at a deep epistemological level, regardless of where these disciplines are sited (and they are certainly set out in the school curriculum):

> It is not that the existence of multiple cultures of knowledge is itself problematic ... but rather that the way the field is partitioned dualistically into the particular gender—and nature—coded forms I have identified hides from us certain hybrid possibilities and inhibits the development of certain mixed forms that are crucial for an ethically integrated science and an ecologically integrated humanities knowledge field. (Plumwood 2002: 51)

The eco-feminist critique is sufficient to challenge the representation of environment in the curriculum discussed here, but Bonnett (2004; see also Bonnett 2007) has offered arguments rooted in similarly hegemonic terms (scientific objectification of nature is 'a powerful expression of its masterful instrumental motives' (p. 92)). Bonnett, too, is aiming to show that the kind of 'knowing' of nature needed for right 'dwelling' in it (as with Plumwood's 'hybrid possibilities') *includes* 'a direct, intimate, tacit knowledge that *affects* us' (p. 93; emphases in the original), drawing on the kind of embodied experience that both he and Plumwood argue is missing. Such a 'knowing' ('knowledge by acquaintance') does not exclude scientific approaches to nature, but neither does it privilege them. Orr (2004: 95), too, concludes that there has been a deep 'failure in the educational process to join intellect with affection and loyalty to the ecologies of particular places'. Finally, and crucially for our discussion here, Bonnett (2004: 93) notes that 'it follows that such knowledge may not be fully articulable because of its intense particularity and therefore non-generalizable features'.

Conclusions

In respect of 'environment', what should the official curriculum be aiming to achieve? This is much contested, and I have attempted to polarize some of that spectrum of contest in the ideal types used in this study. I have attempted to analyse in more detail how an outline story—that the representation of environment in extant curricula is dominated by one end of this spectrum—is realized in a specific curricular text. The text embodies a view of environment in which the environment is not a whole, but a sum of object parts, separate from society and self, and therefore most readily understood to be of instrumental value. I have tried to show quite precisely how this representation lies embedded in curriculum text (as background and between subjects; as a set of nameable parts or generalized models; as a set of shallow causal relations). And I have briefly attempted to show that what drives this representation is locatable. At least one interpretation lies in the official curriculum as a repository and reproducer of a public world-view and that the desire for such a world-view is driven by ideals of social equality. However, its public-ness is limiting: denying depth of enquiry, denying any enquiry of personal relation with environment, and denying alternatives. If we are to hope that the contested possibilities for representing environment in the curriculum might shift or be resolved, then perhaps exploring these kinds of curricular stories will create scope for the progressive, incremental and detailed curricular responses that are suited to processes of official curricular change.

Acknowledgements

I would like to thank various anonymous referees, George Meldrum, Morwenna Griffiths, Pamela Munn, and Justin Kenrick for their contributions to this project. Also the Scottish Qualifications Authority for permission to reproduce figure 1.

Notes

1. This study is part of a wider project (Ross and Munn 2007) and will focus largely on science, geography, and the related sections of Environmental Studies.
2. I needed to ask to see the examiner marking-instructions but all the rest of the source text is public domain.
3. Standard Grades remain the dominant examination syllabuses for Scottish 14–16-year-olds, although there are increasingly prevalent alternatives.
4. It is important to note that the above description is a general pattern in secondary schooling. There is a range of alternative qualifications that are growing in popularity (for an overview, see Scottish Qualifications and Credit Framework 2006, Scottish Qualifications Authority [SQA] 2006) and age-stage restrictions have been removed, allowing younger secondary pupils to sit exams earlier in their school career—an opportunity that some schools are taking advantage of for some of their pupils.
5. A further sub-division is also made as, for example, 'atmosphere' and 'earth-surface' processes are separated throughout.

References

Bernstein, B. (1971) On the classification and framing of educational knowledge. In M. F. D. Young (ed.), *Knowledge and Control: New Directions for the Sociology of Education* (London: Collier Macmillan), 47–69.

Bonnett, M. (2004) *Retrieving Nature: Education for a Post-Humanist Age* (Oxford: Blackwell).

Bonnett, M. (2007) Environmental education and the issue of nature. *Journal of Curriculum Studies,* 39(6), 707–721.

Bookchin, M. (1981) The concept of social ecology. *CoEvolution Quarterly,* 32 (Winter), 14–22.

Carr, D., Allison, P. and Meldrum, G. (2006) In search of excellence: towards a more coherent Scottish common school curriculum for the twenty-first century. *Scottish Educational Review,* 38(1), 13–24.

Chenhansa, S. and Schleppegrell, M. (1998) Linguistic features of middle school environmental education texts. *Environmental Education Research,* 4(1), 53–66.

Clark, M. E. (1991) Rethinking ecological and economic education: a gestalt shift. In R. Costanza (ed.), *Ecological Economics: The Science and Management of Sustainability* (New York: Columbia University Press), 400–415.

Committee to Review Assessment (Dunning Report) (1977) *Assessment for All. Report of the Committee to Review Assessment in the Third and Fourth Years of Secondary Education in Scotland* (Edinburgh: Her Majesty's Stationary Office).

Consultative Committee on the Curriculum (Munn Report) (1977) *The Structure of the Curriculum in the Third and Fourth Years of the Scottish Secondary School* (Edinburgh: Her Majesty's Stationary Office).

Cotton, D. R. E. (2006) Implementing curriculum guidance on environmental education: the importance of teachers' beliefs. *Journal of Curriculum Studies*, 38(1), 67–83.

Covert, D. C. (1969) Toward a curriculum in environmental education. *Journal of Environmental Education*, 1(1), 11–12.

Fien, J. and Trainer, T. (1993) Education for sustainability. In J. Fien (ed.), *Environmental Education: A Pathway to Sustainability* (Geelong, Australia: Deakin University Press), 11–23.

Giddens, A. (1971) *Capitalism and Modern Social Theory: An Analysis of the Writings of Marx, Durkheim and Max Weber* (Cambridge: Cambridge University Press).

Goodson, I. F. (1994) *Studying Curriculum: Cases and Methods* (Buckingham, UK: Open University Press).

Hirst, P. H. (1969) The logic of the curriculum. *Journal of Curriculum Studies*, 1(2), 142–158.

Huckle, J. (1996) Realizing sustainability in changing times. In J. Huckle and S. Sterling (eds), *Education for Sustainability* (London: Earthscan), 3–17.

Learning and Teaching Scotland (LTS) (2000) *Environmental Studies: Society, Science and Technology: 5–14 National Guidelines* (Dundee, Scotland: Learning and Teaching Scotland). Available online at: http://www.ltscotland.org.uk/5to14/guidelines/environmentalstudies.asp, accessed 8 May 2007.

MacGregor, J. (2003) Science education. In: T. G. K. Bryce and W. M. Humes (eds), *Scottish Education: Post-Devolution*, 2nd edn (Edinburgh: Edinburgh University Press), 606–616.

Moore, T. W. (1982) *Philosophy of Education: An Introduction* (London: Routledge & Kegan Paul).

Munn, P., Clark, G., Dargie, R., Meldrum, G. and Ross, H. (2004) *Education for Citizenship: Mapping the Social Subjects Curriculum: A Research Report to The Social Subjects Citizenship Liaison Group* (Dundee, UK: Learning and Teaching Scotland).

Naess, A. (1973) The shallow and the deep, long-range ecology movement: a summary. *Inquiry*, 16(1), 95–100.

O'Riordan, T. (2000) *Environmental Science for Environmental Management* (London: Prentice-Hall).

Orr, D. W. (2004) *Earth in Mind: On Education, Environment, and the Human Prospect* (London: Island Press).

Ozga, J. (2000) *Policy Research in Education Settings: Contested Terrain* (Buckingham, UK: Open University Press).

Palmer, J. A. (1998) *Environmental Education in the 21st Century: Theory, Practice, Progress and Promise* (London: RoutledgeFalmer).

Payne, P. (1999) The significance of experience in SLE research. *Environmental Education Research*, 5(4), 365–381.

Plumwood, V. (2002) *Environmental Culture: The Ecological Crisis of Reason* (Abingdon, UK: Routledge).

Pring, R. (2000) *Philosophy of Educational Research* (London: Continuum).

Ross, H. and Munn, P. (2007) Representing self-in-society: education for citizenship and the social subjects curriculum in Scotland. *Journal of Curriculum Studies*, in press.

Scott, W. and Gough, S. (2003) *Sustainable Development and Learning: Framing the Issues* (London: RoutledgeFalmer).

Scottish Executive (2004) *A Curriculum for Excellence: The Curriculum Review Group* (Edinburgh: Scottish Executive). Available online at: http://www.scotland.gov.uk/Publications/2004/11/20178/45862, accessed 8 May 2007.

Scottish Executive (2006) *A Curriculum for Excellence: Progress and Proposals* (Edinburgh: Scottish Executive). Available online at: http://www.scotland.gov.uk/Publications/2006/03/22090015/0, accessed 8 May 2007.

Scottish Qualifications and Credit Framework (2006) *SCQF Information Card 2006* Available online at: http://www.scqf.org.uk/downloads/SCQF%20Information%20Card%202006.pdf, accessed 8 May 2007.

Scottish Qualifications Authority (SQA) (2001a) *2001 Geography SG General: Finalised Marking Instructions* (Glasgow: Scottish Qualifications Authority).

Scottish Qualifications Authority (SQA) (2001b) *2001 Science SG Foundation: Finalised Marking Instructions* (Glasgow: Scottish Qualifications Authority).

Scottish Qualifications Authority (SQA) (2001c) *National Qualifications 2001: Contemporary Social Studies Standard Grade: Foundation Level* [0580/401](Glasgow: Scottish Qualifications Authority).

Scottish Qualifications Authority (SQA) (2001d) *National Qualifications 2001: Contemporary Social Studies Standard Grade: General Level* [0580/402] (Glasgow: Scottish Qualifications Authority).

Scottish Qualifications Authority (SQA) (2001e) *National Qualifications 2001: Geography Standard Grade: Credit Level* [1260/405] (Glasgow: Scottish Qualifications Authority).

Scottish Qualifications Authority (SQA) (2001f) *National Qualifications 2001: Science Standard Grade: Credit Level* [3700/403] (Glasgow: Scottish Qualifications Authority).

Scottish Qualifications Authority (SQA) (2001g) *National Qualifications 2001: Science Standard Grade: Foundation Level* [3700/401] (Glasgow: Scottish Qualifications Authority).

Scottish Qualifications Authority (SQA) (2001h) *National Qualifications 2001: Technological Studies Standard Grade: Foundation Level* [4020/401] (Glasgow, UK: Scottish Qualifications Authority).

Scottish Qualifications Authority (SQA) (2001i) *Scottish Certificate of Education: Standard Grade Revised Arrangements in Geography: Foundation, General and Credit Levels in and after 1999* (Glasgow: Scottish Qualifications Authority). Available online at: http://www.sqa.org.uk/files_ccc/geography_sg.pdf, accessed 8 May 2007.

Scottish Qualifications Authority (SQA) (2001j) *Scottish Certificate of Education: Standard Grade Revised Arrangements in Science. Foundation, General and Credit Levels in and after 1999* (Glasgow: Scottish Qualifications Authority). Available online at: http://www.sqa.org.uk/files/nq/science.pdf, accessed 8 May 2007.

Scottish Qualifications Authority (SQA) (2002a) *2002 Geography SG General: Finalised Marking Instructions* (Glasgow: Scottish Qualifications Authority).

Scottish Qualifications Authority (SQA) (2002b) *National Qualifications 2002: Geography Standard Grade: Credit Level* [1260/405] (Glasgow: Scottish Qualifications Authority).

Scottish Qualifications Authority (SQA) (2002c) *National Qualifications 2002: Geography Standard Grade: Foundation Level* [1260/401] (Glasgow: Scottish Qualifications Authority).

Scottish Qualifications Authority (SQA) (2002d) *National Qualifications 2002: Geography Standard Grade: General Level* [1260/403] (Glasgow: Scottish Qualifications Authority).

Scottish Qualifications Authority (SQA) (2006) *Conditions and Arrangements for National Qualifications 2006/2007*(Glasgow: Scottish Qualifications Authority). Available online at: http://www.sqa.org.uk/files_ccc/NQConditionsArrangements20062007.pdf, accessed 8 May 2007.

Simpson, M. (2006) *Assessment* (Edinburgh: Dunedin Academic Press).

Sterling, S. (1993) Environmental education and sustainability: a view from holistic ethics. In J. Fien (ed.), *Environmental Education: A Pathway to Sustainability* (Geelong, Australia: Deakin University Press), 69–98.

Sterling, S. (1996a) Developing strategy. In: J. Huckle and S. Sterling (eds), *Education for Sustainability* (London: Earthscan), 197–211.

Sterling, S. (1996b) Education in change. In: J. Huckle and S. Sterling (eds), *Education for Sustainability* (London: Earthscan), 18–39.

Strauss, A. and Corbin, J. (1998) *Basics of Qualitative Research: Techniques and Procedures for Developing Grounded Theory* (London: Sage).

Weber, M. (1949) *The Methodology of the Social sciences,* ed. and trans. E. A. Shils and H. A. Finch (Glencoe, IL: Free Press).

Ziman, J. (2000) *Real Science: What It Is and What It Means* (Cambridge: Cambridge University Press).

Environmental and health education viewed from an action-oriented perspective: a case from Denmark

BJARNE BRUUN JENSEN

The paper describes the action-competence approach used in environmental and heath education in Denmark. This approach implies students' genuine participation and actions, as well as interdisciplinarity. The concept of action is often described in vague terms with the implication that the action concept is ambiguous in educational practice and discussion. Here I distinguish 'action' from 'behavioural change' and 'activity'. I outline different forms of action and explore the issue of knowledge about environmental and health issues from an action-oriented perspective. Three case studies illustrate the approach.

In this paper, I present an action-oriented and interdisciplinary approach to environmental and health education,[1] although most of the examples and illustrations I will be presenting are from environmental education. First, I outline the challenges as schools work with environmental problems. I then introduce a number of key concepts and present three cases from the Danish context to illustrate the overall approach. The paper ends by addressing some of the overall conclusions and main challenges lying ahead.

My point of departure is that, if environmental problems are to be solved in the long run, teaching is needed that contributes to the development of students' abilities to influence local and global environmental problems, i.e. teaching should develop students' abilities to take action themselves, that is their *action-competence*. This educational challenge stands faced with two different trends in modern societies. Concern about environmental problems has never been greater; but, at the same time, there is an increasing action-paralysis when confronted with pervasive technological and social developments, a powerlessness that manifests itself in introverted, narcissistic activities. The most important task for environmental education in schools is in the area of tension between these two trends: starting with the views, concerns and anxieties of students and working systematically towards transforming the sense of powerlessness into the desire and ability to act. Passing on the contemporary action-paralysis to the coming generation must be averted!

Many claim that people in the western world are environmentally-conscious. Here, I will make a sharp distinction between concern and consciousness. Consciousness includes the sense that—over and above being concerned—one can contribute to solving the problems in question. The overall purpose of teaching towards environmental and health education can, thus, be characterized as up-grading—or qualifying—the pupils' anxieties to a '*real*' consciousness in the fields of the environment and health.

The heightened contemporary concern *and* action-paralysis around environmental issues derives from the status that scientific viewpoints have been given in our culture and ways of thought. This is also the case in health education where a vast amount of work has been done, often on a biological or scientific basis, to illustrate, and prove, that we have problems. Subsequent scientific investigations have sought to make plain how extensive these problems are. This scientific imperialism, which pervades our whole culture, bears a significant part of the blame for the anxiety and for the feeling of powerlessness. In the following, I will develop a framework for an alternative to this 'scientific' vision of environmental education: this alternative vision can be characterized by the keywords *interdisciplinarity*, *participation*, and *action-orientation*.

Interdisciplinarity

Concern, feelings of powerlessness, and individualization of the causes

In many countries in Europe, the natural science subjects of biology, chemistry, and physics have been responsible for environmental education in schools. In biology, efforts have been made to increase understanding of how levels of concentration of pollutants affect animal and human life and balances in ecosystems. Efforts have been made to provide predictions of the consequences of future increases of nitrates in surface waters, CO_2 in the atmosphere, the devastation of the rain forests, and the consumption of drinking water, etc. Laboratory testing illustrates and documents the seriousness of environmental problems, e.g. nitrates in water, acidity of rain water, oxygen content in lakes, pollutants in the air, and organisms in polluted water, etc. Such practical and theoretical work in the natural sciences has undoubtedly provided students with a sound and well-founded understanding of, and insight into, the effects of environmental problems on individuals and societies, and the natural world. And students have been provided with insight into the worsening of environmental conditions, which increases the seriousness of future effects.

In short, the studies have been concerned with the effect-levels, in theory and practice. The reason given for this approach is that, if students can acquire an understanding of how pollution is threatening human life and natural environments, they will become involved, and motivated to solve the problems. However, environmental problems are closely related to our technologies and societies, and revealing their complex causes must be a necessary pre-requisite for any contribution to their solution. This demands educational approaches which go beyond effect-level to include causes and actions.

In other words, an approach to environmental education based on the sciences can contribute towards creating concern, worry, and anxiety among students about environmental problems and environmental futures. However, there is a danger of not progressing towards the level of action. A science-oriented environmental education can contribute towards developing both worry and action-paralysis.

Thus, even when time and energy are spent measuring the oxygen content in a stream or measuring the acidity in rain in order to illustrate environmental problems, the result is knowledge of the existence and extent of problems, but *not* an increased competence to take action. Furthermore, the scientific approach results in the individualization of the causes of problems and, consequently, in an individualization of the responsibility for them. When, for example, the concern with energy problems and the concentration of pollutants is centred only on measuring exhaust gases from cars, students can be left with the belief that problems connected with increased energy consumption can be blamed on motorists—and not on the conditions in our society that have made the car an indispensable means of transport. And, when the focus is on connections between smoking and lung cancer, or between eating habits and heart diseases, the effects on human behaviour from the advertising campaigns of major industries are neglected.

When environmental and health education is exclusively based on the sciences, environmental and health problems are not placed within their cultural and economic reality. The implicit viewpoint, offered students, is that the blame and cause of the problems lies with the individual. The complex of 'causes' is not clarified, and the action level is never reached. More importantly, this dominant approach to environmental and health education provides a basically erroneous, individualistic conception of society. The challenge is to break environmental and health education out of the science framework. In this spirit, we can ask how much science is really needed in environmental education? I frame the issue this way not necessarily because little science is what is needed, but rather to ensure that perspectives from the humanities and social sciences are given attention, time, and energy in environmental and health education.

The interdisciplinary approach

The natural sciences have a role to play in describing the nature and extent of environmental problems. However, the humanities must also be drawn into the work as we consider desirable changes in the society of the future. The social sciences must be introduced to elucidate the whole spectrum of action-possibilities (collective and individual) that exist in a democratic society—and to clarify the barriers around these action-possibilities.

This interdisciplinary approach is illustrated by the 'eight dimensions' or steps, and the related IVAC model[2] (see table 1), that have emerged from our Danish development projects in the areas of environmental and health education (Jensen *et al.* 1995). We use the 'eight dimensions' to avoid the 'individualization trap' and to help us, and students, see environmental

Table 1. The IVAC-approach (investigations, visions, actions, and changes)

A. Investigation of a theme
- Why is this important to us?
- What is its significance to us/others?–now/in the future?
- What influence do life style and living conditions have?
- What influences are we exposed to and why?
- How were things before and why have they changed?

B. Development of visions
- What alternatives are imaginable?
- How are the conditions in other schools, countries, and cultures?
- What alternatives do we prefer and why?

C. Action and change
- What changes will bring us closer to the visions? Changes within ourselves? In the classroom? in society?
- What action possibilities exist for realizing these changes?
- What barriers might prevent the undertaking of these actions?
- What barriers might prevent actions from resulting in change?
- What actions will *we* initiate?
- How will we evaluate those actions?

problems as the structural and interdisciplinary problems they are. Work with the eight dimensions aims at developing students' action-competence:

(1) Which topic or theme should be worked on?
(2) Which problem within the topic in question should we work with?
(3) What are the causes of this problem?
(4) Why did it become a problem?
(5) What alternatives can we imagine?
(6) What action possibilities exist to secure these alternatives?
(7) What barriers will be brought to light through these actions?
(8) What actions will be initiated?

The first and second of these steps deal with the need to reach a common perception of what the problem actually is that I/we are working with. In other words, what conditions would I/we like to change, and what should be kept? It is not enough to simply decide that I/we want to work on a particular topic (for instance 'water'); we must go further to specify, for example, whether the issue is increased water-consumption or increased pollution of drinking water.

In connection with the second step, it must be made clear exactly why I/we feel that the actual conditions present a problem, and for whom they are a problem. These questions are important whether one works with conditions within a school, the local, or the global environment. 'Science' plays an important role in resolving these issues as a problem, and in their identification and description.

The third step deals with the need to reach a common understanding of the underlying causes of any problem that is selected as a focus of work.

These causes must be surfaced as completely as possible: if, for example, the problem is increased water-consumption, we must estimate the relative consumption on the part of households, agriculture, and industry. The factors influencing each sector's consumption must also be considered. The work must be systematic in order to clearly define the underlying causes. Even if the problem manifests itself in the classroom or the school (whether in relation to alcohol, bullying, or the quality of drinking water), the underlying causes will often turn out to be outside the school. Sociological approaches, whereby health and environmental problems are set in the economic, cultural, and social structures, and, thus, in the contexts in which they develop, are important here.

The fourth step deals with history. In order to reach an evaluation of how present-day conditions, or a 'development', can be influenced, it is important to understand what has contributed (over time) to the development. In short, an 'alteration perspective' makes it necessary to look at the conditions from a 'development perspective'.

The fifth step deals with the development of visions about how the conditions under which I/we have decided to work with, and would like to change, could look in the future. Students need to develop ideas, dreams, and perceptions about their future lives, and a view on the societies in which they will be living. It is here that imagination sprouts and a wealth of possible actions blossom as we seek to reach one or another of the visions that have been offered. All proposals must be brought to the discussion!

And, finally, the different actions must be explored in relation to their effects, and the barriers that might arise as we decide on the one or more actions that we might undertake.

I must note that, while this framework of 'eight dimensions' or steps follows a logical sequence, project work very rarely starts with step 1 and ends with step 8. Sometimes, a faster process is enacted wherein some steps are skipped—only for it to be discovered that the actions that have been initiated don't actually alter the situation very much. Is this due to not having been clear on what problems we are actually working with?—with the implication that we have to go back to step 2 and work on from here. Or do we have to go back to step 6 to develop and evaluate afresh possible actions?

Instead of looking at the eight dimensions as goals to be worked on in a specific sequence, the process can be seen as a spiral in which we keep going back to the different steps to explore them more deeply. Furthermore, it should be pointed out that students obtain insight and commitment best if they themselves experience the various dimensions by working on projects. This also means that the teacher becomes, to a large extent, a guide and consultant for students rather than a dispenser of cold, hard facts.[3]

The action-orientation

The concept of action is a consequence of the challenge of developing the action-competence I have been emphasizing.[4] There are many, very different reasons for this emphasis, but I will highlight only four of these reasons:

- The 'scientific' focus on giving students knowledge about the seriousness and extent of environmental problems has not been able to incorporate the social and societal perspectives involved in questions about action-possibilities, for society and for the individual.
- Moralizing, behaviour-modifying teaching never—or only very rarely—leads to the intended behavioural changes. This has brought about a new focus on 'student action'.
- The growing criticism that schools give priority to the 'academic' at the expense of the more practical has led to increased interest in the 'action-oriented'.
- Criticism of the schools' work with artificial 'as if' situations, e.g. role-playing, has led to demands for authenticity and for participation in the reality of society as part of teaching.

Put in this way, the 'reasons' for 'action-oriented teaching' are taken from completely different places and, as a consequence, the concept does not have an unequivocal content in educational practice and discussion. In what follows, I will venture a definition of the 'action' concept. I will in particular distinguish between 'action', 'behavioural change', and 'activity'.

Action vs behaviour change

In many contexts we encounter tendencies to equate actions with behavioural changes. It is often admitted that knowledge does not necessarily lead to 'action', i.e. 'changed behaviour' (see, for instance, Kollmuss and Agyeman (2002) for a review) and other approaches must be attempted. However, this admission does not necessarily lead to a reconsideration of the actual goal of behavioural modification. Instead, efforts are concentrated on research and development of 'new' and more efficient strategies for influencing students' behaviour.[5]

We can characterize all these approaches as efforts to influence students directly—outside, as it were, the 'knowledge component'. Such approaches do not seek to have students make up their own minds and decide on the behavioural change *they* might intend. The goal is to bring students to behavioural change in a previously determined direction—using nearly all of the available means.

This goal exactly defines the important difference between behavioural change and action, as well as the two fundamentally—or paradigmatically—different goals for environmental education: behaviour modification and action-competence. Before any *action* there will always be a conscious making up of one's mind; this is not necessarily the case with a behavioural change. The notion of participation is, so to speak, built into the action concept.

When a teacher in an interview, e.g. about environmental education, says that 'I do really try to change the pupils' behaviour', she is not seeking to have her students act, even if their behaviour changes (Jensen and Nielsen 1996: 122). And we can question whether such teaching can contribute to the development of her students' environmental action-competence. *Here it is the teacher who is the actor* (as she acts upon her students). The first element in the definition of action is that students themselves are involved in taking the decision to do something, whether it is a question of a change in their behaviour or an attempt to influence their broader environments.

Action vs activity

Another strong tendency in environmental education—often emerging as a reaction to the academically-oriented content—has practical activities incorporated in the teaching. These activities can consist of excursions to more or less untouched natural areas, of physical, chemical, and biological investigations of a polluted lake or stream, etc. In many places, such units, courses, and programmes are described as action-oriented. However, while such activities are valuable and productive to the extent that they help motivation and the acquisition of knowledge, to characterize them as actions they must be addressed towards solutions of the problem which is being worked with.

If, for example, work is being done with problems connected with agricultural fertilizer-consumption, investigating the amount of oxygen in a nearby lake cannot be characterized as an action, but as an *activity* (which, as I have suggested, can be valuable). An action-perspective around such an issue as fertilizer-consumption leads to work on, e.g. how, boycotting products from conventional agriculture might promote opportunities for products from ecological agriculture, and help solve the problems of nitrate pollution. In other words, an action must be targeted towards solutions of the problem that is being worked with. This is the second element in the definition of action.

Figure 1 summarizes these criteria for actions. The horizontal dimension marks the boundary between *behaviour* and *action* and, thus, the question of whether the students themselves decide to 'do something'. The vertical

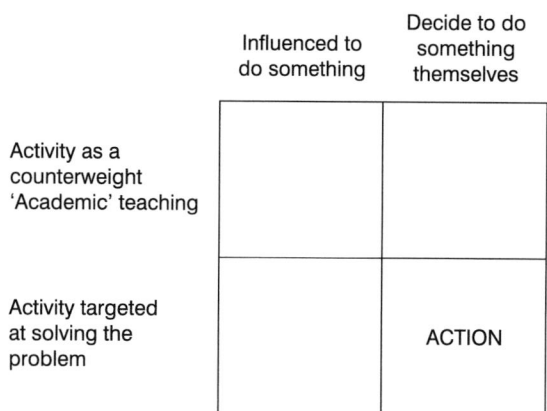

Figure 1. Defining the concept of action (from Jensen and Schnack 1997).

dimension marks the difference between *activity* and *action* and, thus, focuses on whether or not what 'is done' is addressed to a solution of the actual problem.

Different kinds of actions

There is a need to further differentiate and define the terms 'action' and 'activity' with a view to defining the concept of 'action' within its problem-solving aspect. For example, if students decide of their own accord to examine the degree of pollution of a stream and, as a result, initiate chemical and biological analyses, we can ask whether this is an action. In this case, the students made their own decision to carry out the analyses, and they are, thus, directed towards solving *their* problems by acquiring more knowledge about the extent and prevalence of pollution.

Thus, we must distinguish between actual *environmental actions* and *investigative actions*. To test the content of oxygen in water is a 'scientific investigative action'; to interview community members about their opinions on a potential environmental hazard is a 'social investigative action'. These activities can each be characterized as actions, but it is only those actions geared toward solving a specific environmental problem that can be described as 'environmental actions'.

Figure 2 illustrates the relation among the three concepts of *activity*, *action*, and *environmental action* (Jensen and Nielsen 1996). The most general concept is 'activity' in that an activity is defined as just 'doing something'. Some activities will also bear the character of actions—when students have an influence on what is to be done and why it should be done. All actions will also be activities, but all activities, however significant they might be from a pedagogical point of view, are not actions. Environmental actions not only aim at solving a problem but aim at solving the specific environmental problem being worked on.

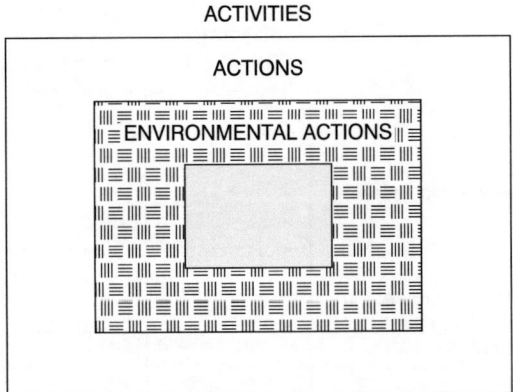

Figure 2. Relations between activities, actions, and environmental actions.

Different kinds of environmental actions

Environmental actions can also be grouped into two main classes:

- actions which directly contribute to solving an environmental problem, and
- actions whose purpose is to influence others to do something to contribute to solving an environmental problem, i.e. indirect environmental actions.

In other words, *indirect actions* involve 'people-to-people' relations; *direct actions* involve relations between people and their environment.

One example of a direct action could be a farmer who decides to halve his or her consumption of fertilizers; the laws, and taxes legislated by politicians in order to influence the farmer to do this or that are indirect actions. Politicians' actions can also be seen as a outcome of the indirect actions of public groups, i.e. their protest letters, demonstrations, lobbying, voting, etc. In other words, a direct action will typically be caused by a web of indirect actions.[6]

We can also distinguish individual and collective actions, which can be direct as well as indirect. 'Social investigative actions', such as interviews and the like, represent one class of indirect actions, and it could be argued that interviews can be environmental actions. When students decide to test farmers' knowledge and attitudes towards the environmental problems around agriculture with the help of questionnaires and interviews, these activities in themselves may influence the farmers. Such social investigative actions can be characterized as indirect environmental actions. In short, an action-oriented environmental education should be defined as an education in which working on developing students' environmental actions is an essential and integrated element.

Action-competence

I have argued that an action-perspective must be brought into environmental education. However, at the same time I should stress that too great a focus on an action-perspective and specific actions has its own problems. Thus, if actions deal only with the individual or school level, e.g. building a compost-heap for the use of the school, or turning out lights on leaving the classroom, there is the risk of presenting students with only an *individualistic* approach to environmental problems and their causes. The question is how do we ensure that a specific action contributes to the development of *understanding* of this or that environmental problem? If action-competence means that insight into the solution of environmental problems requires social and structural changes, major demands are placed on the teacher's ability to put individual actions, and their potential, into perspective—both locally and globally. Such teaching must ask: Which environmental problem does this action help to solve? Does a solution to this problem require many to act in the same way? Are there conditions that make many choose *not* to act in this way? What can be done to make it possible for more people to act?

And, finally, Are there other sources of the problem or conditions in society which are more important as we think about this problem?

The same need for perspective arises as we consider the projects that target their indirect actions toward e.g. politicians, companies, or other institutions within the local and global community. Such actions will often come across barriers in the form of, for example, insufficient response or no response at all. For such indirect actions to result in increased action-competence, demands must be made on teachers to put these barriers into perspective. If teachers lack the skills to deal with how to overcome or tackle barriers—such as non-response—teaching leads to incompetence and indifference among students.

Furthermore, it is important that a particular action not be viewed as an end-product of an environmental education project. Students must have the opportunity to evaluate, reflect, and restructure their actions—within their project and with their teachers—in order to develop their action-competence. Environmental actions must be integrated into the larger educational process.

To sum up, two conditions must be provided in order that 'actions' within environmental education programmes contribute to the development of action-competence. First, prior to an action, *students* must develop a critical starting point; and, second, actions must not only be directed at but also put into the perspective of the problem being worked on.

Acting by students in the field of the environment embraces issues that are different in principle. Students need to develop coherent knowledge about what the problems are, how they arose, and what possibilities there are for solving these problems. In addition, we need to promote students' sense of satisfaction and accomplishment, their commitment, and their drive. Knowledge about problems is not transformed into action if courage and commitment are not present; and commitment does not lead to actions without an associated insight into the problem. Put in another way, knowledge without commitment is empty and commitment without knowledge is blind.

The links between knowledge and commitment are complex. Teaching that is full of facts about how serious environmental problems are can appear onerous and labourious to the students, and risks destroying their commitment. Thus, we must ask which forms of knowledge contribute to the development of students' action-competence, and how should such a knowledge be acquired.

From action-paralysis to action-competence—a coherent landscape of knowledge

Working with students as active partners in an action-oriented environmental education does not make environmental 'content' superfluous. Instead, it has to be re-thought from an action perspective. The Danish psychologist, Larsen (1998: 22; my translation), argues for the necessity of:

> the professional, experienced teacher being in natural control of the substance. And what does that mean? That means that the content substance is controlled

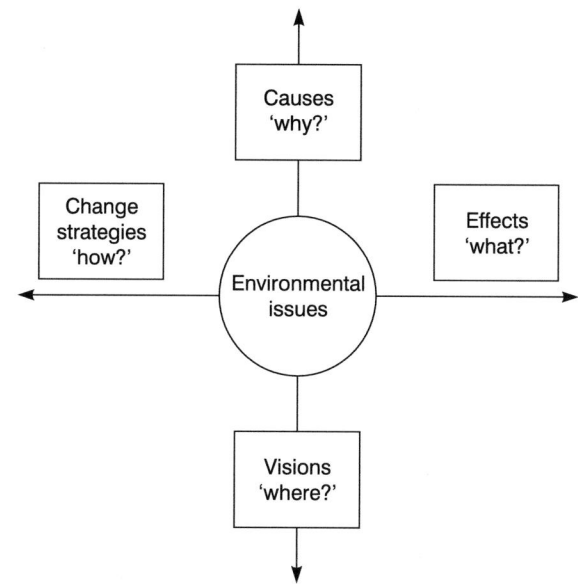

Figure 3. Four dimensions of action-oriented knowledge.

at a level such that it becomes an integral part of the teacher's personality. He does not need to use attention and resources on the professional side, but can concentrate all his energy on choreographing the educational process.

However, this leaves us with the question 'What should this "substance" contain?' In the following, I contend that an action-oriented substance involves interdisciplinary connections between environment, health, people, culture, and society. This point of departure has significant consequences for the kind of knowledge that will be the focus of planning, and implementing, and evaluation of the teaching and learning.

Four aspects of action-oriented knowledge are set out in figure 3 (Jensen 2000, 2002). These four dimensions become questions which illustrate different perspectives through which any environmental or health topic can be viewed and analysed.

What kind of problem is it?—Knowledge about effects

We must have knowledge about the existence and spread of health and environmental problems, about, e.g. the consequences of a behaviour such as drug-abuse or too much fat in a diet, acid rain or bad air-quality in a city or workplace. Such knowledge awakens concern and attention, and creates the starting point for a willingness to act. It is one of the pre-requisites for developing an action-competence. However, such knowledge is typically 'scientific' and, on its own, risks contributing to the development of concern on the part of students, but also an action-paralysis. It provides no explanation for why we have this or that problem, and no insight into how we can contribute to solving it.

Why do we have the problems we have?—Knowledge about root causes

We must deal with the 'cause' dimension of health and environmental problems, including the social factors behind behaviour. Why do so many people use private cars instead of public transport? Why is smoking more common in certain social groups? Which aspects of living conditions influence whether the use of alcohol leads to abuse? Why is unemployment associated with increased illness and greater risk of death? Such knowledge belongs mainly in the sociological, cultural, and economic areas.

How do we change things?—Knowledge about change strategies

We must deal with knowledge about both how to control one's own life and how to contribute to changing living conditions in society. How do we change the structures in, e.g. a school, a workplace, or a local community? Who do we turn to, and who could we ally ourselves with? Such knowledge also includes knowing how to encourage co-operation, how to analyse power relations, and so on. Such issues are explored in psychological, political, and sociological studies, and they are central to an action-oriented health and environmental education.

Where do we want to go?—Knowledge about alternatives and visions

We must deal with the necessity of developing one's own visions. Seeing real possibilities for forming and developing dreams and ideas for the future in relation to one's own life, work, family, and society, and having the support and surplus energy to realize them, is a pre-requisite to any motivation and ability to act and change. Such visions include knowing how people go about things in other cultures and other places: knowledge about other possibilities is a source of inspiration for personal visions.

Traditional environmental information finds its place in the first of these questions (see figure 3): it offers knowledge of the effects associated with environmental and health problems. However, we must also include both causal analyses and understandings of the ways to produce changes around health and the environment—and this is particularly important at a time when increasing globalization and individualization is leading to action-paralysis. In Joergensen's (1999:3; my translation) words, it seems as if:

> we have lost our eye for noticing that certain problems arise and appear more frequently, and how this can be related to cultural and societal factors—and that the solution of such factors should, therefore, be found in taking a starting point in how we live our lives and organize our society, which could be potentially different. The point is of course not that it should be simple and straightforward to agree on carrying out greater societal changes—but that we think and behave as if we live in the only possible world and that this can hardly be different, which is clearly wrong.

The model presented in figure 3 has been used in several environmental and health education projects in Denmark, and both students and teachers have used it to analyse their work. For example, a 9th-grade student commented on how using the model influenced her thinking processes:

> The material which I had collected was given more structure. Now I was able to sort the material based on the four areas in the model. And this 'forced' me to not only look at the effects and causes in relation to my subject—CO_2 pollution and the greenhouse effect. I was also forced to deal with alternatives and possibilities for change. The model is also good when you're 'stuck in a rut'. The model shows other ways of moving along. (Schmidt 1999: 12; my translation)

In other words, a participatory, action-oriented health and/or environmental education is not without basic knowledge and insight; on the contrary, it demands a new 'landscape' of extensive and coherent knowledge and insight. This creates important demands and challenges for teachers: they should be in a position to fulfil the consultant role *and* from their own experience and talent perceive today's health and environmental conditions from an inter-subject and action-oriented point of view.

Furthermore, students' personal influence on the whole process maintains and improves commitment and ownership (Jensen 2000). The dialectical process between the individual and his or her world, in which action-competence is developed, finds its optimal conditions when the person is actively involved in authentic problem-solving situations. The principles of student participation as well as authenticity are, therefore, important pre-requisites for the development of students' action-competence around environmental issues. Let me offer three examples from the Danish context to illustrate how the approach works in practice.

Cases

Traffic conditions around the school

This project was carried out as part of an environmental education initiative involving all students and teachers in three schools (Jensen *et al.* 1995, Jensen 2002, Uzzell *et al.* 1997). In a preparatory phase lasting for 8 months, teachers and some other key people from the community participated in various in-service activities. The project itself lasted for 6 full days (of which 3 were consecutive), with follow-ups covering a number of weeks. The agreed-upon criteria for the projects were:

- The environmental problems and the proposed actions must be chosen jointly by the students, teachers, and other involved adults.
- The environmental problem to be selected must provide scope for a local solution.
- All projects must involve concrete actions on the part of students as integrated elements of the learning process.
- The changes in the community should be permanent and continue after the school's initiatives come to an end.

- All projects must involve the development of new human relationships, i.e. social capital, in the community as a consequence of the project.
- All projects must involve the strengthening of insight, commitment, and visions, i.e. action-competence, on the part of the participating teachers and students.
- All projects must include highly-involved participants to support and facilitate the students. These facilitators (municipal civil servants, teachers, and other adults) should participate in the development prior to the students' projects in order to develop a shared language, including a common understanding of the processes and aims of environmental education.
- All projects must involve political and budgetry support from the school boards and the municipality.

The guidelines also saw teachers, students, parents, and other community members working and acting together.

As the beginning phase of the actual project, the students participated in both various brainstorming exercises and walking-tours in the local area in order to identify the environmental issues they would be working with. Subsequently, and in accordance with their interests, they were placed in groups of 15–20, each with one or two teacher-facilitators, to explore a set of broad topics. These groups, which contained students from different age-groups (7–14 years old), further specified the environmental problems they wanted to address. The majority of the groups chose to work on traffic conditions in the neighbourhood.[7]

During the project, students were involved both in measuring traffic intensity, speeds, etc., and in discussions of various mechanisms by which both traffic and its speed might be diminished. They developed their own visions with regard to collective transport, bicycle paths, etc., in the local area. One of the groups decided to organize a demonstration as part of their work:[8] they erected a transportable homemade zebra-crossing where they felt that zebra-crossings should be established; they made signs indicating speed limits and other reminders to motorists. They also gathered signatures from 160 local inhabitants on a petition supporting their proposals. The demonstration ended with the students arriving at the town hall, where their petition for a re-organization of the traffic patterns was acknowledged by the municipality.

Although the idea and the topic for this project originated with the students, their teachers, police, and community members as well as the students were responsible for the actions that were undertaken. Later, several other groups worked on the traffic problems and, as a consequence, a number of concrete changes appeared in the months after the project.[9]

The evaluation of the project addressed the various changes that emerged from the project as a whole. These changes can be divided into three groups.

First, there were the changes that came about as a result of individual groups' *direct actions* to attempt to change environmental conditions, e.g. the setting-up of compost containers and various embellishment-like changes. Each of these changes could be traced back to the actions of one group.

Secondly, there were a number of *indirect actions*, whereby the groups approached the appropriate authorities (i.e. politicians, the city government, etc.) with the aim of getting them to act, or to widen, the action-space for community members, by, for example, establishing public places, a children's village board, playgrounds, etc.. These changes also came from single groups.

Finally, there were changes concerning traffic. These changes could not be traced back to a single group's actions; rather, the many different projects around traffic-patterns contributed to intensifying the attention given to the issue, and the debate, at all levels within the municipality. Moreover, there were a number of changes already being discussed when the project was initiated. However, there is no doubt that the local groups' projects and actions contributed to the changes, in particular by hastening both the decision-making and the content of the decisions. The set of *indirect actions* formed a web that contributed to the effort towards the same solutions.

The evaluation study involved interviews and questionnaires—to students, teachers, and key leaders from the community. The findings were very positive about the students' experiences, and in favour of the approach. Many of the students spoke warmly of the authentic and action-oriented aspects of the project (Jensen *et al.* 1995: 89):

> When you're living in your own area, you begin to realize there are problems. Then you are more likely to go to the authorities, go to people's houses, and get signatures. And you don't give up just because they say 'But I'm sorry, it just can't be done'. You just keep on.

> Now I just don't worry; now I'll phone and complain . . . if there are any traffic problems again. I've learned a lot from this. Like giving them a kick in the backside.

The teachers considered the project to be very demanding, but at the same time many concluded that the approach was superior to traditional teaching. In the words of the teacher planning-group (Jensen *et al.* 1995: 118),

> It is our opinion that the majority of children have experienced that the school can function as catalyst in the local community. They have experienced being part of a process in which, during the projects, ideas became a reality. We hope such projects will flourish and maybe spread like rings on water with the contributions of parents, children, and other citizens.

Strengthening social capital in the local community

In this project, grades 1–6 (5–11 years-old) students in a small town worked for one full day on developing their ideas around the question 'What makes me feel good?'. At the end of the day, the classes presented their ideas in posters. The students from grades 2 and 3 decided, with their teachers, to continue their work in the following week. During this week, they worked with the more focused question 'What could make me feel better at school?'

In this work, the classes used the IVAC approach, implying that they should address the following perspectives:

- *investigation* of what 'feeling good' is for them;
- the development of their own *visions* about how they would like the school to change; and
- the initiation of concrete *actions* in order to facilitate the necessary health-promoting *changes*.

The students discussed a number of ideas about improving their school and, after 2 days in which they presented and discussed their ideas, they finally reached an agreement. They would work for (1) more lessons on food and cooking, and (2) more lessons related to physical exercise and movement. They then went through a 'who-has-the-power' exercise in which they invited the head to clarify how they could develop a strategy to reach their goals. The head explained that the person in charge of the lessons in schools was the city's director of education, and showed them organizational pictures to illustrate the municipal organization and the distribution of power.

The head also tried to convince the students to include cooking and working with food in their normal Danish language lessons. However, the students did not accept this suggestion and decided to approach the director of education with their proposals. They prepared an interview schedule, and one of their main questions was 'If we want extra lessons in home economics and physical exercise, how would you advise us to proceed?' His response gave them a number of ideas about how they might reach their vision.

After the interview they worked hard—in collaboration with their teacher—to draw up a formal proposal to submit to the director of education. A few weeks later, they received a response from the municipality allocating resources for two extra lessons per week in home economics during the first half-year and two lessons in physical education during the second half-year.

The following year, the teachers decided to carry out a project using the same basic approach. This time the students—now 9–10 years old—worked with the question 'What will make our community better?' Again the IVAC approach was used as a guiding framework.

The students developed the vision that they wanted their community to be livelier and provide more leisure activities. During the investigation phase, for instance, they had discovered that, in the past, a circus had visited the town at least once a year. In addition, they had discovered that there had once been many small shops and various social activities in the town, but that all this had gone.

The teachers introduced them to the ideas of networks and social capital (although in other terms), which the students discussed from many different angles and perspectives. Finally, they agreed to try to initiate actions with the aim of improving the community's 'richness' and 'liveliness'. One idea was to 'get the circus back in town!' Another was to establish a play area for children and young people. A third was to plan a charity run for all citizens in the area aiming at, among other things, raising money for their planned activities.

The students soon realized that they needed to address the local politicians if they wanted to make progress with their ideas. They then contacted other adults, including representatives from different sports associations, to present their ideas. Jointly, the adults and the children then initiated a public meeting where the students themselves introduced their ideas. Some of the politicians who attended the meeting tried to convince the students that the play area should be located at a remote site on the outskirts of the community, a site that was unsuitable for other uses. However, the students argued their case convincingly, and were supported by their parents and other adults. In the end, the municipal council agreed to establish a 500 square metre playground in a central location and set aside funds so that a circus could perform in the local community on Constitution Day the following year.

This project can be evaluated using two different sets of criteria: first, in terms of the changes the project facilitated; and second—and perhaps more important—in terms of the development of the students' empowerment and action-competence. It was clear that the project did make a difference in the school as well as in the local community. In the school setting, students succeeded in getting the lessons they had asked for. In the local community, students were actually able to bring about changes that improved the social capital of all citizens. As we evaluated the consequences of the programme in terms of the students' empowerment and action-competence, we saw them expressing a strong commitment to future actions in influencing and developing the community.

One of the teachers involved with this project said:[10]

> It surprised us, as teachers, that the students were so clear about what they wanted to get out of the projects ... When we prepared the project we discussed what to do if we ended up with 26 students who didn't say anything and didn't have any ideas. However, we were impressed by the students' ideas and their commitment, and with the fact that they were able to present such qualified arguments. (Lund 2000: 224; my translation)

Thus, one of the interesting findings of the evaluation was that even the so-called 'marginalized' or 'weak' students derived great benefit from participating in the projects. This is perhaps due to the fact that they—for once—had a genuine opportunity to influence the agenda (the topics to work with, the actions to undertake, etc.). In the words of one of the teachers,

> Even those students who are usually hesitant about participating in class have been very good at arguing for their views and ideas. At the same time they have gained more respect from the other students in the class during the project. (Lund 2000: 226; my translation)

Students examine the attitudes of their parents in relation to drinking

The third project took place at one of the schools in the Danish Network of Health-promoting Schools (Jensen 1997). Students from a 7th-grade class (12–13 years old) decided to work with the issue of alcohol, something that was beginning to interest more and more of the students, and, thus, their parents. Inspired by the IVAC approach, they worked with both the examination and the vision phases, and then in the action phase where they investigated structures of power, actions, and strategy.

In the investigation phase, they analysed alcohol from biological, sociological, cultural, and historical perspectives. From materials developed by the Danish National Board for Health, they learned about alcohol-related diseases and accidents. They also discovered that the consumption of alcohol is dependent on economic conditions and, among other thing, analysed the consumption of alcohol during the last century, making comparisons with the economic development, unemployment rates, etc. By way of interviews with people from different countries living in the local area, they explored how alcohol is used differently within cultures and religions. The students also worked with the factors, i.e. peer-pressure, advertising, loneliness, unhappiness, social events, which might influence their own consumption of alcohol, using role-playing to clarify the mechanisms behind these causes.

Their vision centred on creating a society, and a school, where there was balanced and responsible behaviour in relation to drinking, where alcohol could be viewed as a positive contribution to the culture, and where no one harmed themselves or others as a consequence of alcohol-consumption. They discussed what such a society would look like, how to make sure that all citizens (children as well as adults) would have equal opportunities for stimulating and exiting lives, and what a society (and a school and a family) would look like if no one was left out or felt lonesome.

As one of their concrete actions, the students decided to present some advice to their parents about how they—according to the students—should act towards their own children in connection with alcohol. According to the students, their parents had misinformed and wrong ideas about even their own children. They, therefore, decided to offer seven pieces of advice to their parents:

- Young people have views on alcohol.
- Have faith in young people, and in your own children.
- It is not your problem if other people's children drink.
- Allow your children to go to parties; nothing will happen.
- Allow your children to taste alcohol.
- Allow your children to have pre-parties at home.
- Allow children to drink only low-alcohol drinks.

Of course, the content of these seven statements can be questioned (and was!), but this was not the point. Instead, the point was to illustrate how students' ideas might be used as a stimulating and fruitful starting-point for a general discussion involving the students, parents, and the teacher. The students' tangible action consisted in the planning of a parent evening to present their work, their ideas and advice, and to discuss the whole topic. All of the parents attended the meeting. The students presented their findings about alcohol, drinking and health statistics, etc. and, as a conclusion, introduced the seven suggestions to their parents.

The students developed the seven suggestions as a starting point for discussion in small student–parent groups. They had formed the groups beforehand to make sure that parents were not in the same group as their children. Following the group-work, it was the parents' turn to comment on the students' presentation and a stimulating discussion resulted.

After this evening event, the parents indicated that they had learned a lot about their children's attitude to alcohol, and several parents indicated that they had altered their perception, both on the basis of the students' presentation and on the other parents' comments. The evening did not end with specific agreement on 'when, how, and how much', but with a much greater mutual understanding.

Final remarks and conclusions

Although the cases deal with different topics, were carried out in different settings, and were evaluated in very different ways, they shared the same basic educational approach—action-orientation, participation, and interdisciplinarity. A few general conclusions emerge.

The school and students—under certain conditions—can act as catalysts for environmental and health changes on many levels: in the community, in the school, in students' own lives, and in their families.

The development of students' action-competence and empowerment benefits from working with authentic problems from the base of their participation and actions. The key word is *ownership* as a pre-condition for any actions and changes. In order to stimulate the development of ownership, students' genuine participation in selecting the topics to be addressed, in the development of visions, and of the actions to be carried out is crucial.

Students' concrete actions must be dealt with as integrated elements of teaching (and not as end-products). Students acquire action-competence by taking action and trying to influence 'real life' as part of their education.

The genuine participation of students does not imply that teachers play a passive role. Teachers need to take responsibility and assume an active role in projects; passive teachers who wait for pupils to consult them wreck the process and demotivate students. The core element is genuine dialogue between teachers and students, with the teachers playing the crucially-important role of supervisor and facilitator—asking provocative questions, coming up with suggestions and ideas for action strategies, putting barriers in perspective, pointing out possible collaboration-partners, etc.. The challenge is to find a balance between involving students as active partners who are taken seriously and the partnership of the teachers as they play an important role in the dialogue and the process. Without feedback, the students cannot develop their own attitudes and understandings.

Teachers need high-quality skills and pedagogical competencies to support, stimulate, and challenge students and their ideas. Many teachers are not used to this approach; although they find it stimulating and useful for students' learning, they also find it very demanding. Consequently, there is a need for developing flexible educational models and materials, like the IVAC approach, to provide a focus for teachers' work in projects of this kind. In addition, in the three cases described above, the teachers were supported by an external consultant acting as a 'critical friend'. This specific form of in-service activity, which took place within the school context, was a necessary complement to adequate materials, models, etc.[11]

A genuine commitment and a positive approach on the part of local government stakeholders are also crucial. The key players from the local community have to take students and their interests seriously.[12]

Notes

1. This 'action-competence' approach has been developed in relation to a number of projects within the areas of environmental and health education at the Research Programme for Environmental Education, the Danish University of Education (Jensen 1997, 2002, Jensen and Schnack 1997, Schnack 2003).

2. The IVAC (Investigation, Visions, Actions, and Change) approach (Jensen 1997) is a simplified model of the eight steps. The IVAC approach has been integrated into the national health education curriculum in Denmark, and a number of schools have taken it up.

3. This does not, however, mean that demands on teachers become lessened, or that the teacher's professional knowledge should play a less important role. The opposite is true: The teacher should both be in a position to fulfil the role of consultant and draw on his or her own experience and talent to perceive environmental and health problems from an inter-disciplinary point of view.

4. In practice, there is a strong trend towards an action-perspective—or work in an 'action-oriented way'—in both environmental and health education in many Danish schools.

5. The preparation of smart 'fashionable advertisements' about the 'right' behaviour and using the person of the teacher more consciously as a role model can be cited as examples of this.

6. *Direct environmental actions* could include, e.g. sorting garbage, constructing compost heaps, economizing on water- and energy-consumption, etc. In the health area, it could be changes in behaviour related to eating, drinking, smoking, etc. *Indirect environmental actions* could include, e.g. the preparation and distribution of a newsletter concerning the environment, writing letters to politicians and companies or developing editorials for local papers, organizing debates on environmental conditions, etc. in the health area, developing rules about how to avoid bullying in the school, preparing a healthy meal, etc.

7. It came as a surprise to many teachers that traffic was chosen as the first priority by many students. They had assumed that the students would select garbage as *the* issue because it has been a priority issue in the municipality for a number of years. In fact, a number of teachers had prepared materials on garbage prior to the actual project period—these materials were not used in the project.

8. A demonstration involving 150 students against traffic in the town's main street was an example of the broad co-operation, involving teachers, students, parents, the police, and politicians, with each group pulling its weight concerning actions. As part of their participation in the project, the police had carried out speed tests and given 27 fines to motorists who had been driving too fast the day before the demonstration.

9. These included an intensified debate in the local media on traffic involving local politicians and citizens; the city council's earmarking of DKK1m. (~ US$150 000) for the re-organization of traffic in one local area (roundabout, etc.); the establishment of a flash-and zebra-crossing by one of the schools; the reduction of the speed limit to 30 mph at another school; speed-reduction measures on the town's main shopping street; and the planting of trees along the bicycle paths between two neighbourhoods.

10. The data for the evaluations were obtained using a number of different methods: observation, interviews with teachers and students, analysis of documents (e.g. meeting announcements, posters, recordings of student interviews about the project on local radio, etc.).

11. There are many examples of external partners approaching schools in order to use students as a means to reach *their* pre-determined goals. Such an attitude to the role of the school in the community runs counter to the approach advocated here. Schools and students should be viewed, and approached, as partners and resources in their own right.

12. However, if the school and the local community are to profit from such collaboration, traditional in-service training activities are out of step with this educational approach. Key persons from the school and community together must come together at in-service courses to develop a shared language and common concepts, and to reach a mutual understanding of the overall aim(s) of any collaboration between the school and the community. The traffic project illustrates very well the potentials in such an inter-'sector' approach.

References

JENSEN, B. B. (1997) A case of two paradigms within health education. *Health Education Research*, 12 (4), 419–428.

JENSEN, B. B. (2000) Participation, commitment and knowledge as components of students' action competence. In B. B. JENSEN, K. SCHNACK, and V. SIMOVSKA (eds), *Critical Environmental and Health Education—Research Issues and Challenges* (Copenhagen: Danish University of Education), 219–238.

JENSEN, B. B. (2002) Knowledge, action and pro-environmental behaviour. *Environmental Education Research*, 8 (3), 325–334.

JENSEN, B. B. and NIELSEN, K. (1996) Pupils' activities, actions and action competence. In S. BREITING and K. NIELSEN (eds), *Environmental Education Research in the Nordic Countries* (Copenhagen: The Royal Danish School of Educational Studies), 120–143.

JENSEN, B. B. and SCHNACK, K. (1997) The action competence approach in environmental education. *Environmental Education Research*, 3 (2), 163–178.

JENSEN, B. B., KOEFOED, J., UHRENHOLDT, G. and VOGNSEN, C. (1995) *Environmental Education in Denmark—The Jaegerspris Project* (Copenhagen: The Royal Danish School of Educational Studies).

JOERGENSEN, C. R. (1999) Globaliseringens identitet [Identity of globalization]. *Politiken*, 16 May, 3.

KOLMUSS, A. and AGYEMAN, J. (2002) Mind the gap: why do people act environmentally and what are the barriers to pro-environmental behaviour? *Environmental Education Research*, 8 (3), 239–260.

LARSEN, S. (1998) *Den ultimative formel for effective læreprocesser* [The ultimate formula for efficient teaching processes] (Hellerup, Denmark: Steen Larsen).

LUND, J. H. (2000) Hvad gør dig glad?—et projekt om handling og sundhedsundervisning [What makes you feel good?—a project on actions and health education]. In B. B. JENSEN (ed.), *Handling, læring og forandring—beretninger fra Den Sundhedsfremmende Skole* [Action, learning and change—experiences from the Health-promoting School] (Copenhagen: Danish Committee for Health Promotion and Education), 210–227.

SCHMIDT, S. E. (1999) Erfaringer med at bruge det udvidede faglighedsbegreb [Experiences using the extended concept of professionalism]. *Newsletter for the Danish Network of Health-Promoting Schools*, 3 (Copenhagen: The Royal Danish School of Educational Studies).

SCHNACK, K. (2003) Action competence as an educational ideal. In D. TRUEIT, W. E. DOLL, H. WANG, and W. F. PINAR (eds), *The Internationalization of Curriculum Studies* (New York: Peter Lang Publishing), 271–291.

UZZELL, D., DAVALLON, J., FONTES, P. F., GOTTESDINER, H., JENSEN, B. B., KOFOED, J., UHRENHOLDT, G. and VOGNSEN, C. (1997) *Children as Catalysts of Environmental Change* (Lisbon: Ministerio de Ambiente, Instituto de PromoÁao Ambiental).

Implementing curriculum guidance on environmental education: the importance of teachers' beliefs

D. R. E. COTTON

Many observers have commented on disparities between the theoretical understandings of environmental education portrayed in academic literature and the environmental education that takes place in schools. In much of the literature and in curriculum documents there has been an increasing emphasis on promoting positive attitudes towards the environment, and the results of several surveys suggest that many teachers support this aim. This paper explores the beliefs of three geography teachers teaching controversial environmental issues in UK secondary schools. In contrast to the findings of prior studies, the teachers in this study feel strongly that they should try to avoid influencing students' attitudes, or imposing any kind of pro-environmental agenda. There is a substantial divergence between the teachers' beliefs and the espoused aims of much environmental education literature and the geography syllabus they were following. This suggests that, unless curriculum developers take account of teachers' beliefs in designing new curriculum materials, those materials are unlikely to be implemented in their intended format.

Environmental education has been increasingly incorporated into the UK school curriculum, and has aroused substantial research interest. Several journals now exist which are devoted entirely to this subject; a large number of contributors over recent years have analysed different forms of environmental education, and advocates of particular approaches have provided cogent arguments in support of their favoured form (e.g. Huckle 1985, Fien 1993, Tilbury 1995, Aldrich-Moodie and Kwong 1997). In this debate, there has been a growing emphasis on 'education *for* the environment': socially critical environmental education which includes 'an overt agenda of values education and social change' (Fien 1993: 16). Proponents of education *for* the environment contend that teaching can never be a value-free activity, and that the attitudes generally promoted are those of the dominant culture and of powerful groups in society. The role of socially critical environmental education is, therefore, to challenge the dominant ideology, and put forward an alternative world-view, promoting both personal and structural transformation (Fien 1993).

Many aspects of education *for* the environment have been adopted by curriculum developers (e.g. Naish *et al.* 1987, Curriculum Council for Wales 1992, 1993), including the National Curriculum guidance document on environmental education (National Curriculum Council [NCC] 1990).[1] At the time this paper was written, environmental education was a statutory, i.e. required, cross-curricular theme, and the accompanying curriculum guidance for teachers stated that:

> [Teachers should be] promoting positive attitudes towards the environment by encouraging: appreciation of and concern for the environment; independence of thought; respect for others' beliefs ... (NCC 1990:6)

It is possible to criticize this statement on several grounds, not least its use to teachers as guidance. There is little evidence available about appropriate strategies for encouraging 'appreciation of and concern for the environment'. Many studies, particularly in the USA, have considered the impact of educational interventions on students' environmental attitudes (e.g. Gross and Pizzini 1979, Yount and Horton 1992, Showers and Shrigley 1995, Uzzell 1999), but these studies have frequently produced contradictory results, and largely failed to draw any substantial conclusions. In a review of the relevant literature, Rickinson (2001) suggests that there is evidence that some interventions can lead to attitude change, at least in the short-term, but little is known about the characteristics of successful programmes. Moreover, Elliott (1993) maintains that attempting to influence students' attitudes and behaviour is of little educational value, and is likely to prove limiting and counter-productive.

Even if it were possible to identify an agreed set of desirable environmental attitudes, the NCC document does not make clear *how* teachers are to promote certain specified attitudes, while at the same time encouraging their students to develop 'independence of thought'. Various critics have argued strongly that the role of education should be to encourage independent thought, not to promote a specific world-view (e.g. Jickling 1992), and that the teacher should impart knowledge rather than attempt to act as an agent of change (Aldrich-Moodie and Kwong 1997). However, despite these criticisms, in much of the environmental education literature there is a growing expectation that teaching environmental education should be about changing attitudes, or even engaging students in taking action on environmental problems (Breiting and Mogensen 1999, Posch 1999). As Scott and Oulton (1999) note, the consequence of this exclusive focus on education *for* the environment has been that alternative forms of environmental education have been dismissed as being irrelevant or even damaging, thereby excluding much of the work currently taking place in schools.

This restricted focus of much of the environmental education literature may go some way to explaining what Grace and Sharp (2000: 333) describe as the 'rhetoric-reality gap'. The differences between environmental education as advocated by many theorists and the environmental education which takes place in schools have been widely recognized and discussed (Lee 1993, Walker 1997, Grace and Sharp 2000). Various studies have identified constraints on implementing environmental education in schools including lack of time, lack of unbiased resources, lack of school support, and lack of staff expertise and motivation (Tomlins and Froud 1994, Ballantyne 1999, Lee 2000). Grace and Sharp (2000) also found that student teachers trying to implement an approach to environmental education as advocated in the literature faced sustained opposition from the teachers in the schools in which they were working. Walker's (1997) research in Australia suggested that the requirements for implementing socially critical environmental education—particularly in terms of taking action on environmental problems—were simply too great for many teachers to take on board.

Despite these constraints, there is some research evidence that teachers are strongly in favour of the more affective side of environmental education, with 'personal responsibility for the environment' being rated as the most essential aspect in two separate surveys (Tomlins and Froud 1994: 17, Grace and Sharp 2000: 337), and 'future attitudes to the environment' also being rated as very important in the latter survey. However, other research has provided more ambiguous results. Ballantyne's (1999) international survey reported a similar stated commitment to promoting environmental attitudes and values among geography teachers and teacher educators, but found that UK teachers were least committed to promoting a particular environmental ethic (though the majority still apparently favoured this over a neutral approach). That survey also showed that investigation of local controversial issues, and teaching strategies for exploring attitudes and values, were used only occasionally by the teachers. Lee (1993), in a comparative study of England and Hong Kong, also noted espoused support for teaching attitudes of concern for the environment in both countries, but very little support for those teaching strategies that might enable teachers to achieve this aim. These findings may suggest that teachers are keen to promote positive attitudes towards the environment in their teaching but are limited in their delivery of such aims by constraints on time and resources. However, they may indicate that teachers' commitment to promoting environmental attitudes is less wholehearted than the surveys initially suggested.

The study

Methods

Given the ambiguities in the environmental education literature, it is possible that teachers' pedagogical and environmental beliefs are more important in guiding their teaching about controversial environmental issues than have previously been recognized. Much existing research in this area has involved either large-scale surveys of teachers' attitudes (e.g. Lee 1993, Tomlins and Froud 1994, Ballantyne 1999, Grace and Sharp 2000), students' attitudes (e.g. Richmond and Morgan 1977, Hausbeck et al. 1992, Scott and Willits 1994, Morris and Schagen 1995), or classroom interventions that attempt to change students' attitudes (e.g. Gross and Pizzini 1979, Yount and Horton 1992, Showers and Shrigley 1995, Uzzell 1999). These studies have provided little insight into teachers' beliefs and normal classroom behaviours. This study, therefore, aimed to provide an alternative perspective by producing a detailed account of teachers' beliefs and practices in relation to teaching controversial environmental issues, collecting extensive data, and undertaking a detailed analysis of those data in relation to the literature. The aspect of this research reported herein concerns the beliefs and aims of three experienced geography teachers, and the ways in which they approached teaching about controversial environmental issues, taking into account the realities of the classroom and the practical constraints under which they work.

The teaching of controversial issues was chosen to provide a suitable context for eliciting teachers' beliefs about the role of environmental attitudes in the curriculum (see table 1 for some examples of issues studied). Many environmental issues are controversial at least in part because of the differing attitudes and values held by interest groups, and geography is considered by some to be the ideal place for environmental education because of its long-standing focus on incorporating different attitudes and values (Bailey 1974, Naish *et al.* 1987). Despite the brief attempt to introduce environmental education as cross-curricular theme (NCC 1990), the majority of environmental education in UK schools is still taught in geography and science disciplines (Tomlins and Froud 1994, Grace and Sharp 2000). Furthermore, the 'Geography (16–19 Project) Syllabus' which these teachers were following explicitly recommended, as part of its approach, that the course should 'enable candidates to acquire ... an attitude of concern for the quality of environments, for the condition of human life, and for the biosphere as a life support system' (University of London Examinations and Assessment Council 1996:3). This syllabus and subject matter would seem to provide plentiful opportunities for investigating teachers' beliefs and practices regarding teaching controversial environmental issues.

The findings presented in this paper are based on detailed interviews undertaken with three experienced geography teachers, Mary, Sam, and Chris,[2] in three English secondary schools as part of the research study described above. The three cases were studied sequentially over the course of 2 years, and involved spending a total of 5–6 weeks at each school. (Table 1 provides some brief biographical details about the teachers involved.) A series of semi-structured interviews based around classroom observations, following the style of Cooper and McIntyre (1996), were undertaken with each teacher, using an interview framework which ensured that the research

Table 1.   Details of research participants, and controversial issue being taught.

	Teacher 1: Mary	Teacher 2: Sam	Teacher 3: Chris
Position in school	Head of Geography	Geography teacher	Geography teacher
Years teaching	24	4	25
Gender	Female	Female	Male
School size	2000	1000	1000
School situation	Rural	Urban	Rural
Year group studied	12	13	12
Example of controversial issue	Indigenous people's land rights in the rainforest	The role of NGOs in governing Antarctica	Reconciling the needs of conservation and tourism in National Parks

agenda was covered but also allowed the interviewee to be expansive and to lead the interview to a certain extent. The interviews covered a range of topics including specific lesson-events (aided by stimulated recall using sections of the lesson which had been recorded on audio-tape), and more general questions. By grounding interviews, at least in part, in shared classroom events, it was hoped that the discussion would focus on the reality of teaching about controversial environmental issues, rather than upon an ideal view held by the teacher, i.e. 'theory-in-use' rather than 'espoused theory' (Argyris and Schön 1974). The final interview with each teacher covered a broader range of topics, including explicit discussion of the research focus, in order to probe the teachers' beliefs in greater depth. Selected students from each class were also interviewed, and the data were used to provide an alternative account of classroom activities and to add depth to the analysis. All interviews were undertaken by the same person, tape-recorded, and transcribed.

It was considered particularly important to interpret the meaning of events in terms of the teachers' own beliefs and aims, rather than any preconceived ideas about how the topic should be taught, especially in the light of the literature that contains a large number of prescriptions about what teachers *should* be aiming for in their teaching about environmental issues. However, while the aim of the interviews was to provide an understanding of how each teacher viewed teaching about environmental issues, the final goal was a cross-case comparison that attempted to identify commonalities among the three accounts. In this sense, these were 'instrumental' case studies—they had an aim over and beyond that of understanding each particular teacher (Stake 1995). The case-study approach was chosen on the basis of its strong grounding in reality, accessibility to teachers, and the ability to generate a rich, detailed account. Generalization, in this study, takes the form of 'theoretical inference' (Hammersley 1998), in which the conclusions move beyond the claims made about the individual cases to a more general, theoretical level that is potentially of wider interest. Data arising from the individual cases in this study are, therefore, used to theorize about the possibilities and problems of teaching controversial environmental issues in contemporary schooling. Any theoretical understanding thus produced should be considered provisional in nature and would benefit from further investigation.

Analysis

Data collected during this study were analysed, using a detailed framework of principles and procedures (Fido 1999). Two over-arching principles guided the analysis:

- *Principle 1*: Data analysis and interpretation should start from, and be based upon, the teacher's own understanding of events.
- *Principle 2*: All propositions and outcomes should be supported by the evidence, to the extent that there are no negative cases.

The approach chosen for the first stage of data analysis was that of analytic induction, involving the development and testing of propositional statements (Silverman 1993). This procedure involves examining small sections of data to see if they fit the initial hypothesis or proposition, then reformulating the hypothesis to enable the inclusion of all relevant data. Using the techniques of constant comparison and negative-case analysis, analytic induction 'offers a powerful tool through which to overcome the danger of purely "anecdotal field" research' (Silverman 1993: 170). In this stage, each case study was analysed in turn to develop a set of propositional statements about that teacher's beliefs (i.e. 'This teacher believes that ...'). These propositions represented second-order constructs (Schütz 1973) that sought to encapsulate each teacher's beliefs about teaching controversial environmental issues.

The second stage of analysis involved a comparison of the 'beliefs propositions' of the three teachers in order to identify commonalities among them. This involved reading and re-reading the 'beliefs propositions' in conjunction with each other, to identify commonalities in wording or meaning among the three teachers.[3] The rationale for the cross-case comparison was that the results of this study would be of most benefit if they had the potential to be generalized to other teachers. This suggests that 'the analysis of the data must attempt to establish generalizations by seeking common concepts across teachers' (Brown and McIntyre 1993:50). The areas of concern identified by the analysis as being common to all three teachers were then considered in greater depth, by re-examining the interview data in detail to investigate the teachers' understandings of the key concepts and to develop an account of the teachers' beliefs (Fido 1999).

Findings

The study revealed a remarkable degree of similarity among the beliefs of the three geography teachers—and a very substantial divergence from the aims of environmental education discussed earlier. Data from all three cases suggest that one of the teachers' main aims in their teaching was to provide a 'balanced' picture of environmental issues by ensuring that students were exposed to a range of different viewpoints, and that the teachers' own views were not imposed on the students. All of the teachers made great efforts to provide access to a variety of viewpoints in their lessons by means of their selection of resources, their use of role-plays, and their management of class discussions. It was clear that these teachers were not in favour of promoting particular 'attitudes of concern for the environment', as the syllabus and much of the environmental education literature suggests:

> I think they're quite capable of thinking for themselves ... you can't have independence of thought if you then want to determine and define their attitudes. So, I think I would go for the independence of thought first, and if you are going to really respect that and you mean that, then you can't say that you're going to promote positive attitudes. (Mary, third interview)

However, despite their apparent agreement about the teacher's role, the picture which emerged from detailed analysis of the interview data is that the teachers' beliefs about balance were highly complex, and in some ways problematic in terms of their potential for translation into practice in the classroom.

While the teachers were agreed that balance was desirable, they differed in their reasons for valuing this approach, and their methods of implementation. Mary appeared to view her teaching as providing an alternative to widely-held views about the environment, in one example aiming to 'debunk the myth that ... indigenous people always know best about the forest'. This teacher was critical of the simplistic ideas that can be transmitted through some textbooks, especially in the lower secondary school, and was keen to provide a balanced view that delved more into the complexity of the issues. As well as explaining this issue to the students herself, the teacher also selected textbooks which covered the issue from a different angle:

> They've obviously been so taken aback by this idea because if you read ... the standard A-level texts for 16–19, you quite happily go away saying, 'Oh we'll leave the forest to indigenous peoples'. Then you read this thing about Papua-New Guinea and ... you start revising your view. And I think that's perhaps why I put that in at that point, to make them stop and think. (Mary, third interview)

Mary perceived that students held strong views about environmental issues prior to studying the subject; therefore, she thought that to implement a balanced approach she needed to challenge the views they held already, taking on a form of 'devil's advocate' strategy (Stradling *et al.* 1984). While Sam and Chris used this strategy less, they also noted that students should sometimes be pushed to extend their ideas about environmental issues, and to reconsider their initial views. So, although they did not want to impose their own views, they acknowledged that exposing students to a diversity of views could include challenging students' current understanding about an issue.

Sam's concern to adopt a balanced approach appeared to be more strongly influenced by the threat of pressure from parents or school authorities. She was keen on using role plays, she strongly encouraged students to express their own opinions in class discussions, and she even took care to arrange her classroom displays to reflect a diversity of viewpoints. Sam's thinking about her classroom environment appeared to reflect a concern with how outsiders might view her teaching, and she noted that the response from other teachers, parents, or governors was likely to be highly critical if a one-sided approach were adopted:

> You could get into serious trouble ... with their parents because, like I say, 'It's not your job to say how they live. It's your job just to make them more aware and to give them ideas, and to give them knowledge of what is going on'. (Sam, third interview)

Sam's understanding of balance involved exposure to as wide a variety of different views as possible, using external resources, and classroom discussions to enable them to be aired. Her use of 'balance' appeared to function as a defensive measure to protect against possible charges of indoctrination, illustrating the politically-sensitive nature of such teaching.

Chris cited a different rationale again for the use of a balanced approach, placing greater emphasis on the need for students to be able to assess and evaluate differing viewpoints:

> What I specifically wanted them to do was to think how they could evaluate opinions numerically so that they could come up with some kind of evaluation ... (Chris, second interview)

From his perspective, it was important for students to be able to consider different points of view, and then come to a decision about 'the best way forward' based on some kind of objective evaluation of the arguments. His emphasis was much more on the requirements of the examination, and he saw analysis of different viewpoints as a useful skill for the students to acquire because it would be needed for the decision-making paper and their individual study, both central parts of the course.

It can be seen that, while the three teachers, at a superficial level, shared a conception of 'balance' as requiring the consideration of more than one viewpoint, the function that balance served in each of their teaching was distinctly different. However, despite these differences, they were all firmly agreed that they, as teachers, should not impose their own views on their students. They thought that students might be overly influenced if the teacher supported one particular point of view, and all three teachers were wary of even stating their own views in lessons, apparently because of an underlying fear of indoctrination:

> I think it's important to give a balanced view and I'm not out there to turn them into green, banner-waving, fundamentalist environmentalists. ... They've got to come to it themselves, I don't think we should be imposing our views on them. I don't think that's my role. (Mary, second interview)

> That's not your job! I don't think you should be telling them what to think. I think because you are a teacher then you may have some influence on them, and therefore you shouldn't be using it to say, 'This is what I think. Do it this way'. (Sam, second interview)

> I don't want to prejudge the situation. I want them to come to their own viewpoints. And if they've exchanged their viewpoints, I might well say 'Well that's interesting, now my viewpoint is ...'. But no, I haven't stood up and banged a gong and said, 'This is what I firmly believe!'. (Chris, second interview)

All of the teachers thought that students should be encouraged to make their own decisions about controversial issues, and they believed that stating their own views might limit the range of responses that students thought would be appropriate. It was noted that some students seemed deliberately to express attitudes with which they thought the teacher would agree—and that others could be 'particularly reactionary', in an attempt to elicit a response!

However, the teachers attempted to avoid open disagreement with the students, irrespective of the attitude voiced:

> I'd never say, for example—well I hope I'd never say—well, you know, 'That's wrong!'. I'd say, 'Well, I don't have much sympathy', or 'I do not myself agree necessarily with what you're saying, we may beg to differ on that'. (Chris, second interview)

It seems that these teachers were extremely conscious of the uncertainty surrounding much environmental information. As Gough (1993) notes, it is especially important in environmental education to be aware that educators cannot give 'one true story', and this was reflected in the teachers' approaches to the subject. However, this was not an unproblematic stance in terms of student understanding and engagement with the issues. Several comments made by the students suggested that they were confused by the multitude of different opinions considered, and were unable to resolve these issues themselves. For example, several of the students mentioned an increased feeling of powerlessness in the face of their growing knowledge about the complexity of environmental problems:

> I think it's so difficult to form a total opinion from the amount of information we're given. I mean we're given a lot, but because it's so impartial either side, you know, there's an equal amount either side, it's really difficult to decide ... you see there are problems either way, and they can't, in the end, be really fixed. (School 2: Lara)

This echoes the findings of Connell *et al.* (1999), who found that a sense of pessimism, frustration, and action-paralysis prevailed among the young people they interviewed, and illustrates the potential for alienation if students are not provided with guidance on how such issues may be resolved.

As well as a concern to avoid expressing their own views in lessons, two teachers also found students with strong views problematic, and were suspicious of those they considered to have 'fixed' viewpoints and who would not consider other points of view:

> [The only difficulties] are the ones [i.e. individual students] that are completely intransigent and not really able to consider any other viewpoint other than their own. And that's sometimes very difficult to break down. And it's not often but every now and then you get somebody who says, 'Well I still think this should happen', and that's fair enough. The majority of people can see that there's this viewpoint and there's that viewpoint. (Chris, second interview)

Chris's students proved hard to manage in the light of the teacher's desire not to state his own view or express disagreement in any way. Both Sam and Chris suggested that their ideal role in lessons would be to do nothing, and let the students discuss the issues, especially in a role-play where it should be possible for the students themselves to present a range of opinions and discuss them together. Their aim was to enable the students to talk among themselves and express different opinions, arguing among themselves. The stance chosen by these teachers appears similar to that of the 'neutral teacher' advocated by the Humanities Curriculum Project (Schools Council/Nuffield Humanities Project 1970), and one of the teachers explicitly stated that he hoped he had 'demonstrated neutrality' in his previous lesson. This is a considerable step away from the environmental education literature in which teachers are encouraged to take a 'committed' approach to their teaching about the environment, in order to encourage students to take action on environmental issues (e.g. Fien 1999).

In practice, however, the role of the neutral teacher proved very difficult to maintain, especially when students expressed strong views that were not effectively challenged by others in the class, or if the teacher had strong opinions on the subject:

> I felt he was trying to say, 'Well, OK then, winner takes all', which is an attitude that I come across quite a lot in this school ... and I just thought, this is

so trite, this is so out of order, and that's how people have become disaffected. (Mary, first interview)

In this example, the teacher described her role as 'imposing balance', but she clearly noted that this student's view of the issue was not one which deserved serious attention, and that the 'prejudiced' views of some of the students should be challenged.

While this was an unusually strong expression of the teacher's view, the limits to all of the teachers' acceptance of the role of the neutral teacher are made clear as they refer to challenging students' initial views to enable a wider variety of arguments to be aired in the lesson and to promote empathy for other cultures or viewpoints. This was achieved by questioning students to encourage them to reconsider their position:

I would probably just keep questioning them, and questioning them ... and if they just kept coming back at me, I'd just be like, 'Well, OK, fair enough, what do the rest of you think?', and open it up to the group, you know. So they would at least be aware that other people felt differently. (Sam, third interview)

While sharing with it a desire to avoid expressing their own values, this view of the teacher's role as involving questioning or challenging students is substantially different from that of the neutral teacher.

The teachers' emphasis on using questioning to challenge students— rather than making explicit statements—appeared to be linked to other pedagogic beliefs, in particular about use of the 'enquiry approach', which is a prevailing theme in the Geography 16–19 syllabus. A key aspect of their understanding of the inquiry approach involved questioning students to encourage them to think about solutions to problems rather than simply providing all the answers:

Through inquiry ... I think there's a lot of value in what students find out for themselves rather than just being told by us. If we tell them too much, it just goes in one ear and out the other and they're sitting thinking about something else, they're not actively involved. (Mary, third interview)

The inquiry approach values *asking*, rather than *telling*, students about issues; therefore, on the face of it, it appears a useful tool with which to engage students in thinking about controversial environmental issues. However, although the teachers intended to allow students with differing views equal time, this strategy proved difficult to enact in the reality of the classroom, where debates could become quite heated:

Sometimes I get more involved than I should be and I need just to step back and to let them take over. It's their ideas, not my ideas, and I'm only prompt-ing discussion like I did with that question—I prompted a line of debate—it's not, for me, necessary to lead that. (Chris, second interview)

I think [Greenpeace] have got a very very important role to play, definitely. ... And I think perhaps that came out a bit, in the questioning as well. (Sam, second interview)

These teachers were conscious that in such situations it was very easy for questions to become 'biased' in some way, offering implicit or explicit support for one side of the argument.

Conclusions and discussion

The findings of this study reveal that these teachers' beliefs are at odds with much published discourse on environmental education. Although much of the literature (including the Geography 16–19 syllabus which the teachers were following) advocates promotion of positive attitudes towards the environment, this agenda is not shared by these teachers. Instead, the teachers aimed at offering a 'balanced' picture of controversial environmental issues, leaving students to develop their own attitudes based on the information provided. The concept of balance which they articulate is also somewhat problematic: Although the teachers all spoke in terms of 'balance', their stance of non-involvement is clearly influenced by the 'neutral teacher' role, and many of the problems they identify with this approach are similar to those observed by teachers implementing the Humanities Curriculum Project (Rudduck 1983). Furthermore, accounts of 'challenging students' views' are indicative of a type of 'devil's advocate' strategy, perhaps suggesting that the distinctions between the various roles are not as clear-cut as the literature on controversial issues may suggest.[4] However, notably, the one role the teachers do *not* support is that of the 'committed teacher'—despite the fact that this is arguably closest to the expressed aims of socially critical environmental education.

Socially critical theorists in environmental education, such as Fien (1999), have argued that the liberal ideology of balance and neutrality disregards the intrinsic value-orientation of the education system itself, and ignores ideological bias within the overt and hidden curriculum. In the wider critical pedagogy literature, theorists have argued that education is not a neutral enterprise, and neither, therefore, can teachers be neutral:

> There is ... always a *politics* of official knowledge, a politics that embodies conflict over what some regard as simply neutral descriptions of the world and what others regard as elite conceptions that empower some groups whilst disempowering others. (Apple 1996: 23; emphasis in original)

From this perspective, the idea of maintaining a neutral position is portrayed as an illusion, and these teachers' attempts at neutrality are, therefore, destined to fail. However, the alternatives—including 'committed impartiality' (Kelly 1986), which involves taking a committed stance while remaining open to alternative views, and avoiding imposing values on the students—are also problematic, given the teachers' concerns about influence and indoctrination.

There are a number of possible reasons why the teachers may not implement the syllabus quite as the curriculum developers intended. Carr and Kemmis (1986) discuss the importance of 'subjective' and 'objective' constraints on action:

> [T]hey recognize that there are 'objective' aspects of social situations which are beyond the power of some particular individuals to influence at a particular time and that to change the way people act it may be necessary to change the way these constraints limit their action. At the same time, they recognize that people's 'subjective' understandings of situations can also act as constraints on their action, and that these understandings can be changed. (Carr and Kemmis 1986: 183)

While the distinction between subjective and objective constraints is often not clear-cut, this analysis may be useful in considering the potential problems of attempting to implement a more radical environmental education in schools. Objective constraints, such as (in the UK) external examinations, parental pressure, and school structure may reduce the possibilities for introducing socially critical environmental education into the curriculum (it has long been argued that schools are a poor place for developing skills such as democratic action, because of the unequal power-relations enshrined in the teacher–pupil relationship).

Scott and Oulton (1999) note the limitations of the environmental education, which may be developed without major changes to the structure and functioning of the school system:

> [Teachers must] attempt to find their own path through a bewildering mixture of often contradictory instruction, guidance and advice, amid doubts about what approaches are effective and what purposes are appropriate, mostly operating within educational systems where school success continues to be measured in terms of traditional academic, rather than more-environmental, criteria. (Scott and Oulton 1999: 90)

In the wider literature, constructivist views of teaching (e.g. Doyle 1992) suggest that teachers' interpretation of the curriculum will also be constrained by what Doyle (1992) describes as the 'co-construction' of the 'classroom curriculum' by pupils and their teachers. Pupils' responses to the teachers' strategies will, therefore, help shape the nature of classroom interaction, and may constrain the teachers' ability to make radical changes.

Clearly, there are several constraints acting upon teachers which militate against a more radical agenda being pursued. However, what this study demonstrates is that subjective constraints are also at work, limiting the opportunities for change where these go beyond the traditional role of the teacher. Beliefs about tolerance of different viewpoints and avoiding imposition of values are deeply embedded in liberal educational thinking while the origins of environmental education are rooted in the political movement of environmentalism, a movement based on a desire for specific behavioural outcomes. Given the dominant status of liberal pedagogy among teachers, the current emphasis on critical environmental education can only increase the gap between what environmental education is supposed to achieve, and what either teachers or schools are willing to take on board. This echoes the reviews of many curricular innovations of the 1950s and 1960s (Sosniak et al. 1994), which suggest that success or failure of these innovations was influenced to a large extent by the compatibility between their content and the practical demands of classroom teaching. Sosniak et al. (1994) point out that these 'new' curricula presented views of the subject matter which did not match the teachers' views, required teaching strategies that were unfamiliar, and expected students to behave in ways inconsistent with usual practices. This may explain to some degree these teachers' rejection of the more radical aims of environmental education.

In a similar vein, Olson (1992) notes how what he terms 'visionary' projects' are difficult to integrate into schools because they conflict with the traditional functions of the school, and with the ideas and beliefs of at least some of the teachers. He also identifies tensions between the goals of the institution and the goals of the practice itself, which teachers must attempt to resolve. Olson recommends that innovators work in conjunction with teachers to develop new projects: 'The new practice is, thus, carefully mapped onto the actual working lives of teachers—neither the adequacy of the new ideas, nor the inadequacy of the old are assumed. Dialectically each is used to assist the other' (Olson 1992: 69).

Such an approach enables curriculum developers to take account of subjective and objective constraints on teachers' action, and moves away from the deficit view of teaching whereby the failure of developments is seen as a failing on the part of teachers to implement them successfully. In this context, consideration of the teachers' views would have revealed a lack of support for the transformational view of environmental education which underpins much of the writing in this field. The teachers' aims when teaching about environmental issues simply did not include 'fostering', 'reinforcing', or 'developing' students' environmental attitudes.

The findings of this study are in sharp contrast to much of the existing quantitative research evidence that suggests widespread agreement among teachers for promoting environmental attitudes (Lee 1993, Tomlins and Froud 1994, Ballantyne 1999, Grace and Sharp 2000). It is possible that these large-scale surveys failed to uncover the complexity of teachers' thinking about these issues—and there are hints in some of the studies that agreement was not as straightforward as the initial results suggested (Lee 1993, Ballantyne 1999). Certainly, the data from the study reported herein are derived much more directly from teachers talking about their own teaching of specific environmental issues in the classroom, and may, therefore, have come closer to accessing the thinking that guides their actual teaching, as opposed to their ideal views of teaching. Furthermore, the nature of the research leaves space for participants to define the subject in their own terms, rather than being constrained by pre-conceived ideas. This suggestion may go some way towards explaining the findings of previous surveys that, while teachers expressed a general commitment to promoting environmental attitudes at an abstract level, their reporting of actual teaching strategies used revealed much less strong support (Lee 1993, Ballantyne 1999). Similarly, previous research in mathematics education (Sosniak et al. 1994) has indicated that teachers' views vary, even to the extent of appearing internally inconsistent, depending upon how intimately connected the issues are with the everyday process of teaching and learning. The research of Sosniak et al. suggests that teachers' views shift towards a more conservative position, the closer the issue is to classroom teaching.

However, it could also be argued that the teachers in this study were simply a particularly conservative group—perhaps a different group of teachers would show stronger support for promoting environmental attitudes? Clearly, further research would be required to refute this hypothesis with absolute confidence. However, the fact that the views held by the teachers in this study converged to such a great degree on the key issue of promoting environmental attitudes suggests that they are unlikely to be isolated examples of extreme views. Moreover, the teachers were working in a context in which they were explicitly encouraged to engage with the more controversial aspects of environmental issues, and to consider attitudes and values in their teaching. Students on the Geography 16–19 course were required to consider different attitudes and values and the implications of holding them as a key part of one of the examinations. Thus, compared to teachers of other subjects, these teachers have every opportunity to implement the more radical aims of environmental education. Indeed, all three teachers did encourage students to consider different attitudes and values in a variety of ways via their use of resources, discussions, role-play, etc. (Fido 1999), and they allowed open, and frequently lengthy, discussion and argument in their lessons. They simply did not want to promote specific pro-environmental attitudes. It can be hypothesized that, for teachers of other syllabuses (or other disciplines) in which the role of attitudes and values is given a lower emphasis, the situation would be even more constrained, although this speculation merits further research.

In other words, I suggest that any attempt to introduce a more radical environmental agenda into schools by making changes to the curriculum will not succeed unless teachers can be convinced that it is desirable. The critical pedagogy for environmental education advocated by Fien (1993) and Breiting and Mogensen (1999), among others, seems far removed from these teachers' aims where their understandings of environmental education seem in conflict with their beliefs about the teacher's role in general. Their beliefs about the teacher's role take precedence over the more radical aims of environmental education, even where these aims are specifically advocated by the syllabus. In this way, the teachers' beliefs act as a critical mediating factor between the syllabus and the classroom. To this degree, this study adds support to the considerable body of literature across different areas of the curriculum which suggests that unless curriculum developers take account of teachers' beliefs in designing new curriculum materials, these materials are unlikely to be implemented in their intended format (e.g. Hargreaves 1989, Ball and Bowe 1992, Brown and McIntyre 1993).

Acknowledgements

This paper reports research undertaken for a doctoral dissertation at the University of Oxford (Fido 1999). This research was made possible by an Economic and Social Research Council grant, together with the support of many staff and students in the Department of Educational Studies. I thank Ann Childs, Graham Corney, and Geoff Hayward for their indispensable advice and guidance, and all the teachers and students who participated in this research and who devoted great amounts of their time and energy to the study.

Notes

1. Curriculum guidance, in this context, consists of non-statutory advice for teachers regarding environmental education.
2. The names of teachers and students are pseudonyms.

3. Similar to the form of 'recursive comparative analysis' described by Brown and McIntyre (1993) and Cooper and McIntyre (1996).
4. See Stradling *et al.* (1984) for a detailed discussion of these teaching approaches.

References

Aldrich-Moodie, B. and Kwong, J. (1997) *Environmental Education*, IEA Studies on the Environment, No. 9 (London: The Institute of Economic Affairs).

Apple, M. W. (1996) *Cultural Politics and Education* (Buckingham, UK: Open University Press).

Argyris, C. and Schön, D. A. (1974) *Theory in Practice: Increasing Professional Effectiveness* (London: Jossey-Bass).

Bailey, P. (1974) *Teaching Geography* (Newton Abbott, UK: David & Charles).

Ball, S. J. and Bowe, R. (1992) Subject departments and the 'implementation' of National Curriculum policy: an overview of the issues. *Journal of Curriculum Studies*, 24(2), 97–115.

Ballantyne, R. (1999) Teaching environmental concepts, attitudes and behaviour through geography education: findings of an international survey. *International Research in Geographical and Environmental Education*, 8(1), 40–55.

Breiting, S. and Mogensen, F. (1999) Action competence and environmental education. *Cambridge Journal of Education*, 29(3), 349–353.

Brown, S. and McIntyre, D. (1993) *Making Sense of Teaching* (Buckingham, UK: Open University Press).

Carr, W. and Kemmis, S. (1986) *Becoming Critical: Education, Knowledge and Action Research* (London: Falmer).

Connell, S., Fien, J., Lee, J., Sykes, H. and Yencken, D. (1999) 'If it doesn't directly affect you, you don't think about it': a qualitative study of young people's environmental attitudes in two Australian cities. *Environmental Education Research*, 5(1), 95–113.

Cooper, P. and McIntyre, D. (1996) *Effective Teaching and Learning: Teachers' and Students' Perspectives* (Buckingham, UK: Open University Press).

Curriculum Council for Wales (1992) *Environmental Education: A Framework for the Development of a Cross-curricular Theme in Wales*, CWW Advisory Paper No. 7 (Cardiff, UK: Curriculum Council for Wales).

Curriculum Council for Wales (1993) *INSET Activities for Environmental Education* (Cardiff, UK: Curriculum Council for Wales).

Doyle, W. (1992) Curriculum and pedagogy. In P. W. Jackson (ed.), *Handbook of Research on Curriculum* (New York: Macmillan), 486–516.

Elliott, J. (1993) Handling values in environmental education. Paper presented at the OECD Values in Environmental Education Conference, 5–7 May, Stirling, Scotland (Norwich, UK: University of East Anglia).

Fido, D. (1999) Teaching controversial environmental isues in 16–19 A-level Geography: possibilities and problems. Doctoral thesis, University of Oxford.

Fien, J. (1993) *Education for the Environment: Critical Curriculum Theorising and Environmental Education* (Geelong: Deakin University Press).

Fien, J. (1999) Towards a map of commitment: a socially critical approach to geographical education. *International Research in Geographical and Environmental Education*, 8(2), 140–158.

Gough, N. (1993) Environmental education, narrative complexity and postmodern science/fiction. *International Journal of Science Education*, 15(5), 607–625.

Grace, M. and Sharp, J. (2000) Exploring the actual and potential rhetoric-reality gaps in environmental education and their implications for pre-service teacher training. *Environmental Education Research*, 6(4), 331 345.

Gross, M. P. and Pizzini, E. L. (1979) The effects of combined advance organizers and field experience on environmental orientations of elementary school children. *Journal of Research in Science Teaching*, 16(4), 325–331.

Hammersley, M. (1998) *Reading Ethnographic Research: A Critical Guide*, 2nd edn (London: Longman).

Hargreaves, A. (1989) Curriculum policy and the culture of teaching. In G. Milburn, I. F. Goodson and R. J. Clark (eds), *Re-interpreting Curriculum Research: Images and Arguments* (Lewes, UK: Falmer Press), 26–40.

Hausbeck, K., Milbrath, L. and Enright, S. (1992) Environmental knowledge, awareness and concern among 11th-grade students: New York State. *Journal of Environmental Education*, 24(1), 27–34.

Huckle, J. F. (1985) Values education through geography: a radical critique. In D. Boardman (ed.), *New Directions in Geographical Education* (London: Falmer Press), 187–197.

Jickling, B. (1992) Why I don't want my children to be educated for sustainable development. *Journal of Environmental Education*, 23(4), 5–8.

Kelly, T. E. (1986) Discussing controversial issues: four perspectives on the teacher's role. *Theory and Research in Social Education*, 14(2), 113–138.

Lee, J. C.-K. (1993) Geography teaching in England and Hong Kong: contributions towards environmental education. *International Research in Geographical and Environmental Education*, 2(1), 25–40.

Lee, J. C.-K. (2000) Teacher receptivity to curriculum change in the implementation stage: the case of environmental education in Hong Kong. *Journal of Curriculum Studies*, 32(1), 95–115.

Morris, M. and Schagen, I. (1995) *Global Environmental Education: Green Attitudes or Learned Responses?* (Slough, UK: National Foundation for Educational Research).

Naish, M., Rawling, E. and Hart, C. (1987) *Geography 16–19: The Contribution of a Curriculum Project to 16–19 Education* (Harlow: Longman).

National Curriculum Council (NCC) (1990) *Curriculum Guidance 7: Environmental Education* (York: National Curriculum Council).

Olson, J. (1992) *Understanding Teaching: Beyond Expertise* (Buckingham, UK: Open University Press).

Posch, P. (1999) The ecologisation of schools and its implications for educational policy. *Cambridge Journal of Education*, 29(3), 341–348.

Richmond, J. M. and Morgan, R. F. (1977) A national survey of the environmental knowledge and attitudes of fifth year pupils in England. ERIC ED 134 478.

Rickinson, M. (2001) Learners and learning in environmental education: a critical review of the evidence. *Environmental Education Research*, 7(3), 207–320.

Rudduck, J. (1983) *The Humanities Curriculum Project: An Introduction*, revised edn. (Norwich, UK: School of Education, University of East Anglia).

Schools Council/Nuffield Humanities Project (1970) *The Humanities Project: An Introduction* (London: Heinemann).

Schütz, A. (1973) *Collected Papers: The Problem of Social Reality*, ed. and intro. M. Natanson (The Hague, The Netherlands: Martinus Nijhoff).

Scott, D. and Willits, F. K. (1994) Environmental attitudes and behaviour: Pennsylvania Survey. *Environment and Behaviour*, 26(2), 239–260.

Scott, W. and Oulton, C. (1999) Environmental education: arguing the case for multiple approaches. *Educational Studies*, 25(1), 89–97.

Showers, D. E. and Shrigley, R. L. (1995) Effects of knowledge and persuasion on high-school students' attitudes toward nuclear power plants. *Journal of Research in Science Teaching*, 32(1), 29–43.

Silverman, D. (1993) *Interpreting Qualitative Data: Methods for Analysing Talk, Text and Interaction* (London: Sage).

Sosniak, L. A., Ethington, C. A. and Varelas, M. (1994) The myth of progressive and traditional orientations: teaching mathematics without a coherent point of view. In I. Westbury, C. A. Ethington, L. A. Sosniak and D. P. Baker (eds), *In Search of More Effective Mathematics Education: Examining Data from the IEA Second International Mathematics Study* (Norwood, NJ: Ablex), 95–112.

Stake, R. E. (1995) *The Art of Case Study Research* (London: Sage).

Stradling, R., Noctor, M. and Baines, B. (1984) *Teaching Controversial Issues* (London: Edward Arnold).

Tilbury, D. (1995) Environmental education for sustainability: defining the new focus of environmental education in the 1990s. *Environmental Education Research*, 1(2), 195–212.

Tomlins, B. and Froud, K. (1994) *Environmental education: teaching approaches and students' attitudes: a briefing paper* (Slough, UK: National Foundation for Educational Research).

University of London Examinations and Assessment Council (1996) *Geography (16–19 Project) Syllabus 9219* (London: University of London Examinations and Assessment).

Uzzell, D. (1999) Education for environmental action in the community: new roles and relationships. *Cambridge Journal of Education*, 29(3), 397–413.

Walker, K. E. (1997) Challenging critical theory in environmental education. *Environmental Education Research*, 3(2), 155–162.

Yount, J. R. and Horton, P. B. (1992) Factors influencing environmental attitude: the relationship between environmental attitude defensibility and cognitive reasoning level. *Journal of Research in Science Teaching*, 29(10), 1059–1078.

Curriculum change and climate change: Inside outside pressures in higher education

SHIREEN J. FAHEY

In higher education today, institutions are facing a number of challenges—including the challenge to create future-proof graduates. Higher education institutions have a particular mandate to develop future leaders and decision-makers capable of understanding and providing solutions to complex, global issues. Education programmes that focus on multi-disciplinary thinking are required to prepare future leaders to solve problems not yet known to be problems. Using a case study of a postgraduate climate change programme, this study illustrates the challenges addressed and resulting rewards when reforming the curriculum. Two theoretical curriculum models informed the re-imagination of the programme: objectives-based and action research following the process inquiry model. The reformation was undertaken by the programme teachers as researchers of their practice. To future-proof graduates, this study discusses how curricular intentions are aligned with the institution's capacity for action towards change. Avoiding a business-as-usual scenario when faced with complex, politicized and global issues such as climate change requires both programme and course curricula continuous evaluation and revision. Alignment with internal (university and teacher-level) goals and external directives is required.

Setting the scene: The role of higher education and climate change

The role of higher education in general and specifically how a university approaches that role varies (see for example Kemmis *et al.* 1983, UNESCO 1998, McDonald and Van der Horst 2007, Saleh *et al.* 2010). Institutional roles, in addition to educating future leaders, should produce 'graduates for the social and economic development of a country' (McDonald and Van der Horst 2007). Higher education should also create opportunities for experimentation, and enable life-long-learning.

Other external factors that mediate a university's role include workforce expectations and meeting professional standards. In Australia standards include Australian University Quality Assessment (AUQA),[1] Tertiary Education Quality Standards Agency standards (DEEWR 2011,[2] TEQSA[3] 2011); potential implications resulting from the Bradley Review[4] (2008) and government funding requirements. Roles are often expressed through university goals, mission statements, and graduate attributes. The fulfillment of a university's chosen role or roles in learning, teaching, and research can be accomplished in part through specific programme curricula. For the present study, a curriculum is defined as a programme of study leading to the award of a university degree. The definition of *curriculum* follows Stenhouse (1975), who expressed the term as essential principles and features of an educational programme that are effectively translated and able to be put into practice.

This article explores the subject of role by describing the curriculum renewal process that the academics adopted at one university in Australia to provide education for future societies facing the global issue of climate change. The academics were motivated by personal direction and initiative, while pro-actively addressing future, externally originating factors. The case study is a 5-year young, science-based, multi-disciplinary postgraduate curriculum: the Master of Climate Change Adaptation (MCCA). When the programme was first conceived in the mid-2000s, the academics and the university recognized that changes in societal values and behaviours are needed to avoid a 'business-as-usual' approach (also known as 'doing nothing' (Stern 2006)) to face global climate change impacts and actions.

Human-induced, global climate change is a reality, has entered mainstream discourse, and is highly politicized (UN 1992, IPCC 2001, 2007, Wigley 2001, Oreskes 2004, McMichael et al 2006, Ramaswamy et al 2006, Whetton 2007, Ritter 2009, NOAA 2011, 2012). Despite the overwhelming scientific evidence, skeptics remain (Cater 2007, Fenton 2007, Wight 2010) and are provided media coverage which influences public perception (UN 2012). Nonetheless, climate change education is considered essential on global scales and it is now taught in schools and universities.

Higher education has a particular mandate to educate students for individual needs such as approaching career decisions and also for the potential for students' societal influence. For example, an estimate of the need for climate change professionals comes from the Australian Office of Climate Change (Australian Government 2009).[5] This body estimates the need for greater than 75 000 'green skills' workers. Workers will plan for, mitigate, and adapt the environment and society to human-induced climate change (DEEWR 2008, Hatfield-Dodds et al. 2008, Connection Research and DECC NSW 2009). Current graduates with expertise in climate change adaptation number in the low 1000s through the southern hemisphere. A programme to up-skill and enrich the capabilities of the climate change profession is required.

One factor that sets climate change education aside from other curricula is that current and future societies are already, and will continue to be in the future, impacted by the climatic changes underway. Education for the future to cope with climate change must therefore include the relevant science content (Leal Filho 2010, along with skills such as futures thinking, critical analysis, and deepened understanding of the impacts of climate change (DEEWR 2009, DEWHA 2010). To refine the expertise of postgraduate students at the higher education level not only on the issue of climate change, inter-disciplinary training for other global issues such as socio-economic, governance, and complex problem-solving is also required (Leal Filho 2010). The climate change adaptation education programme is one example of a university's internationalized curriculum that 'prepares graduates for defined international professions, is inter-disciplinary and covers more than one country' (IDP Australia 1995). Producing multidisciplinary science graduates with a holistic world view is significantly aided by multi-national student cohorts, such as those who complete the programme at the case study institution. These students enhance both the cultural flexibility and understandings necessary to approach a global issue—within the context of the needs of local peoples. The curriculum renewal process was used by the academics at the case study university in Australia to evaluate whether their multidisciplinary programme would provide relevant education for future societies facing the global issue of climate change.

The present article examines the higher education curriculum evaluation process, the pressures, the rewards, and lessons learned for curriculum developers. The findings are, therefore, relevant to the wider, non-specialist audience. Two curriculum development models are applied to evaluate the case study curriculum and to compare and contrast the models' applicability for any review:

(1) the outcomes-led or objectives-based model (Tyler 1949), and
(2) action research (based on the process-inquiry model) (Stenhouse 1975, McKernan 2007).

Methodology: Choosing curriculum review models

The steps to develop, and the preferred model to review programme curricula, have been argued for decades (see for example Tyler 1949, Stenhouse 1975, McKernan 2007).

There are three reasons why the two curriculum review models were chosen for this study. First, Tyler's model presents four fundamental questions that underpin any curriculum. These four questions represent guidance to structure the experience for learners. The questions are: (1) What are the (curriculum) educational purposes? (2) What are the educational experiences to attain the purposes? (3) What is the organization of the learning experiences? and (4) Have the purposes of the experience been met? (Tyler 1949). At the core of these questions is objectives-based learning, which Lawton (1983) simplified to: aims/objectives, content, organization, and evaluation. Tyler's questions aligned with the present case study objectives, as well as with this university's Program Review Policy (university 2010).

In contrast to an objectives-based model, Stenhouse's (1975) process-inquiry model specifically excludes the use of objectives as necessary for curriculum planning, evaluation, and education generally. Rather, it is the process of learning, as opposed to the products of learning (measureable outcomes) that lie at the centre of the process-inquiry model. Stenhouse (1975) wrote that the objectives-based model defeats the educational process by leading students (and teachers) to focus on end-points or views. McKernan (2007) states that the educational process is further defeated by setting arbitrary solutions to problems of knowledge, as opposed to treating (and designing) education for its own value.

The second reason for selecting the two evaluative frameworks is the researcher's interest in the distinction between the role of the teacher in both models: the Tyler (objectives-based) and the Stenhouse (process-inquiry) model. The distinction between the role of teachers is recognizable in empowerment and autocratic evaluations, further described in the next section. The case study was conducted as an empowerment evaluation, a strategy that can be used for any curriculum revision.

The third reason that the two models were chosen for the present review is the contrasting view of teacher quality and context. Stenhouse (1975) identified teacher quality as the strength and potential weakness of the process-inquiry model. This is in contrast to the objectives-based model, which focuses on students' behaviours, not on the teacher's actions or skills (Biggs 1999). Admittedly, the objectives-based model specifies that students' behavioural changes occur if the teaching is successful (Lawton 1983). However, ultimately, the objectives-based model emphasizes the role of the learner.

The study assessed a core precept in the educative process; that of teachers as researchers, integrating their knowledge with skills and values. Lewin's (1946) and Stenhouse's (1975, 1981) focus on teachers as researchers of their own practice is central to action research.

The selected models are employed in the context of two challenges facing universities that are developing and renewing curricula (Stenhouse 1975, McKernan 2007). These challenges are described in detail below. In short, applying Tyler's four questions along with comparing and contrasting the two curriculum review models is a strategy applicable for any curriculum review. This strategy led to the recommendation for model selection that achieves the desired curriculum renewal outcomes discussed in the conclusion section.

The challenges in higher education of curriculum revision

Academics face a multitude of challenges when undertaking any curriculum evaluation. They respond to internal and external direction, pressures, and evaluations. At the case study university, as is common with most universities, internal directions come from sources including:

- Academic incentive, i.e. professional development;
- University policy, plan, and strategy documents;
- University quality assurance; and
- Academic Committees and Support Units.

External pressures or directives also come from several sources including:

- global societal issues, including in the present case, that of climate change,
- employer requirements,
- industry expectations,
- professional norms, and
- external-to-the-university quality reviews and teaching standards.

Irrespective of the pressures or drivers for change, the degree programme, or which curricular development model is selected, there are two additional, common challenges faced by higher education institutions and academics (Stenhouse 1975, McKernan 2007):

- Challenge #1 is the difficulty of putting curricular intentions into practice; and
- Challenge #2 is the institution's capacity and support for change (Stenhouse 1975).

Challenge #1

The key to address the difficulty of putting curricular intentions (i.e. aims, purposes, values) into practice is to unite the curriculum with evaluation (McKernan 2007). When the teacher undertakes the curriculum evaluation as part of an internal professional mandate (an empowerment evaluation) the activity becomes part of her professional development as a researcher (McKernan 2007). This is an important concept, when one considers that teachers, as researchers, integrate their knowledge with skills and values, all core precepts in the educative process. Stenhouse's (1975) focus on the teacher-as-researcher of their own practice is central to the process-inquiry model. The present internal evaluation was undertaken voluntarily through the incentive of the academic staff as part of their professional interests and development as researchers. They are scientists by training and practice, and they undertook the curriculum review as an empowerment evaluation. The present case study explores the level of difficulty faced by these researchers as well as the rewards perceived as they progressed through the programme renewal process.

In contrast to internally-motivated evaluations, an external evaluation is one conducted by an evaluator in response to an external central authority that sets the policies and recommendations for the curriculum. The external evaluator, as opposed to the teacher, takes on the role of expert and judge. McDonald (cited in Hamilton 1976) labelled this as an 'autocratic evaluation'.

Challenge #2

The capacity and support for change, in general terms, can be defined as 'the ability to perform functions, solve problems and set and achieve objectives' (Fukuda-Parr *et al.* 2002: 8). The thesis explored herein is that a curriculum should be considered as content, form, and design that can be assessed by the teacher in a professional role as a practitioner and researcher. The curriculum should be changed, revised, and modified (following McKernan 2007: 56, Stenhouse 1981: 109) as internal and external-to-the-university demands dictate such as changing needs of society or students. Furthermore, a curriculum should not be fully formed or set in place in advance (McWilliam 2005).

In the context of the case study, the capacity and support for change is examined through examples including teachers taking the initiative as curriculum developers and researchers and supported by the institution to do so. McKernan (2007: 18) asserts that allowing teachers these roles is the university's admission of the power to change at a local level.

The results section provides examples of the above two challenges at the case study university in the context of responses to and by internal university sources.

Case study

The selected case study is an example of a multidisciplinary degree programme (MCCA) at an Australian university. The programme was the first postgraduate level, fully articulated climate change adaptation-focused programme in Australia. No other university in the world had yet offered an accredited climate change degree programme. The programme sits within the School of Science, Education, and Engineering in the Faculty of Science, Health, Education, and Engineering. Students are predominantly drawn from the ranks of professionals having an undergraduate degree. Some of these students enrolled in and completed courses to meet their 'just-in-time' professional development needs. Some are career changers and some are returning to education to up-skill. The MCCA is the sum of 12 subjects or courses or 144 units, completed over a minimum of 18 months. Topics presented in the cross-disciplinary courses include climate and hydrology, climate change adaptation and mitigation, ecosystem-based environmental management, impact assessment, governance-engagement-capacity-building, coastal systems dynamics, environment and resource economics, and marine resources. This case study was chosen for four key reasons to demonstrate generic relevance to curriculum theorists:

(1) The programme is an example of an internationalized curriculum, preparing students for international professions.
(2) The programme is multidisciplinary, involving education on global issues that require complex problem-solving skills.

(3) The curriculum renewal is being undertaken voluntarily by the programme leader and course instructors as part of their professional development.

(4) The curriculum was originally designed using the two development models presented above, and the renewal process also applied the same two models.

The results section presents the finding from the first stage (hereafter Stage 1) of the curriculum renewal process: aligning course intended learning outcomes (ILOs), course content, and assessment to the MCCA programme ILOs (figure 1). Stage 2, in which the MCCA programme structure was aligned to external directives, is discussed in a forthcoming paper.

Results

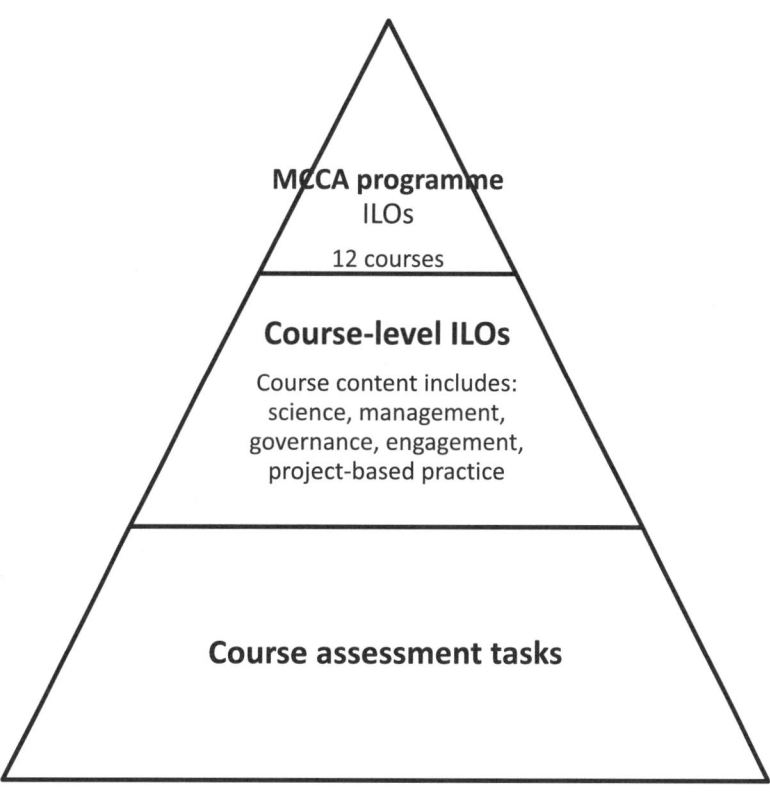

Figure 1. The first stage (Stage 1) of the curriculum alignment process. The case study involved a parallel process of aligning specific course' intended learning outcomes (ILOs), course content, and assessment tasks to the MCCA programme ILOs. Course content includes theory, science, governance, and management; for example, weather and climate, integrated environmental management, governance, engagement and capacity building, and associated problem-based, independent projects.

Challenge #1: The difficulty of putting curricular intentions into practice—Evidence that evaluating a curriculum can achieve programme goals

Evidence for a process-inquiry framework to couple theory and practice

Evidence for coupling theory and practice within a process-inquiry framework is found in the original, and current individual courses' content delivery and assessment methods (documented in the course outlines). Content includes both theory (lecture-based teaching activities) and practical (independent, project-based) applications. Examinations are not included in the assessment repertoire. Instead practical applications of theory are taught using problem-based learning (Biggs 1999) and assessed for understanding using multi-modal methods rather than memorization exercises as befitting a postgraduate course (McKernan 2007). Multi-modal activities include oral and written critiques of literature, production of professional-standard manuscripts, and web-friendly summary reports.

Further evidence of the curriculum having a process-inquiry focus is found in the teaching methods. For example, the review of the course outlines indicates that many techniques are used that focus on the process of learning, rather than the outcomes (after Stenhouse 1975: 92). These techniques include conducting classroom open discussions that encourage students to exchange ideas, question their own views, and develop the ability to use evidence to support their views.

Outcomes-based framework: Evidence for and challenges ofpre-selected behavioural and non-behavioural outcomes

Although there is evidence of a process-inquiry framework, the programme also contains a set of pre-selected outcomes following an objectives/outcomes-based model such as that of Tyler (1949) and Taba (1962). The objectives include both non-behavioural/non-measurable and behavioural/directly measurable components.

Behavioural objectives are specific behaviours students must demonstrate to indicate that learning has occurred; what the students are meant to do, and how that behaviour will be assessed. Non-behavioural objectives are not easy to align to assessment tasks, nor assess. Examples from the course outlines include:

- 'Recognize problems and pitfalls ... assess a range of models ... recognize and understand the key concepts ...
- Demonstrate relevant knowledge of the theory and key concepts ... critically appraise a diverse array of data ...'. (University 2005 programme documents)

Examples of non-behavioural objectives from the original programme's stated educational goals (hereafter intended learning outcomes (ILOs)) included:

- 'Ensure that professional development in the field of climate change provides practitioners with practical skills in problem-solving as related

to adaptive environmental assessment and integrated management systems. Such skills will be acquired through a combination of face-to-face teaching and practical projects. Training sessions utilize environmental simulation models to practice students in: problem-defining; data and information manipulation, management, and application.
• Equip environmental and natural resources managers with the theoretical understanding and practical tools to enable them to more effectively integrate best practice vulnerability assessment tools and techniques ...'. (University 2005 programme documents)

It remains unclear how the teacher would 'ensure professional development' or 'equip a student with theoretical understanding'. Hence, the challenge remained of putting the curricular intentions into practice.

In addition to problematic programme-level ILOs, the original course-level ILOs were difficult to assess (refer to figure 1). Some examples of the awkward course-level ILOs included: 'students will attain theoretical understanding and practical tools; manage project outcomes to ensure sustainability of natural resources; examine types of governance; and practical problem-solving skills' (University 2005 programme documents).

Aligning programme-level and course-level intended learning outcomes
While undertaking the revision of programme-level ILOs, it became evident that each course within the programme must introduce and establish course-specific ILOs, which would also require methods to assess ILO achievement. Thus, the programme-level review proceeded in parallel with the course-level review (figure 1). While this was a challenge due to workload issues, the result of this parallel revision is that the academics were able to align their individual course ILOs to the MCCA programme ILOs. This alignment was not evident in the original programme structure. The results of the parallel review are discussed in more detail below.

Evidence for alignment: ILOs and assessment
When attempting to align course-level ILOs with the assessment tasks, gaps became apparent, particularly for the non-behavioural ILOs noted above. In short, alignment of ILOs and assessment should form the basis upon which student learning could more readily be inferred (Biggs and Tang 2009). There were multiple examples of assessments that were either misaligned with the ILOs or the validity or fairness of the assessment tasks remained unclear. Validity is defined in the context of outcomes-based education as relevance and ILO coverage (Killen 2005: 114). I based the evaluation of fairness of the assessment practice on 'fairness of what we are asking students to do to demonstrate their learning' (Killen 2005: 230), and also on the fairness of our 'judgement of the quality or adequacy of the task'. Fair assessment tasks include a variety of opportunities for students to demonstrate evidence of their understanding (Brown 2004–2005). In fact, it was the inability to align the course-level

ILOs to the assessment tasks that first set in motion the programme ILO revision.

In order to align the assessment tasks to the ILOs, I re-formatted the course assessment tasks following Biggs' (2003) model of learning and teaching. Meyers and Nulty (2009) illustrated how curriculum design and assessment can be structured to guide deep learning and engagement in science education in higher education. So, for the climate change science degree, the assessment tasks were restructured to provide students with authentic, constructive learning experiences, and aligned to ILOs. The tasks also allow delivery in more flexible formats, to accept assessment in a range of contexts (White and Blythe 1992, Brown 2004–2005). Assessments are now accepted using any delivery medium, as opposed to only written or PowerPoint© delivery format. Suggested media include poster format, oral, video, or written presentation, combinations of these, or other creative means to demonstrate understanding of key concepts. The alignment of teaching and learning activities to assessment, ILOs, to content delivery methods and activities has been completed. The delivery methods include using more problem-based learning contexts, more use by the teachers of visual media (YouTube), and facilitating active rather than passive roles for students, with more open discussion and group activities.

Evidence for programme structure alignment to the stated goals or outcomes
In addition to the issues of assessment validity and fairness, and course-level ILOs and assessment task misalignment, the overall structure of the MCCA programme was not aligned to the stated programme goals or outcomes. Evidence for this is that the original programme structure required only four compulsory courses and allowed students to select eight from 24 elective courses to design their own programme. It is not clear how the programme outcomes would be achieved by each student, given that any combination of courses could be selected. The present programme renewal included reducing and re-structuring the number and ratio of compulsory vs elective courses. For example, the total number of courses on offer was reduced from 24 to 13 and required six compulsory and six elective courses with revised ILOs aligned to the programme-level goals. As an aside, further restructuring is in progress to require all 12 courses to be compulsory, thus assuring that each student will be able to achieve the stated programme-level ILOs.

Seven of the 13 courses offered within the MCCA programme were reviewed for ILO alignment and assessment tasks during Stage 1. The re-design of the new assessment tasks for these courses is based upon the research findings of authors including Brown (2004–2005) and the Australian Learning and Teaching Council (2009). Their research has indicated that assessment, both summative and formative, must be based not only on assessment criteria, but also aligned to the course and programme ILOs. Research has emphasized the need for constructive alignment (Biggs 2003): to align ILOs and assessment tasks with teaching and learning activities In this context, the present study considered the learning outcomes as originated from three sources: those established by the University's Learning and Teaching Plan (2009a), the MCCA programme-level ILOs, and course-level ILOs.

As noted previously, working in parallel at all levels of the MCCA allowed the programmatic 'big picture' to appear. Originally, each course was structured to attempt delivery of all programme ILOs, in isolation of the other courses. For example, one of the first courses to be reviewed listed 20 programme-level ILOs as the basis for that single course. After the review, the course ILOs were reduced to four, to which each assessment item was then aligned. The same situation of multiple ILOs was encountered with the next six courses reviewed. The big picture that appeared was that the context of how each course contributed to the overall programme was likely not considered when each course was first developed. Further, the programme ILOs in general were too many, too broad, and the wording was such that in most cases they were not assessable.

To summarize Challenge #1, the difficulty of putting the curriculum intentions into practice at the case study university is evidenced by (1) the number of elective courses that were offered, (2) the number of ILOs stated for each course, (3) the attempt to construct and deliver each course in isolation of the others, (4) the missing alignment between the courses and the programme ILOs, and (4) the validity and fairness of assessment. The number of elective courses, overlapping ILOS, and overlapping content between the courses would indicate that the developers missed the notion of alignment (course/programme/assessment tasks) to achieve the curriculum aims.

Challenge #2: The capacity for action towards change—Curriculum flexibility, alignment, and renewal

Capacity for action and potential to make changes
There are two components with regard to a university's and academics' capacity for action towards change. The first is the *short-term* capacity to change course and programme ILOs as needs, i.e. student, employer, and society demands. The second is the potential for the teacher and responsible academics to make these changes.

I noted previously that the case study programme and component courses were developed using a hybrid curriculum development model, involving process-inquiry and outcomes/objectives-based. Further, the evidence suggests that primarily the objectives-based model was followed, although aspects of a process-inquiry framework were also incorporated. The case study (Stage 1 programme evaluation) was undertaken also using a hybrid model, but weighted primarily towards a process-inquiry framework. This is important when considering Challenge #2.

Because of the hybrid process, the strengths of each model were able to be drawn upon. For example, the founding academics established a programme and courses as student needs and societal values presented. The process-inquiry model advocates that value issues must be crucial ingredients in curriculum and a central component of content (McKernan 2007). Evidence for these value issues in the MCCA programme curriculum is that it was established by practitioners in the field of climate change in view of societal values. The accreditation documents stated:

> Heightened awareness of the implications of global warming and the potential effects of climate change and associated environmental changes in the coastal zone are now being accepted as a reality by national, regional, and local governments; especially with regard to impacts on physical, biological, social, economic, and cultural conditions of coastal environments. Many of the existing and emerging issues can be addressed by focused research and research training activities that can lead to the award of the Master's degree. This gives rise to the need for professional development, through research training, for all spheres of government, business, industry, and non-government organizations to build capacity in order to better assess and manage the effects of changing conditions in natural and human affected coastal zones. (University 2005; supporting documents to accredit the MCCA programme)

Five years later (during the present study), the academics responsible for the programme were able to make changes and adjustments to the structure as needs, aims, and purposes changed. This is evidence to support the finding that, at the case study university, the academic staff are encouraged, supported, and have the capacity for action to change curricula and courses in response to changing student and societal needs and values.

The university's (and academic staff's) capacity for action towards change is also reinforced within an objectives-based framework. However, this capacity comes with the proviso that the ILOs set will be flexible and not subjective or have biased limits or solutions to the uncertain problems associated with climate change. In response to this proviso, the researchers revised the MCCA programme ILOs to be flexible. Examples of revised, flexible ILOs that could be readily assessed include:

Upon completion of the MCCA programme, students will be able to:

• Explain and evaluate, by making and defending their judgements, the natural and human-induced factors responsible for rates of environmental and climate change;
• Identify the elements of complex systems such as integrated environmental management, ecosystems, and governance; and
• Produce practical environmental assessment proposals and plans using problem-solving skills that include the identification and analysis of relevant stakeholders and on-going monitoring and plan improvement components. (University programme documents 2011)

Further evidence of the capacity for action towards change at this university is that the MCCA programme and courses have been continuously evolving over nearly 5 years voluntarily, by internal direction of the academic staff. This course of action illustrates the concept of the teacher having the potential and capacity to effect change; an element of both process-based and objectives-based models.

The capacity for action towards change is further evidenced at this university by stated professional development indicators. For example, the Learning and Teaching Plan (L&TP) document (2011) Key Performance Indicators (KPI) 2.2 and 2.3 focus on professional development for the teaching staff. If this definition is taken to include professional development as researchers, then the university has likely considered this KPI through the lens of an inquiry-based framework. The remaining KPIs in this document focus on student outcomes and objectives, thus presenting a ready alignment to using an objectives-based model as a programme evaluation tool at this university.

Discussion

The present evaluation of the MCCA curriculum is evidence to support McKernan's (2007) view that, when teachers undertake such an evaluation, it becomes part of their professional development, particularly their development as researchers. The investigation demonstrates that teachers who act as researchers are integrating their knowledge with skills and values; all core precepts in the educative process. Stenhouse's (1981) focus on the teacher-as-researcher of her own practice is central to the process-inquiry model.

The results of the present study show insights into teacher quality at the case study university. The process-inquiry model's focus on the teacher is one aspect of the model that contributes as an evaluative tool for teacher quality. During Stage 1, the researchers reflected on their practice as teachers, against one particular internal imperative; the university's Learning and Teaching–Academic Policy (L&TP). This Policy contains the key elements of learning and teaching that address teacher quality. The key elements include:

5.2 Lecturers and tutors who:

- take a scholarly approach to their teaching;
- display expert knowledge of and enthusiasm for their discipline;
- plan, design, manage, deliver, reflect upon, and improve their teaching and curricula to enhance student learning;
- engage with current research and creative outputs to inform their teaching;
- respect contributions from, and encourage participation by all students;
- provide advice to students regarding their academic work and academic choices;
- provide fair, critical, helpful, and timely feedback on student work;
- make use of appropriate technologies and media to support and enhance learning, teaching, assessment, and evaluation;
- comply with relevant legal, ethical, and policy responsibilities; and
- strive to be excellent teachers. (University 2010–2014)

In summary, at the case study university, academic quality and development are forefront in the policies and plans (University 2009a, 2009b). Participation of academic staff accountable for the MCCA programme for the present evaluation and renewal is evidence of the on-going application of both the process-inquiry and objectives-based models of curriculum development using an action research framework. The process of curriculum evaluation and re-structuring has contributed to the professional growth of the teachers at this university, a principle argued quite rigorously by Stenhouse (1975) and McKernan (2007).

The present evaluation allowed the programme to advance by virtue of the internal evaluators' (programme leader and teachers) abilities and experience to create solutions to problems identified (Fetterman 2001: 147). The evaluation also enabled the academics to continue with the good practices established. The academics, in their roles as researchers, identified challenges rather than problems with the curriculum. Some remaining challenges include renewal of the learning outcomes for all the courses within the MCCA programme, and final course alignment to the proposed new programme outcomes. The MCCA programme curriculum, while responding proactively to societal issues confronted by on-going climate change, is also expected to align to the university policies, goals, graduate attributes, and key elements of the L&TP. Stage 2 will complete the MCCA programme renewal. The Stage 2 process will be documented in preparation for the expected future internal quality audits and for the external-to-the-programme evaluations such as the TEQSA audit in 2012 (see note 3). The outcomes will be described in a subsequent paper.

Key findings

There are four key findings from the case study; first, the original MCCA programme designers used pre-selected, course-level outcomes (objectives-based model) as the basis for the curriculum design. This model appears to have achieved course-level outcomes, but curriculum-level outcomes could not be assured for all students. The original programme comprised 12 courses; four required 'core' courses and eight electives selected from a list of 24 courses (University student handbook semester 1 2007). All students in the programme, therefore, did not necessarily learn the same content, nor achieve the same programme learning outcomes. Each course represented a specific and independent learning experience for students. Students of the programme were and still are predominantly drawn from the ranks of professionals with an undergraduate degree. Some of these students enrolled in and completed courses to meet their 'just-in-time' professional development needs. The programme structure allowed these students to enrol in and complete individual courses. On accepting the role of programme co-ordinator, the MCCA team and I commenced re-designing the courses so that they were cumulative in effect to ensure the courses each contributed to programme-level student learning outcomes (Meyers and Nulty 2009). The original design of individual, stand-alone courses necessitated course and programme-level re-design to allow students to achieve specific programme-level learning outcomes.

The second finding is that revising the programme-level ILOs allowed for coupling teaching with practice. The MCCA programme is delivered by practitioners who unite their teaching with their research and practice. This feature results in training provided by teachers who have practical and technical expertise. For this case study, the teachers who are scientists by discipline and practice were empowered by the programme, and discipline leaders and the head of school to make on-going programmatic adaptations. It would appear that this is a primary reason for the success of the evaluation, and the continued interest in educational research by teachers whose primary interests lie in science. These academics expressed enthusiasm and increased enjoyment of their teaching practice having participated in the Stage 1 evaluation. Rather than expressing the renewal process as a constraint, they conveyed the opposite to be true.

The third finding is that the curriculum re-design used a foundation of ILOs with improved pedagogy to achieve learning outcomes with greater emphasis on problem-based learning (Barrows and Tamblyn 1980). Postgraduate students given open-ended problems to solve which are aligned with their interests are challenged to develop critical thinking and creative skills. They are motivated to transfer their existing and newly-developing knowledge to new situations. This style of learning is particularly suited to a complex discipline such as climate change.

The fourth main finding from the case study is from the course-level review. The original course ILOs claimed to address all programme-level ILOs, and all seven of this university's selected graduate attributes. The course-level outcomes could number up to 27. Practically, academics and students found challenging the prospect of addressing 27 course-level outcomes in only four assessment tasks. Additionally, there were too many vaguely worded ILOs that assessment tasks failed to require students to address. Examples of the vaguely worded learning outcomes are: 'To understand the principles and skills for assessment' and 'read and comprehend'. The first component of the re-design process refined and reduced both course-level and programme-level ILO's. Assessment tasks were then re-designed to align with the newly-refined ILOs.

Like many studies (Biggs 1999, Brown 2004–2005, Race 2005, Baume 2009) my work demonstrated that having multiple ILOs in a single course produced one or more of the following challenges:

(1) the teaching focuses on content rather than process and outcomes,
(2) students may express disappointment because the full range of ILOs were not addressed in the course,
(3) the course may be trying to address too many ILOs in a short time frame,
(4) the course contributes little to students' staged and sequenced acquisition of programme-level outcomes,
(5) the course is not aligned with other courses in the programme, and contributes little to the achievement of programme-level learning outcomes, and

(6) the course ILOs may be peripheral to the assessment tasks and vice versa.

Conclusions

The results from this case study may be helpful for academics undertaking development of new programme curricula, or evaluating, re-imagining, and re-designing an existing programme. For example, when curriculum development, evaluation, and renewal are undertaken for multi-disciplinary programmes in higher education such as climate change adaptation, applying an objectives-based model proved effective. However, curriculum objective-setting should be flexible to also permit alignment to learning activities, desired learning outcomes, and to assessment, and established in collaboration between the institution and the stakeholders. Further, coupling an objectives-based model with a process-inquiry framework that ensures an empowered evaluation by the stakeholders can enhance a programme curriculum and result in higher job satisfaction for the academic staff. Applying aspects of two curriculum development/evaluation models has several strengths. One of these is the ability for academic staff to respond quickly to emerging trends in student, employer, and societal needs, skills, and values. The academics have the benefit of being researchers of their own practice and can quickly integrate their knowledge with skills and values. This is a core principle in the educative process.

Although most curriculum developers and evaluators strive to achieve multiple purposes and desired outcomes to satisfy internal and external stakeholders, this is especially true for internationalized and politicized curricula such as climate change. Developers and evaluators can respond to external pressures with proactive, on-going internal review that invites and enlists the support of programme stakeholders. Academics in higher education who research actively in areas that change as rapidly as the daily tides are well-placed to integrate their evolving knowledge quickly into programme and course curricula. When teachers are researchers, they exhibit a self-directed focus (Stenhouse 1975, 1981) to undertake programmatic adjustments in response to the demands of internal and external circumstances. Active researchers focus on educational principles to embed social issues into students' learning in ways that prepare them to resolve multi-dimensional, international issues and problems such as those associated with climate change.

Notes

1. AUQA: Australian University Quality Assessment, Established early in 2000 following agreement by Australia's Education Ministers. It is 'responsible for auditing the quality of Australian universities. It also audits those Commonwealth, State and Territory authorities responsible for accrediting universities and higher education courses offered by other providers' (http://www.deewr.gov.au/Higher-Education/Programs/Quality/QualityAssurance/Pages/TheAusUniQualityAgency. aspx, accessed 23 January 2012).

2. DEEWR: The Department of Education, Employment and Workplace Relations. The 'lead government agency providing national leadership in education and workplace training, transition to work and conditions and values in the workplace' (http://www.deewr.gov.au/Department/Pages/About.aspx, accessed 23 January 2012).
3. TEQSA: Tertiary Education Quality Standards Agency. TEQSA is 'Australia's regulatory and quality agency for higher education. TEQSA's primary aim is to ensure that students receive a high quality education at any Australian higher education provider (TEQSA 2011)' (http://www.teqsa.gov.au/, accessed 23 January 2012).
4. The Bradley Review (2008): 'In March 2008, the Government initiated a Review of Higher Education to examine the future direction of the higher education sector, its fitness for purpose in meeting the needs of the Australian community and economy, and the options for ongoing reform. The Review was conducted by an independent expert panel, led by Emeritus Professor Denise Bradley AC' (http://www.deewr.gov.au/HigherEducation/Review/Pages/default.aspx, accessed 23 January 2012).
5. Australian Office of Climate Change and Energy Efficiency: The AOCCEE is the key Commonwealth agency on greenhouse matters, and is responsible for both the co-ordination of domestic climate change policy and for administering the Climate Change Science Programme. The Department of Climate Change was established on 3 December 2007. On 8 March 2010, as a result of Machinery of Government changes, a new Department of Climate Change and Energy Efficiency was established (http://www.climatechange.gov.au/about.aspx, accessed 24 January 2012).

References

Australian Government (2009) Transforming Australia's Higher Education System. P 28 $250 m for VET, HE and research infrastructure related to cc and sustainability activities. P 55 $40m for CC and energy research at UWS. Available online at: http://www.deewr.gov.au/HigherEducation/Documents/PDF/Additional%20Report%20-%20Transforming%20Aus%20Higher%20ED_webaw.pdf, accessed 23 January 2012.

Australian Government Department of Education, Employment and Workplace Relations (DEEWR) (2008) Climate Change and Skills for Sustainability. Available online at: http://www.deewr.gov.au/Skills/Programs/WorkDevelop/ClimateChangeSustainability/Pages/GreenSkillsAgreement.aspx, accessed 26 August 2011.

Australian Government Department of Education, Employment and Workplace Relations (DEEWR) (2009) Transforming Australia's Higher Education System. Available online at: http://www.deewr.gov.au/HigherEducation/Documents/PDF/Additional%20Report%20-%20Transforming%20Aus%20Higher%20ED_webaw.pdf, accessed 23 January 2012.

Australian Government Department of Education, Employment and Workplace Relations (DEEWR) (2011) Tertiary Education Quality Standards Agency (TEQSA) Draft Provider Standards, Bradley Review (2008). Available online at: http://www.deewr.gov.au/HigherEducation/Policy/teqsa/Pages/DraftProviderStandards.aspx, http://www.deewr.gov.au/highereducation/review/Pages/default.aspx, http://www.deewr.gov.au/HigherEducation/Review/Pages/ReviewofAustralianHigherEducationReport.aspx, accessed 30 May 2011.

Australian Government Department of Sustainability, Environment, Water, Population and Communities (DEWHA) (2010) Sustainability Curriculum Framework: A guide for curriculum developers and policy makers. Available online at: http://www.environment.gov.au/education/publications/curriculum-framework.html, accessed 23 January 2012.

Australian Learning and Teaching Council (ALTC) (2009) Assessment 2020 (Sydney: University of Technology).

Barrows, H.S. and Tamblyn, R.M. (1980) *Problem-based learning: An approach to medical education* (New York: Springer).

Baume, D. (2009) *Writing and using good learning outcomes* (Leeds, UK: Leeds Metropolitan University).

Biggs, J. (1999) *Teaching for Quality Learning at University: What the Student Does* (Buckingham, UK: Society for Research into Higher Education/Open University Press).

Biggs, J. (2003) *Teaching for Quality Learning at University*, 2nd ed. (Buckingham, UK: Society for Research into Higher Education/Open University Press).

Biggs, J. and Tang, C. (2009) *Teaching for Quality Learning at University*, 3rd ed. (Buckingham, UK: Society for Research into Higher Education/Open University Press).

Brown, S. (2004–2005) Assessment for learning. *Learning and, Teaching in Higher Education*, 1, 81–89.

Cater, R. M. (2007) The myth of dangerous human-caused climate change. *Paper presented at The AusIMM New Leaders' Conference*, 2007, Brisbane, Queensland.

Connection Research and Department of Environment and Climate Change New South Wales (2009) Who are the Green Collar Workers? Available online at: http://www.eianz.org/sb/modules/news/attachments/71/Green%20Collar%20Worker%20report%20Final.pdf, accessed 26 August 2011.

Fenton, L., Geissler, P. and Haberle, R. (2007) Global warming and climate forcing by recent albedo changes on Mars. *Nature*, 446, 646–649.

Fetterman, D. (2001) *Foundations of Empowerment Evaluation* (Thousand Oaks, CA: Sage).

Fukuda-Parr, S., Lopes, C. and Malik, K. (2002) Overview: Institutional innovations for capacity development. In S. Fukuda-Parr, C. Lopes and K. Malik (eds), *Capacity for Development, New Solutions to Old Problems* (London and Sterling, VA: UNDP-Earthscan), 284.

Hamilton, D. (1976) *Curriculum Evaluation* (London: Open Books).

Hatfield-Dodds, S., Turner, G., Schandl, H. and Doss, T. (2008) Growing the green collar economy: Skills and labour challenges in reducing our greenhouse emissions and national environmental footprint. *Report to the Dusseldorp Skills Forum*, June 2008. CSIRO Sustainable Ecosystems, Canberra.

Intergovernmental Panel on Climate Change (IPCC) (2001) Climate change 2001: The scientific basis. In J. Houghton, Y. Ding, D. Griggs, M. Noguer, P. van der Linden, X. Dai, K. Maskell and C. Johnson (eds), *Contribution of Working Group I to the Third Assessment Report of the Intergovernmental Panel on Climate Change* (Cambridge, UK and New York, NY: Cambridge University Press).

Intergovernmental Panel on Climate Change (2007) Climate change 2007: The physical science basis. In S. Solomon, M. Qin, Z. Manning, M. Chen, K. Marquis, M. Averyt, H. Tignor and R. Miller (eds), *Contribution of Working Group I to the Fourth Assessment Report of the Intergovernmental Panel on Climate Change* (Cambridge, UK and New York, NY: Cambridge University Press).

International Development Partners (IDP) (1995) *Curriculum development for internationalisation* (Canberra: Department of Education and Training).

Kemmis, S., Cole, P. and Suggett, D. (1983) *Orientations to curriculum and transition: towards the socially-critical school* (Melbourne: Victorian Institute of Secondary Education).

Killen, R. (2005) *Programming and assessment for quality teaching and learning* (Southbank, Victoria: Thomson Learning).

Lawton, D. (1983) *Curriculum Studies and Educational Planning* (London: Hodder & Stoughton).

Leal Filho, W. (2010) Climate change at universities: Results of a World survey. In: W. Leal Filho (ed.), Universities and Climate Change (Berlin: Springer-Verlag), v–vi.

Lewin, K. (1946) Action research and minority problems. *Journal of Social Issues*, 2(4), 34–46.

McDonald, R. and Van Der Horst, H. (2007) Curriculum alignment, globalization, and quality assurance in South African higher education. *Journal of Curriculum Studies*, 39(1), 1–9.

McKernan, J. (2007) *Curriculum and Imagination. Process theory, pedagogy and action research* (New York: Routledge).

McMichael, A., Woodruff, R. and Hales, S. (2006) Climate change and human health: present and future risks, 367(9513), 859–869. Available online at: www.thelancet. com, accessed 18 January 2012.

McWilliam, E. (2005) Unlearning pedagogy. *Journal of Learning Design*, 1(1), 1–11.

Meyers, N. and Nulty, D. (2009) How to use (five) curriculum design principles to align authentic learning environments, assessment, students' approaches to thinking and learning outcomes. *Assessment & Evaluation in Higher Education*, 34(5), 565–577.

National Oceanic and Atmospheric Administration (NOAA) (2011) Human-caused climate change major factor in more frequent Mediterranean droughts. *Science Daily*, 28 October. Available online at: http://www.sciencedaily.com-/releases/2011/10/ 111028115342.htm, accessed 18 January 201.

National Oceanic and Atmospheric Administration (2012) *Global Climate Change Climate change: how do we know?* Available online at: http://climate.nasa.gov/causes/, accessed 18 January 2012.

Oreskes, N. (2004) The scientific consensus on climate change. *Science*, 306(5702), 1686. DOI: 10.1126/science.1103618.

Race, P. (2005) *Making Learning Happen: A Guide to Post-Compulsory Education* (London: Sage Publications).

Ramaswamy, V., Schwarzkopf, M., Randel, W., Santer, B., Soden, B. and Stenchikov, G. (2006) Anthropogenic and natural influences in the evolution of lower stratospheric cooling. *Science*, 311, 1138–1141.

Ritter, S.K. (2009) *Climate Debate: What's Warming Us UP? Human Activity or Mother nature? Chemical & Engineering News Dec 21 2009* Global Warming and Climate Change. *Chemical & Engineering News*, 87(51), 11–21.

Saleh, A., Lamkin, M. and Cox, D. (2010) *The role of higher education in America: A spa or a smorgasbord?* (Academic Leadership: The Online Journal). Available online at: http://www.academicleadership.org/article/The_Role_of_Higher_Education_in_Amer- ica, accessed 5 May 2011.

Stenhouse, L. (1975) *An Introduction to Curriculum Research and Development* (London: Heinemann).

Stenhouse, L. (1981) What counts as research? *British Journal of Educational Studies*, 29 (2), 103–114.

Stern, N. (2006) The economics of climate change. *Population and Development Review*, 32, 793–798.

Taba, H. (1962) *Curriculum Development: Theory and Practice* (New York: Harcourt Brace and World).

Tyler, R. (1949) *Basic Principles of Curriculum and Instruction* (Chicago: University of Chicago Press).

United Nations (UN) (1992) United Nations Framework Convention on Climate Change (UNFCCC). Available online at: http://unfccc.int/resource/docs/convkp/conveng.pdf, accessed 3 August 2011.

United Nations (UN) (2012) Global Issues: Climate Change. Available online at: http:// www.un.org/en/globalissues/climatechange/, accessed 23 January 2012.

United Nations Education, Scientific and Cultural Organisation (UNESCO) (1998) World Declaration on Higher Education for the Twenty-First Century: Vision and Action. Available online at: http://www.unesco.org/education/educprog/wche/declara- tion_eng.htm, accessed 5 May 2011.

University (2009a) Learning and Teaching Plan 2009-2011. Available online at: http://www. usc.edu.au/University/AbouttheUniversity/Governance/Policies/LearningTeachingPolicy. htm, accessed 22 May 2011.

University (2009b) Learning and Teaching – Academic Policy. Available online at: www. usc.edu.au/University/AbouttheUniversity/Governance/Policies/LearningTeachingPol icy.htm, accessed 22 May 2011.

University (2011) Academic Committees and Support Unit (ACAPSU). Available online at: http://online.usc.edu..au/etc http://online.**.edu.au/webapps/portal/frameset.jsp?tab=community&url=%2Fbin%2Fcommon%2Fcourse.pl%3Fcourse_id%3D_2496_1, accessed 26 May 2011.

Whetton, P. (2007) *Stronger evidence for human-induced climate change* (Melbourne: CSIRO).

White, N. and Blythe, T. (1992) Multiple intelligences theory: Creating the thoughtful classroom. In A. Costa, J. Bellance and R. Fogarty (eds), *If minds matter: A forward to the future* (Palantine, Illinois: Skylight Publishing, Inc), 128–135.

Wight, J. (2010) Skeptical science. Available online at: http://www.skepticalscience.com/climate-change-little-ice-age-medieval-warm-period.htm, accessed 18 January 2012.

Wigley, T. (2001) The science of climate change 2001. In E. Claussen (ed.), *Climate change: science, strategies, & solutions* (Arlington, VA: Pew Center on Global Climate), 6.

Towards a socially critical environmental education: water quality studies in a coastal school

ANNETTE GREENALL GOUGH AND IAN ROBOTTOM

In many schools, studies of water quality in their vicinity might involve little more than students collecting water samples, carrying out standard analytic tests and reporting the results in a conventional 'scientific' manner. But for several schools working with Deakin University in an environmental education project, and for one school in particular, matters were not so straightforward. In this paper we will describe the activities in which these Australian schools were engaged as instances of what might count as socially critical environmental education and explore some of the issues that arise from such attempts to realize a socially critical curriculum and to support its development. The project as a whole is considered in terms of the extent to which it exemplifies socially critical education and, further, the idea of socially critical education itself is appraised in the light of our experiences of participating in the curriculum work undertaken by the schools.

Socially critical schooling is intended to provide students with a map of the existing culture and society and a map of what a better society might be like (Kemmis 1986: 19). We have chosen a socially critical orientation to curriculum (Kemmis, Cole and Suggett 1983) as the conceptual framework for this paper because, like environmental education, such an orientation concerns itself with a critical understanding of, and an informed commitment to, the improvement of society. Socially critical pedagogy is not without its critics (see Kemmis 1991 for a review of some of these), but it also has many supporters (see, for example, Shor 1980, Freire and Shor 1987, Giroux 1989). As a parallel development, there is evidence that environmental education is being characterized increasingly as socially critical in its intent (see, for example, Di Chiro 1987, Greenall Gough 1990, Victoria Ministry of Education 1990, Huckle 1991). Both environmental education and socially critical pedagogy seek to empower students to participate in a democratic transformation of society. Our interest in socially critical education arises largely from our work in environmental education (see, for example, Greenall 1981,1987, Greenall Gough 1990, 1991, Robottom 1985, 1987, 1990) and our recognition of similarities in their respective characteristics. We present here instances from a project which we see as exemplifying this complementarity.

We first describe the context of the project. This is followed by a short discussion of our framework of analysis (socially critical pedagogy and environmental education) and an analysis of the project within the framework of socially critical pedagogy. Finally we raise some questions about the practice of socially critical pedagogy, the answering of which will require further research with the project schools or elsewhere.

Context

The schools described here participated in a project co-ordinated by Deakin University. The project began in late 1987, with five schools in the south-western coastal region of the state of Victoria, Australia engaging in water quality studies and linking their studies with each other and a number of overseas schools by the use of a computer conference (see Preston 1988). The water quality studies were relatively narrowly focused and included the monitoring of bacteria levels and physical variables as well as determining topographical and demographic patterns. The computer conference was accessed by each school's microcomputers, linked by modem with the Deakin University mainframe and, thence, to a water quality conference program housed in a mainframe at the University of Michigan, USA (for further details on the computer conference itself see Robottom and Muhlebach 1989).

The project has three dimensions:

- teachers and students are engaged in scientific/geographical studies of nearby freshwater and marine environments;
- teachers and students are participating in an international computer conference that enables interactions with some 100 overseas schools, all of which are engaged in similar scientific/geographical studies;
- teachers and students are engaged in a form of participant research focused on the educational issues that arise from their investigations of water quality and participation in the computer conference.

Each school adopted a different mode of participation in the project. For example, Lavers Hill Consolidated School is a small school in a relatively isolated rural, forested area of a coastal mountain range. The school participated in the fresh water quality study part of the project in their lower secondary science programme and were particularly interested in the computer conference as a means of compensating for their isolation.

Lorne Higher Elementary School is a small school in a holiday resort town located on the coast. The school participated in the both the fresh and marine water quality study parts of the project as well as the computer conference. There was a high level of public awareness about water quality in the town: students and teachers were interested in investigating this community issue partly as a result of a cryptic memorandum issued to all ratepayers by the local Water Board recommending that all residents in the town boil their drinking water.

While the activities of the Lavers Hill and Lorne schools are interesting in their own right (see Muhlebach and Robottom 1990), this paper focuses on the project-related activities of another participating school. Queenscliff High School, a relatively large school in a coastal holiday resort town, also began its activities with a study of fresh water quality but, in response to local imperatives, changed the focus of its research to a critical study of sewage pollution at their nearby swimming and surfing beaches.

This school's involvement in water quality studies was in part a response to concerns expressed within the student body and within the community at large about the obvious sewage pollution of nearby beaches. These beaches are widely known for their swimming, surfing and fishing, and there was an increasing community concern about the amounts of disposable items (such as plastic syringes and condoms) and sewage finding their way onto the beaches. The school developed a research programme (nominally within Year 11 Marine Studies and Year 12 Biology) beginning with studies of the sand dunes along the foreshore and including tests of bacteria levels in the ocean. The school's investigations showed that coliform bacteria counts in the seawater were far in excess of acceptable Environment Protection Authority (State of Victoria) guidelines for safe body-contact and that such readings were common in locations where thousands of people swim, surf and fish on a regular basis.

Prior to 'going public' with the results of their investigations by providing their data to the mass media, the students and teachers attempted to discuss the issue with the local Water Board but were rebuffed. The Water Board controls a large, new sewage treatment plant which voids its effluent into the ocean via an outfall located offshore from the popular beaches. The Water Board had previously been invited to participate in the school's study in an advisory capacity (for example, by contributing technical expertise via the computer conference), but had also declined. Following these rebuffs the school invoked the state's Freedom of Information Act to gain access to the Water Board's own records of bacteria levels.

Early in 1989 the school published an account of its activities and their outcomes in the local press (*The Geelong Advertiser* 3 February 1989: 1):

> Testing of bacteria levels from Thirteenth Beach to Ocean Grove has revealed readings up to 40 times above a safe limit. The tests, taken by Queenscliff High School students during December, were done before the Geelong and District Water Board started commissioning its $32 million sewage outfall. The teacher said he had been refused access to the water board's figures and had lodged a Freedom of Information request so he could compare the readings.

This disclosure triggered a powerful response from several sections of the community:

- Local, state and national print and electronic media ran stories publicizing the school's activities, the issue itself, and its relationship to other instances of marine pollution elsewhere in the country.[1]
- The local surf-riders' association aligned themselves with the school in criticizing the state of the local beaches, specifically in terms of sewage pollution (Melbourne (Australia) Age 27 February 1989: 15):

> The State Government would insist on further work being done on a new sewage treatment plant, forcing Geelong sewage into Bass Strait, if the plant failed to meet Environmental Protection Authority objectives, the minister for Planning and Environment, Mr Roper, said today . . . .

The president of the Thirteenth Beach Surf Lifesaving Club, Mr Brett Cooper, said last night that, despite the new treatment plant, club members continued to find syringes, condoms and plastic along the beach.

'We've found enough for truckloads', he said. 'And we are concerned about the pollution also because some readings obtained by Queenscliff High School as part of an Education Department programme at the outfall show the *E. coli* count is 40 times higher than permissible levels.'

- The local health centre began to maintain records of complaints about infections and illnesses possibly associated with the bacteria present in the seawater.
- Funding support for the school's project activities was forthcoming from Victoria's Ministry of Education, a regional education/industry collaborative, and the State Rural Water Commission.
- Some of the other schools participating in the project shifted the emphases of their investigations from fresh water quality to seawater bacteria levels at beaches close to their communities.

In short, the school's activities were at least partly responsible for mobilizing a general and sustained community interest in this issue. One outcome of this increased public consciousness was a requirement for the Water Board to justify its environmental actions in regard to this issue. Ultimately, the Water Board was requested by the state's minister for environment and planning to undertake substantial improvements to their sewage treatment facility, at a likely cost of more than $5 million (*The Geelong News* 27 June 1989: 3):

> The State Government last night told the Geelong and District Water Board its new $30 million Black Rock sewage outfall is not good enough.
>
> Planning and Environment Minister, Mr Tom Roper, in Drysdale last night, said the Government would step into the ongoing dispute over the outfall.
>
> While the outfall has undoubtedly improved the situation significant concerns remain. It is clear additional works are required.
>
> Board corporate services manager Mr Rob Jordan said the board was 'bound to disagree' with some of Mr Roper's comments. Responding, Mr Roper said, 'Disagree as much as you like. You will be cleaning up your water.'

A framework of analysis

Kemmis, Cole and Suggett (1983) describe three orientations to education that are currently adopted, in varying degrees, by Australian schools: vocational/neo-classical, liberal/progressive and socially critical. The dominant concerns of these three orientations are summarized in table 1. Kemmis *et al.* (1983: 9) argue for education which engages society, social structures and social issues immediately, rather than merely preparing students for later participation. They believe that education 'must engage social issues and give students experience in working on them – experience in critical reflection, social negotiation and the organization of action'. While there is an emphasis on immediacy in action and reflection, socially critical education also seeks (Kemmis 1986: 13) to locate and interpret current culture within an historical context: 'We need to learn or re-learn what our

Table 1. Dominant concerns of three curriculum orientations.

	Vocational neo-classical	Liberal progressive	Socially critical
School concern	Preparation of students for work	Preparation of students for life	Engagement of students in critical thinking and action
Dominant curriculum concern	Course or subject content	Issue or process	Critical theory and group processes or action oriented
	Structuring learning	Facilitating student learning	
Focus for expression of concern	Faculty and subject organization	Total school teaching team	Teaching team, students and community
Frequency of concern	Episodic (e.g. prior to course construction)	Ongoing	Ongoing
Themes of concern	Student discipline, teaching resourses, curriculum packages, testing procedures	Student participation, group process, small group management, programme review	Student action, group processes and community linkages, negotiating tasks
	Developing diligent obedient students	Developing caring, co-operative environment	Developing working knowledge and critical perspective on society

Source: Stephen Kemmis, Peter Cole and Dahle Suggett (1983) *Orientations to Curriculum and Transition: Towards the Socially-Critical School* (Melbourne: Victorian Institute of Secondary Education).

bureaucratic culture has hidden from us – that we are not only the products of history but also the makers of history'.

A socially critical orientation has implications for the styles of teaching and learning adopted in schools and for the content of the curriculum. A socially critical education (as summarized in Kemmis *et al*. 1983: 11–14) has characteristic views of knowledge, the role of teachers and learners, broad curriculum organization, school–community liaison, and the role of consultants:

- Knowledge is seen as constructed through social interaction and thus as historically, culturally, politically and economically located; it has its meaning in the actions of projects whose significance is in specific historical, political and economic contexts.
- A learning role for students is in using available knowledge through interaction with others in socially significant tasks of critique or collaborative work.
- A proper role for the teacher is as project organizer and resource person, organizing critical and collaborative activities in negotiation with students and community, demanding joint values of autonomy and social responsibility.
- Differentiation of subjects and use of time is seen as being based on negotiation between community, teachers and students about the whole curriculum as a 'project'.

- Schools are seen as taking initiatives in the community as well as 'on behalf of the community' within the school, with reciprocal interaction as the preferred goal rather than procedural 'liaison'.
- Outsiders, where they are called in for consultancy tasks, are seen as contributors to a collaborative school- and community-based process; they are resources.

Kemmis and his colleagues have tended to focus on the theory of socially critical pedagogy; we are attempting the analysis of particular instances of practice (within the project) in terms of the framework their theory provides.

Teachers working within the framework of a socially critical pedagogy attempt to provide learning experiences that give students a historical and critical perspective on society, and give them opportunities to engage in activities that are consistent with building a responsive democratic society. These learning experiences are negotiated with the students, other staff and the wider school community. Learning is seen as a co-operative process and students are seen as agents for producing working knowledge through interaction with others in socially significant tasks. Such tasks include collaborative community projects which are a response to community concerns and which engage students in collaborative reflection and learning from direct experience. In a socially critical school teachers teach less often by didactic approaches such as telling and testing and more often by encouraging inquiry, critical reflection and action. They work with students on topics that the students believe to be important and through tasks they find rewarding and significant.

Being socially critical is a process which must be initiated and sustained. To become socially critical a school simply begins the process and sticks to it. It maintains the process by observing and analysing its progress in being critical. As such, being socially critical in practice is methodologically similar to *reflective deliberation* (Reid 1981: 167) in being concerned 'to improve people's capacity both individually and collectively, to make good decisions'. However, while the socially critical and deliberative orientations are similar in proposing a method for actualizing 'a belief in the possibility of improvement through working with present institutions' (Reid 1981: 167), socially critical writers are more explicit than deliberative scholars in encouraging the questioning of existing social structures.

Much Australian and international literature concerning environmental education asserts that it 'should adopt a critical approach to encourage careful awareness of the various factors involved in the situation' (UNESCO 1980: 27). According to such literature, environmental education should also involve learners in planning their learning experiences, 'utilize diverse learning environments and a broad array of educational approaches to teaching/learning' and 'focus on current and potential environmental situations' (UNESCO 1978: 27).

These characteristics of environmental education, in particular the emphasis on adopting a critical approach and focusing on current and potential situations, suggest a common ground between environmental education and socially critical pedagogy as described by Kemmis *et al.*

(1983) and others (Shor 1980, Freire and Shor 1987, Giroux 1989, Greig, Pike and Selby 1989).

Kemmis (1990: 83) writes of 'Education *for* the environment' initiatives in Australian schools as a demonstration of 'one way in which schools have taken a progressive stand on a pressing matter of public importance, giving what is seen to be a lead in developing new ways of viewing the interconnectedness of nature and human use of the planet'. However, we contend that although the environmental content of school curricula has increased, most schools are not involved in socially transformative environmental education. They are incorporating environmental content into their existing curricula rather than engaging in the kinds of social action that are being undertaken by other community agencies and activists. Much of what seems to be going on in schools under the guise of environmental education appears as 'nature study' or 'doom and gloom' current affairs topics such as the greenhouse effect and the depletion of the ozone layer. Little is done to empower the students to address the issues and resolve the problems. Personal action guides for the earth and green consumer guides[2] are produced for the general public but few action equivalents have been developed by and for students.

In general, the emphasis given to environmental education in schools has been on *environment* rather than *education*, and certainly not on socially critical education. This conclusion is supported by Fien (1990) who notes that the environmental education policies published by state education departments since 1987 'fail to advocate engaging students in a critique of dominant material and ideological forces'[3] and only make lame references to social action for the environment, if such a section is there at all in the policy.

The project within the framework

The project, and perhaps especially the activities of Queenscliff High School, appears to express some of the characteristics of socially critical pedagogy. Our intention here is to analyse the project in terms of those characteristics and to consider the idea of socially critical pedagogy in terms of the practical experience of the project – to expand on some criteria and to comment on some of the silences of the literature on socially critical education.

Some of the apparent correspondences between the project experience and socially critical pedagogy (as summarized in Kemmis *et al.* 1983: 11–14) relate to views of knowledge, the role of teachers and learners, broad curriculum organization, school–community liaison, and the role of consultants, which we have described above. The project will now be analysed in terms of these views.

View of knowledge

In the project described here, the substantive knowledge of the programmes engaged in by the schools was 'working knowledge' (the emerging curriculum content of socially critical education) generated by participants themselves. The 'locatedness' of the 'working knowledge' was evident: from a common starting point each of the schools has implemented a curriculum with a different emphasis.

At Lavers Hill Consolidated School, the emphasis was on involving students in the computer conference aspects of the project: knowledge as water quality data, though gained through students' own enquiries, was perceived primarily as grist for the computer conference mill – as a means of, and justification for, students establishing and exercising their contacts with a larger community of student-scientists, thereby diminishing the isolation of students in this small rural community. At Lorne Higher Elementary School, student involvement in water quality studies arose from an issue of community concern. Their curricular emphasis was the investigation of water quality: an interest in the (unstated) justification of a directive from the local water board (that all drinking water should be boiled) led students and their science teacher to investigate this issue. Students became involved in some local politics through confrontations with water board personnel. Overall, though, the curriculum was translated into a conventional scientific study of a social issue rather than a socially critical study. This is perhaps best summarized by the teacher concerned (Masanauskas 1989): the project 'really opens their eyes to the world of science instead of just getting them into doing small experiments in school'.

As outlined earlier, the curriculum emphasis at Queenscliff High School was different again. We suggest that their orientation was socially critical insofar as the teacher and students treated society as problematic. Through their enquiries they developed their own 'working knowledge' of relevant constituencies in their community (such as the surfers and the medical centre), sought critical understandings of social structures and relationships (such as the powerful influence of the Water Board), appraised this information, and acted with some community agencies (such as the media) to change a social practice (in this case the disposal of sewage wastewater).

One conclusion from this project is that, in socially critical education, social critique is an action-based, community-embedded form of inquiry yielding 'working knowledge' that is transactional rather than transmissional, generative/emergent rather than preordinate, opportunistic rather than systematic, and idiosyncratic rather than generalizable. The project demonstrates that two of the mainstays of traditional curricula – the idea of a 'universal' course content and the idea that 'textbooks' are a prime source of worthwhile knowledge – are inapplicable in a socially critical curriculum.

The project also demonstrates some of the alternative concrete forms that 'working knowledge' can take. The developing understandings of project participants about the environmental issue being investigated – about the power relationships between various constituencies and agencies, about the role of the media, about the technical parameters of water quality, about their own capacity to influence the outcomes of environmental issues – were represented in the computer conference transactions (available as a hard-copy printout), in newspaper articles, national television documentaries, project newsletters, academic papers, workshop reports, and student reports published in environmental newsletters. The teacher's role was to ensure that these diverse forms of emergent 'working knowledge' remained prominent in classroom discourse.

Roles of teachers and learners

The activities of each of the schools exemplify a learning role for students as 'using available knowledge through interaction with others in socially-significant tasks of critique or collaborative work' (Kemmis *et al.* 1983: 11–12). Direct involvement in data collection about water quality entailed collaborative work (collection of water samples; conduct of tests; compiling results); and the focus of the research was perceived as socially significant by the broader community as well as by school participants, as newspaper articles reveal (see note 1). The enacted learning theory was not a behaviourist, 'deficit' one concerned primarily with the transmission of a preordinate, systematically organized body of knowledge. Instead it provided the conditions for learners to appreciate social reality (and, in particular, environmental problems) as socially constructed and subject to reconstruction through historical and social processes – including political processes in which they were themselves involved.

Teachers in this project assumed the responsibility for manifesting a project perspective and, as indicated earlier, the focus of this project perspective differed from school to school under the developing interests of the students and the organizing direction of the teachers. 'Critical and collaborative activities' (Kemmis *et al.* 1983: 11–12) organized by the teachers in the project included modification of the school timetable to enable protracted outdoor fieldwork, establishment of links with community agencies such as print and television media, the local water board, rural water commission, the local water board, rural water commission, the local university and marine sciences laboratory, and encouragement of the writing of reports by students and colleagues.

Broad curriculum organization

Kemmis, Cole and Suggett's (1983: 13–14) suggestion that in a socially critical curriculum 'differentiation of subjects and use of time [should be] based on negotiation between community, teachers and students about the whole curriculum as a "project"' is reflected in curriculum organization at participating schools. In these cases, the schools' 'science curriculum' had no existence independent of the 'project'. The curriculum's use of time and space within and outside the school, and its disciplinary orientation (for example, a 'science' focus *vis-à-vis* a 'social studies' or 'computer/technology' focus) depended as much on project imperatives (for example, tides) and negotiations with community agencies (for example, the media, marine science laboratory and water authorities) as it did on the schools' published subject timetable.

A consequence of the adoption of a 'project perspective' in the curricula of participating schools was that from an initially similar starting point the curricula diverged and evolved differently in the different schools as they followed their respective lines of inquiry. This had implications for the kinds of liaison the schools entered into with support agencies in the community (including consultants).

School–community liaison

In this project, it could be argued that the relationship between schools and community was not an instrumentally dependent one with the curriculum serving to prepare students for participation in society *as it exists*. Rather, the curriculum and its participants (teachers and students and co-operating community agencies) undertook initiatives in the community *for community change* – change that could be perceived as being as much on behalf of the community as on behalf of the schools. School initiatives evolved into a general and sustained community endeavour to resolve a serious environmental problem and, within the context of this endeavour, liaison and communication became interactive and mutually supportive. Community agencies depended on the schools for scientific information about bacteria levels; and the schools were dependent on the community for funding, technical support and visibility.

Liaison between the schools and media was particularly important, and supported the claim made by Aronowitz and Giroux (1985: 72) that schools often appear to be existing 'in a contradictory relation to the dominant society, alternately supporting and challenging its basic assumptions'. The media exploited the opportunity the schools provided for interesting and (in light of similar environmental issues elsewhere in the country) timely copy; for the schools, positive media coverage of their activities provided a kind of legitimation of their involvement in controversial, political activities which otherwise the schools may have avoided. Having said this, it is also apparent that the legitimating effect of mass media coverage was not limited to the schools' activities; other agencies, including the Water Board, benefited from this visibility. For example, on one occasion the Water Board purchased advertising rights to a whole page of the local newspaper to present its own position on the issue in an uncontested fashion. On another occasion it was able to draw a distinction between 'awareness' and 'knowledge' – a distinction aimed at re-establishing the board's own authority in the issue (*Geelong Advertiser* 25 March 1989: 2):

> Ratepayers in Geelong are in for a shock if a major sewage review of the region condemns the controversial Black Rock outfall.
> Geelong and Distict Water Board chief executive, Mr Geoff Vines, has warned that an unfavorable review could be costly. Mr Vines admitted that the board had bowed to the enormous public pressure over Black Rock by ordering a thorough review of its sewage disposal policy.
> 'This review could have financial implications not just for the board but for the people of the region', he said. 'There has been an enormous cranking up of

public awareness on this matter but the level of public knowledge is abysmal.
It is our job to inform the public of the consequences of their aspirations. . .'

In becoming part of the environmental controversy, then, the media acted
more as 'agents of legitimacy' equally available to all parties than independent
pro-environment 'watchdogs'. Thus, while it is likely that media involvement
accelerated the controversy initiated by the schools and provided some
needed legitimacy in a potentially risky pursuit, it is also likely that the media
served largely to reinforce rather than challenge existing power relationships
through their policy of evenhanded reporting.

The role of consultants

Resource people external to both the schools and Deakin University were
involved in this project. For example, on two occasions teacher-participants
expressed a need for a workshop to address methodological problems: on
one occasion, a specialist in telecommunications was invited to advise
teachers on means for reducing their online computer conference costs;
on the other occasion, two specialists in bacteriological testing were invited
to advise on the transfer of techniques of fresh water bacteria testing to
the marine situation. On both occasions the perceived problems were
identified, and the outside expertise was invited, by participants themselves.
These workshops supported a different relationship between teachers and
curriculum support agents from traditional inservice teacher education
activities which tend to be instrumental to the preordinate goals of outside
curriculum change agents rather than responsive to issues of interest and
concern to practising teachers.

While recommending 'reciprocal interaction' as the preferred goal of
school–community liaison, Kemmis *et al.* (1983) are otherwise silent on the
issue of liaison, offering no alternative to conventional constructions of the
'problem of communication'. In this project, it became clear that an
important role for consultants lay in responding to the emerging practical
needs of participants by establishing and supporting a range of forms of
communication. In an account of the OECD's international Environment and
School Initiatives project, Posch (1988) recommends a participatory research
role for teachers in environmental education curriculum development.
In Posch's view (1988: 15), curriculum development in environmental
education requires the creation of conditions for teachers to communicate
extensively with each other and to reflect systematically on personal practices:

> Teachers who take on this [curriculum development in environmental
> education] duty need to communicate with each other and need external
> support. . . . This, however, is not enough. . . . When a teacher no longer
> contents himself [or herself] with imparting systematic knowledge, but
> exceeds the limits set by the school and accepts to cope with unstructured
> situations, he [she] increasingly needs to be aware of what he [or she] does, a
> kind of systematic reflection on his [her] own actions, in order to keep a check
> on the risks connected with environmental projects, and in order to facilitate
> communication on his [or her] actions and further development. Therefore we

want to encourage and help project teachers to evaluate their work with the pupils themselves and to write about it. . .

As this project proceeded and teachers worked their way through a number of technical, teaching and curriculum issues, a number of channels of communication were developed between participants – teachers, students, university personnel, specific community agencies and the general community. These forms of communication were: individual visits to schools by university personnel; the monthly project newsletter with copy from teachers and university personnel; the computer conference accessible through the school's own computers and modems; project workshops called by teachers; and print and electronic media.

Each of these forms of communication offers a different kind of support to participants. The locus of control over these forms of communication varies: the individual visits and project newsletter are largely organized by university personnel and, it could be argued, reinforce external control over the project; the computer conference is an open forum for private and public communication between all participants (including, significantly, students), but technical problems detracted from its potential to serve as a medium for symmetrical communications; the workshops were organized at the request of teachers on the basis of issues identified by them, and constitute a powerful means for influencing the direction of the project; and the print and electronic media enable teachers and students to speak directly to the community about the project. Other ways in which teachers and students were able to communicate with peers beyond the project included publishing accounts of their experiences for the information and interest of their colleagues at other schools (see for example, Gumley 1989, Shepherd 1990, Shepherd, Hunt, Norman and Gray 1990), and presentations at science and technology workshops organized by the State Ministry of Education.

The experience of this project supports Posch's (1988: 15) view that in environmental education it is important to work towards 'the improvement of teacher–teacher communication and the integration of a greater number of teachers/schools into this exchange of experiences, [and] the production of knowledge on environmental project instruction by the teachers themselves'. It seems important to allow these communication structures to develop practically in response to the range of technical, pedagogical, curricular, political and social issues encountered by teachers as they attempt socially critical forms of environmental education. These issues are idiosyncratic and cannot be foreseen, and can only be fully understood by practitioners in whose professional lives they are expressed. There is a need for an alternative to the technocratic conception of the 'massive problem of communication' (Doll 1974) that bureaucratizes communication and reifies technologies for supporting and reinforcing linear monologues rather than enabling interactive discourse, and which sustains a division of labour between those who would produce and sanction knowledge and those who would acquire it. Social critique, as demonstrated by the experience of this project, is a collaborative endeavour involving several different constituencies within a particular community, and participants in social critique exercise a range of forms or channels of communication. The key questions about communication instanced in this example of socially critical environmental education are: Communication between which parties? Through communication channels designed by whom? Who controls the substantive messages that form the communicated discourse?

Summary and conclusions

In the instances described in this paper the curriculum work of participating schools stimulated social and environmental change: the schools played a role in changing community consciousness about an instance of human impact on an important aspect of the environment (the environmental quality of nearby surfing beaches). The changes in community consciousness stimulated by the activities of Queenscliff High School staff and students brought about a major redistribution of resources aimed at environmental protection (the government directive to upgrade the sewage treatment works). There were changes in the ways certain sectors of the community related to other sectors of the community. There was criticism of conventional wisdoms: for example, that the government agency's authority in respect of water management need not and should not be questioned. And there was the realization that individuals can act collectively to shape society in a way which recognizes, but is to some extent independent of, the constraining influences of traditional hierarchical bureaucracies.

We have emphasized the positive aspects of Queenscliff High School's experience from a socially critical perspective. From the same perspective, the project also had some shortcomings. For example,

- although there were some elements of negotiation between teachers and students in the initial selection of the issue for study there was little negotiation once the study was under way and the study appeared at some times and at some schools to be teacher dominated;
- there was no overt attention to the notion of the empowerment of the students through their activities, nor to the notion of subjecting to criticism the 'dominant ideology of the state' expressed in the role of the Water Board;
- only a small proportion of each school's population was engaged in the project and so a socially critical orientation may not have been emphasized elsewhere in the school.

These points notwithstanding, we believe that the instances we have described exemplify the complementarity of socially critical pedagogy and environmental education, and illustrate ways in which students may be empowered to participate in transforming society democratically. In particular we have described correspondences between socially critical pedagogy and environmental education in terms of the views of knowledge, the role of learners and teachers, broad curriculum organization, school-community liaison and the role of consultants as manifested in the project school.

We believe that the experiences described here also raise a number of questions requiring further research. The first is whether students possess a consciousness of empowerment through involvement in curriculum work of this kind – a realization of their ability to influence substantially the direction of proposed social/environmental changes. A second and related question is whether and in what sense a socially critical perspective in the curriculum is dependent on the presence and activities of particular teachers. A third question concerns the role of the community in supporting and, in a sense, 'justifying' a socially critical orientation in the school. A fourth question concerns the challenge to the conventional dependence on texts and other course materials posed by the necessarily contextualized nature of the emergent working knowledge of socially critical education. These questions may be addressed in future research studies with the project schools.

Notes

1. See for example *The Geelong Advertiser* 10 February 1989; 23 February 1989; 23 March 1989; 25 March 1989; 7 April 1989; 22 April 1989; and [Melbourne] *Age* 20 February 1989; 27 February 1989.
2. See for example, Commission for the Future 1989, Elkington and Hailes 1989, *Margaret Gee's Green Buyer's Guide* 1989, Gell and Beeby 1989 and *101 Ways to Protect Our Environment* 1989, to name a few.
3. Similar conclusions regarding environmental education policies prior to 1987 can be found in Greenall 1987.

References

ARONOWITZ, S. and GIROUX, H. (1985) *Education under Siege: The Conservative, Liberal, and Radical Debate over Schooling* (London: Routledge & Kegan Paul).

COMMISSION FOR THE FUTURE (1989) *Personal Action Guide for the Earth* (Melbourne: Australian Government Publishing Service).

DI CHIRO, G. (1987) Environmental education and the question of a gender: a feminist critique. In I. Robottom (ed.) *Environmental Education: Practice and Possibility* (Geelong: Deakin University Press), 23–48.

DOLL, R. (1974) *Curriculum Improvement: Decision Making and Process* (Boston: Allyn and Bacon).

ELKINGTON, J. and HAILES, J. (1989) *The Green Consumer Guide* (Ringwood, Vic: Penguin).

FIEN, J. (1990) Towards school-level curriculum inquiry in environmental education. Unpublished paper (Division of Australian Environmental Studies, Griffth University, Queensland).

FREIRE, P. and SHOR, I. (1987) *Pedagogy for Liberation: Dialogues on Transforming Education* (London: Macmillan).

GELL, R. and BEEBY, R. (1989) *It's Easy Being Green* (Carlton, Vic: McCulloch).

GIROUX, H. A. (1989) *Schooling for Democracy: Critical Pedagogy in the Modern Age* (London: Routledge).

GREENALL, A. (1981) *Environmental Education in Australia: Phenomenon of the Seventies – A Case Study in National Curriculum Development*. Occasional Paper No. 7 (Canberra, ACT: Curriculum Development Centre).

GREENALL, A. (1987) A political history of environmental education in Australia: snakes and ladders. In I. Robottom (ed.) *Environmental Education: Practice and Possibility* (Geelong: Deakin University Press), 3–20.

GREENALL GOUGH, A. (1990) Red and green: two case studies in learning through ecopolitical action. *Curriculum Perspectives*, 10 (2): 60–65.

GREENALL GOUGH, A. (1991) Greening the future for education: changing curriculum content and school organization. *Journal of Curriculum Studies*, 23 (6): 559–571.

GREIG, S., PIKE, G. and SELBY, C. (1989) *Greenprints for Changing Schools* (London: The World Wide Fund For Nature and Kogan Page).

GUMLEY, S. (1989) Monitoring marine pollution – Queenscliff High School. *MESA* 2 (2): 6–7.

HUCKLE, J. (1991) Education for sustainability: asessing pathways to the future. *Australian Journal of Environmental Education*, 7: 43–62.

KEMMIS, S. (1986) Mapping Utopia: towards a socially critical curriculum. *Interaction*, 14 (5): 11–30.

KEMMIS, S. (1990) School curriculum and social context: contemporary issues. In P. Langford and V. d'Cruz (eds) *Issues in Australian Education* (South Melbourne, Vic: Longman Cheshire), 82–139.

KEMMIS, S. (1991) Emancipatory action research and postmodernisms. *Curriculum Perspectives*, 11 (4): 59–65.

KEMMIS, S., COLE, P. and SUGGETT, D. (1983) *Orientations to Curriculum and Transition: Towards the Socially Critical School* (Melbourne: Victorian Institute of Secondary Education).

MARGARET GEE's GREEN BUYER's GUIDE (1989) (Melbourne: S&W Information Guides).

MASANAUSKAS, J. (1989) Science project grows into a water watchdog. [Melbourne] *Age*, 15 March: 3.

MUHLEBACH, R. and ROBOTTOM, I. (1990) *Environmental Education and Computer Conference Project: Supporting Community-based Environmental Education* (Geelong: Deakin Institute for Studies in Education).

101 WAYS TO PROTECT OUR ENVIRONMENT (1989) (Melbourne: Ministry for Planning and Environment and the Victorian Association for Environmental Education).

POSCH, P. (1988) The project 'Environment and school initiatives'. In *International Conference on Environment and School Initiatives Catalogue Linz, Austria, 26–30 September*. (Paris: Organization for Economic Co-operation and Development, Centre for Educational Research and Innovation), 9–16.

PRESTON, S. (1988) A global approach to learning science. *Education Victoria*, April: 10.

REID, W. A. (1981) The deliberative approach to the study of the curriculum and its relation to critical pluralism. In M. Lawn and L. Barton (eds) *Rethinking Curriculum Studies: A Radical Approach* (London: Croom Helm), 160–187.

ROBOTTOM, I. (1985) Evaluation in environmental education: time for a change in perspective? *Journal of Environmental Education*, 17 (1): 31–36.

ROBOTTOM, I. (1987) Two paradigms of professional development in environmental education. *The Environmentalist*, 7 (4): 291–298.

ROBOTTOM, I. (1990) Environmental education: reconstructing the curriculum for social responsibility. *New Education*, 12 (1): 61—77.

ROBOTTOM, I. and MUHLEBACH, R. (1989) Expanding the scientific community in schools: a computer conference in science education. *Australian Science Teachers Journal*, 35 (1): 39–47.

SHEPHERD, J. (1990) Environmental issues at Queenscliff High School. *Lab. Talk*, 34 (1): 28–30.

SHEPHERD, J., HUNT, L., NORMAN, P. and GRAY, C. (1990) Pollution testing: West Coast surf beaches. Unpublished research report (Queenscliff. Vic: Queenscliff High School).

SHOR, I. (1980) *Critical Teaching for Everday Life* (Boston: South End Press).

UNESCO (1978) *Intergovernmental Conference on Environmental Education, Tbilisi (ussn), 14–26 October 1977: Final Report* (Paris: UNESCO).

UNESCO (1980) *Environmental Education in the Light of the Tbilisi Conference* (Paris: UNESCO).

VICTORIA MINISTRY OF EDUCATION (1990) *Ministerial Policy: Environmental Education* (Melbourne: Ministry of Education, Office of Schools Administration).

Teacher receptivity to curriculum change in the implementation stage: the case of environmental education in Hong Kong

JOHN CHI-KIN LEE

This study examines teacher receptivity to the curriculum change embodied in the new environmental education guidelines in Hong Kong. A questionnaire survey, based on a 'receptivity to change' instrument, was distributed and case studies conducted. The analyses revealed that such variables as the perceived non-monetary cost-benefit of implementing the guidelines, perceived practicality, perceived school and other support, and issues of concern were predictors for teachers' behavioural intentions towards promoting environmental education. The qualitative part of the research also found that, in addition to the factor of perceived non-monetary cost-benefit, the dominance of organizational factors may work to shape teachers' receptivity to environmental education.

In many parts of the world in the 1990s, environmental education (EE) has become a recognized area of the curriculum. Hong Kong began to promote EE in schools during the mid-1980s when environmental protection became an increasingly important governmental issue. In the community, non-governmental environmental organizations ('Green' groups) made efforts to disseminate environmental conservation messages. However, Ng's (1991) survey of the general public of Hong Kong indicated that a minority of the respondents had positive attitudes toward the environment.

Despite the pertinence and urgency of resolving environmental problems, EE remains non-statutory in many curricula, e. g. the UK National Curriculum (Gough 1992, Gayford 1996). In Hong Kong, as in the UK, EE has the status of a non-compulsory, cross-curricular initiative. This status was allocated in the 1992 *Guidelines on Environmental Education in Schools* (hereafter *Guidelines*) (Curriculum Development Council 1992); these *Guidelines* stipulate that EE consists of three interrelated components, education about the environment, education in or with the environment, and education for the environment (Robottom 1987a, Gough 1989, Greenall Gough 1990, Fien 1993, Lee 1997). The *Guidelines* also recommend that

teachers should make use of the following principles in designing programmes for EE:

- experiential learning;
- a balanced viewpoint maintained for all issues;
- emphasis on the formation of attitudes; and
- encouragement of individual contribution and participation.

In addition, the *Guidelines* suggest that EE be implemented through both
the formal and informal curricula and in a cross-curricular manner.

Since the publication of the *Guidelines*, EE implementation has been a
matter of discretion by schools (Lee 1993, Morris 1996, Stimpson 1997).
The Hong Kong Education Department tends to adopt a *laissez-faire*
attitude towards the forms of the *Guidelines*, although the Environmental
Protection Department and non-governmental environmental organizations have provided some support for EE in schools (usually in the form
of teaching materials production and extra-curricular activities, such as
talks at schools, public campaigns and competitions). Most schools tend to
teach EE in the formal curriculum through either existing moral, civic and
religious education programmes or such school subjects as general studies
at the primary level and integrated science, geography and biology at the
secondary level. In the informal curriculum, schools tend to organize such
EE activities as visits to nature reserves and urban and country parks, field
trips, and competitions (Lee 1997).

Teachers' receptivity to change

The propensity for adopting EE, however, depends on teachers' attitudes
towards or receptivity to that curriculum innovation. Although positive
attitudes towards a curriculum innovation may not be an accurate predictor
of implementation of an innovation (Morris 1995), teachers' attitudes can
be crucial in determining the success and failure of an innovation (Brown
and McIntyre 1982, Richardson 1991). However, receptivity is innovation-
specific, because innovations differ in their characteristics and pose various
degrees of benefit or threat to individuals (Giacquinta 1975, Katz *et al.*
1994). Age, sex, and experience are also factors affecting teacher receptivity
to change (Bridges and Reynolds 1968, Kelleher 1981). In the case of EE as
a curriculum innovation, studies have indicated that teachers' resistance
may be a barrier to implementation (Ham and Sewing 1987/1988, Stevenson 1987, 1993, Gough 1997).

Waugh and Punch (1985, 1987) propose a model of teacher receptivity
to change and provide empirical support for the following variables
affecting teacher receptivity to a system-wide change:

- beliefs on general issues of education;
- overall feelings towards the previous educational system;
- attitude towards the previous educational system;
- alleviation of fears and uncertainty associated with the change;
- practicality of the new educational system in the classroom;

- perceived expectations and beliefs about some important aspects of the new educational system;
- perceived support for teacher roles at school in respect of the main referents of the new educational system;
- personal cost-appraisal of the change; and
- beliefs on some important aspects of the new educational system in comparison with the previous one.

This model has been modified and applied in studies of curriculum innovation such as economics and industrial understanding in the UK, the unit curriculum system in Australia (e.g. Waugh and Godfrey 1993, Jephcote and Williams 1994), and teaching models in the US (Fleming 1992). However, most of these studies have been conducted in Western countries. As yet, few studies have been conducted in East Asia that focus on teacher receptivity to EE as curriculum change. This study attempts to bridge this gap.

Culture of schooling and context of primary schools in Hong Kong

In the educational and socio-cultural context, the entrenched Chinese culture of passing civil service examinations as a pathway for social mobility is still dominant in Hong Kong society. This cultural influence has great impact on the aims of education, as well as on the role of the teacher and the nature of classroom activities in primary and secondary schools. From the school perspective, the shared East Asian culture of schooling highlights students' hard work, effort and perseverance, and prizes academic excellence. The role of a teacher emphasizes the teacher's authority and students' obedience to, and respect for, their teacher. Classroom activities focus on the maintenance of classroom order and the efficiency of transmission of knowledge, and little time is devoted to group and individual activities (Lee and Gerber 1996, Lee and Dimmock in 1998). These characteristics persist in Hong Kong primary schools, many of which have either a morning or an afternoon session. In these bi-sessional schools, teachers have to teach classes of 35–40 pupils for 5.5 days per week. In primary grade 6, pupils take an academic aptitude test in Chinese and mathematics, the results of which will affect the pupils' placement in the ability bands of the secondary schools to which they are allocated. As a result, teachers of upper primary levels (grades 5 and 6) tend to focus on drilling and preparation for the academic aptitude test. The situation at the lower primary level is still dominated by teacher-centred pedagogy and heavy reliance on textbooks, despite the introduction of activity approaches in some schools (Morris and Marsh 1991).

As early as 1982, the Llewellyn Panel (1982) reported that primary schooling in Hong Kong was characterized by very formal teaching, teacher resistance to innovation, and a subject-centred rather than child-centred curriculum. A recent example of teacher resistance to innovation was exhibited by the negative reaction from the teaching profession towards the Target-Oriented Curriculum (Carless 1997). Comments from teachers on this curriculum innovation have been to the effect that, because of the realities of primary school teaching, proposals for alternative design and implementation have not been followed. In addition, other recent systemic change initiatives in schools, such as the School Management Initiative in 1991, led to an increase in teachers' work. However, the questions remain: Are teachers unreceptive to a curriculum innovation irrespective of its nature and practicality? Are teachers also unreceptive to EE?

Research methods

The aim of this study is to investigate teacher receptivity to EE in the implementation stage in Hong Kong and to test the relationships between the variables in the model of teacher receptivity proposed by Waugh and his colleagues (Waugh and Punch 1985, 1987, Waugh and Godfrey 1993). This study employs both survey research and case studies (involving informant interviewing) to measure and understand teachers' receptivity to EE. The questionnaire used in this study consists of items modified and translated from Waugh and Godfrey's (1993) original 'receptivity to change' instrument. These items, on a 7-point Likert scale, were piloted and checked by a panel of local experts and practitioners before being used in the main study. The dependent variable in this study is teacher receptivity towards EE, defined in two parts:

- attitudes towards the *Guidelines*; and
- behavioural intentions towards promoting EE.

Variation in teacher receptivity towards EE is assumed to be related to a number of independent and demographic variables. The independent variables (see Appendix) consist of:

- perceived non-monetary *cost-benefit* to the teacher: a ratio between the amount of return and the amount of investment (workload) for the teacher in terms of benefit for the teacher and the pupil;
- perceived *practicality* of the *Guidelines*: teachers' views on how practical the principles suggested by the *Guidelines* are for the implementation of EE in schools;
- issues of *concern*: concern about important issues related to the innovation of EE;
- perceived *school support* for promoting EE: alleviation of teachers' concern through meetings, supportive senior teachers, etc.; and
- perceived *other support* for promoting EE: support for the implementation of EE in school from the governmental and non-governmental organizations as perceived by teachers.

The demographic variables include teachers' sex, rank and teaching experience. The reliabilities of the sub-scales in the instrument are reported in table 1.

Table 1. Reliability of receptivity to change instrument by sub-scales.

Receptivity to change sub-scales	n of items per scale	Reliability (Cronbach's α)
Attitudes towards the *Guidelines*	9	0.8050
Behavioural intentions	5	0.9291
Perceived non-monetary cost-benefit	7	0.8942
Perceived practicality of the *Guidelines*	5	0.8101
Perceived school support	7	0.8363
Perceived other support	5	0.8737
Issues of concern	6	0.8081
Total items	44	0.9069

Samples for the questionnaire surveys

There are over 600 primary schools in Hong Kong. The sample of teachers was systematically drawn from schools, and 3000 questionnaires were administered in 132 schools. One hundred and eight schools contributed 1687 questionnaires (cases) for the analyses, a return rate of 81% on a school basis and 56% on a teacher basis. The distribution of the teacher sample by gender and qualification matched quite closely the characteristics of the general population.

Case-study schools for the qualitative study

In the second phase of the study, a number of case-study schools were chosen based on the results of the questionnaire survey. Three types of case-study schools were selected for this study. 'Type I' was those adopting EE programmes; two schools (labelled 'A' and 'B') were chosen as type I because they were considered to be 'reputational cases' (LeCompte and Preissle 1993) after winning awards for their efforts in developing EE. 'Type II' was schools with relatively high mean receptivity scores, and 'Type III' was schools with relatively low mean receptivity scores. Three schools were selected for case-studies in each of types II ('C', 'D' and 'E') and III ('F', 'G' and 'H'). Table 2 illustrates the characteristics of the case-study schools.

In each of the eight case-study schools, the principal and between five and nine teachers were interviewed. The interviews were tape-recorded and transcribed verbatim. The interviews were conducted in Cantonese and the transcriptions, translated into English, were distributed to the interviewees for verification. In total, 62 tape-recordings were obtained from these eight case-study schools.

The qualitative part of this study focuses on perceptions and interpretations of reality as seen through participants' eyes. To elicit those perceptions, semi-structured interviews were used as the main data-collection technique. A list of questions to elicit responses on the topics was prepared prior to the interviews. The interview questions, which focused on teachers' perceptions of promoting EE, were tied in with the questionnaire items:

Table 2. Characteristics of case study schools.

Case study type	School name	Characteristics
I (Adoption of EE programmes)	A	(1) Adoption of EE programmes (2) High means on behavioural intentions (4.30) (3) High mean scores on perceived non-monetary cost-benefit (4.96)
	B	(1) Adoption of EE programmes (2) High mean scores on behavioural intentions (4.72) (3) High means scores on perceived non-monetary cost-benefit (4.92)
II (High receptivity)	C	(1) High mean scores on behavioural intentions (4.73) (2) High mean scores on perceived non-monetary cost-benefit (4.90)
	D	(1) High mean scores on behavioural intentions (4.92) (2) High mean scores on perceived non-monetary cost-benefit (4.89)
	E	(1) High mean scores on behavioural intentions (4.98) (2) High mean scores on perceived non-monetary cost-benefit (5.02)
III (Low receptivity)	F	(1) Low mean scores on behavioural intentions (3.30) (2) Low means scores on perceived non-monetary cost-benefit (3.67)
	G	(1) Low mean scores on behavioural intentions (3.53) (2) Low means scores on perceived non-monetary cost-benefit (3.81)
	H	(1) Low mean scores on behavioural intentions (3.68) (2) Low mean scores on perceived non-monetary cost-benefit (4.33)

(1) If the school were to implement EE in a more extensive and systematic manner, what would be your reactions? What are your concerns and worries?

(2) What are the obstacles to implementing EE at your school?

(3) Do you and your colleagues support the implementation of EE? Does your school authority support the implementation of EE?

(4) What other questions would you ask that I should have asked (or I have not asked)?

Questions (1) and (2) are related to the sub-scales of issues of concern, perceived non-monetary cost-benefit and perceived other support, whereas question (3) is linked with perceived school support and teacher's behavioural intentions towards promoting EE in schools.

The analysis of qualitative data was completed in an evolving process. A coding scheme with operational definitions and examples from transcriptions was developed. The relationship of some selected categories in the coding scheme and the sub-scales of 'receptivity to change' instrument is shown in table 3. The overall analytic process was an inductive, ongoing cyclical process in which categories and patterns emerged from the data and were later cross-checked (Miles and Huberman 1994).

Table 3. **Relationship between selected categories in the coding scheme and sub-scales of the receptivity to change instrument.**

Selected categories in the coding scheme	Example of field data	Subscale of receptivity to change instrument
(1) IP-WORK (Implied change—teacher's workload) IP-BEN (Perceived importance, meaningfulness and need of EE/Perceived benefit of EE to pupils)	'I worry about a corresponding augmentation of workload.	(1) Perceived non-monetary cost-benefit
(2) IP-RES (Implied change—need and adequacy of human, financial and teaching resources; assistance from Education Department, environmental and voluntary organizations)	'The school should provide enough *materials* to the teachers; otherwise it is difficult for the teachers to carry out the programme.' 'The *Education Department* should provide some assistance for teachers...'	(2) Perceived school support Perceived other support
(3) IP-TEACH (Implied change—time available for teaching syllabuses and EE; teaching practices)	'I have to consider whether the implementation will hinder my normal *teaching*.'	(3) Issues of concern

Results

The results of the correlation analyses indicated that the variables of attitudes towards the *Guidelines* and behavioural intentions towards promoting EE were positively correlated ($p < 0.001$) with the variables of perceived non-monetary cost-benefit, perceived practicality of the *Guidelines*, perceived school support and perceived other support (table 4). However, the issues of concern were negatively and significantly correlated with both attitudes towards the *Guidelines* and behavioural intentions towards promoting EE. As can be seen from table 4, teachers who are more likely to have positive behavioural intentions towards the promotion of EE are those:

- with a perception of high non-monetary benefit from the introduction of EE;
- with perceived support from schools and other agencies; and
- with fewer worries about other relevant issues of concern.

Multiple regressions revealed that, taking attitudes towards the *Guidelines* as the dependent variable, the integrated effect of perceived non-monetary cost-benefit and perceived practicality of the *Guidelines*, and issues of concern explained 34% of the variance (table 5); the variable of perceived non-monetary cost-benefit was found to account for 28% of the

Table 4. Results of correlation analysis (*r*): attitudes towards the *Guidelines*, behavioural intentions and factors affecting teacher receptivity.

	Perceived non-monetary cost-benefit	Perceived practicality of the *Guidelines*	Perceived school support	Perceived other support	Issues of concern
Attitudes towards the *Guidelines*	0.5234**	0.4275**	0.2030**	0.1443**	−0.2328**
Behavioural intentions	0.6331**	0.5732**	0.3932**	0.4178**	−0.0701*

$n = 1687$, 2-tailed significance: * $p < 001$; ** $p < 0.001$.

Table 5. Regression analysis: attitudes towards the *Guidelines* and factors affecting teacher receptivity.

Independent variables	β	t	Change in R^2
Perceived non-monetary benefit	0.359	16.055***	0.281
Issues of concern	−0.150	−9.751***	0.046
Perceived practicality of the *Guidelines*	0.127	5.432***	0.012

$R^2 = 0.584$, adjusted $R^2 = 0.339$, $F = 256.68, ***, p < 0.001$.

Table 6. Regression analysis: behavioural intentions and factors affecting teacher receptivity.

Independent variables	β	t	Change in R^2
Perceived non-monetary benefit	0.491	12.270***	0.404
Perceived other support	0.191	9.317***	0.060
Perceived practicality of the *Guidelines*	0.245	8.003***	0.027
Perceived school support	0.128	5.707***	0.010
Issues of concern	−0.060	−3.089*	0.003

$R^2 = 0.711$, adjusted $R^2 = 0.504$, $F = 300.94$, * $p < 0.01$; *** $p < 0.001$.

variance (table 5). As for behavioural intentions towards promoting EE, the results of the analysis revealed that the integrated effect of perceived non-monetary cost-benefit and perceived practicality of the *Guidelines*, perceived school support and other support, and issues of concern accounted for 50% of the variance (table 6); the variable perceived non-monetary cost-benefit accounted for 40% of the variance (table 6).

It should be noted that, in the regression analyses of teacher receptivity to EE, gender, experience and rank were not significant predictors for attitudes towards the *Guidelines* or behavioural intentions towards promoting EE. In addition, the regression coefficients for 'issues of concern' and 'perceived other support' were negative, i. e. when the value for 'issues of concern' is lower (fewer worries), attitudes towards the *Guidelines* and behavioural intentions towards promoting EE tend to be more positive.

Perceived non-monetary cost-benefit

The analyses of the data from the questionnaire found that the perceived non-monetary cost-benefit of the *Guidelines* was a relatively important factor in predicting teachers' attitudes towards the *Guidelines* and their behavioural intentions towards promoting EE.

The interview findings revealed variations in perceived non-monetary cost-benefit across different groups of schools. In type I schools, some teachers perceived more benefits than costs, while others were inclined to support the promotion of EE if they considered this curriculum change meaningful and necessary. For example:

> Teachers really have to put in much time in [promoting EE]. It requires many school resources. However, when we weigh this against its importance, it is worthwhile . . . . Pupils are better off. The idea of [environmental protection] takes root in their hearts as they grow up (Teacher 2, School B, interview).

In type II schools, most teachers supported implementing EE in their schools. In school C, for example, some teachers were supportive of EE, although others saw a conflict between promoting EE and their existing workload. In school D, teachers tended to perceive EE as important and necessary in primary education. In school E, some teachers were supportive of EE: they felt that it would be acceptable if their workload was not enormously increased and their teaching not seriously affected. The following illustrates how teachers valued the importance of (or benefit from) promoting EE and its perceived workload (cost):

> My colleagues will, first of all, consider whether the desired objective is lucid and meaningful. If affirmative, they will support further promotion. If negative, they will think that time will be squandered and prefer normal teaching in the classrooms (Teacher 5, School C, interview).

In type III schools, the principals and many teachers thought that if an EE activity was really good for the pupils and also practicable, they would support it. However, it was notable that in school F there was an ambivalent view, among some teachers, of some of their colleagues' willingness to promote (or receptivity to) EE. In addition, some colleagues were seen as not giving EE a priority over academic performance and other cross-curricular innovations such as moral and civic education. In school G, there was a spectrum of responses towards the EE adoption. Some teachers thought that if the activities were meaningful, they would be willing to promote them; some teachers did not give EE a high priority; and some did not even mention the benefit or importance of EE. School G's teachers saw a dilemma between launching EE activity and facing possible resistance from teachers. In school H, there was also a divergence in the responses of teachers: some were supportive of the EE programme (or activity) but hesitant because of time-constraints; some did not explicitly mention the value or the importance of EE but, rather, worried about the extra workload, because they were already very busy; some even thought the cost of promoting EE activities, such as waste-paper recycling, out-

Table 7. Convergence of quantitative and qualitative findings.

School	Type	Survey findings	Qualitative (interview) findings
A, B	I	(1) high mean score on behavioural intentions.	(1) willingness of individual teachers and their colleagues to support the promotion of EE.
		(2) high mean score on perceived non-monetary cost-benefit.	(2) perceived benefits over cost and willingness to promote EE on the basis of the meaningfulness of the activities.
C, D, E	II	(1) high mean score on behavioural intentions.	(1) willingness of individual teachers and their colleagues to support the promotion of EE.
		(2) high mean score on perceived non-monetary cost-benefit.	(2) perceived benefits over cost, willingness to promote EE on the basis of the meaningfulness of the activities and willingness to promote EE on condition that the workload is not enormously increased and the normal teaching schedule is not seriously affected.
F, G, H	III	(1) low mean score on behavioural intentions	(1) willingness of individual teachers to support the promotion of EE; perception that some of their colleagues might not be willing to bolster the promotion.
		(2) low mean score on perceived non-monetary cost-benefit	(2) willingness to promote EE on the basis of the meaningfulness of the activities; perception of conflict between the benefits of EE activities and the constraints; indifference to the values of EE and even negative reactions towards some EE activities.
			(3) emphasis of existing heavy workload or cost of adopting the innovation.

weighed the benefits. School H's principal did not see EE as having a higher priority than other matters.

It was apparent that there were differences between teachers' attitudes towards the EE adoption in different types of case-study schools. In type I and type II schools, more teachers tended to attribute benefits and importance to EE and expressed a willingness to support it if activities were meaningful and worthwhile, the workload acceptable, or the interference in normal teaching not serious. In type III schools, there was a spectrum of teachers' responses, ranging from willingness to endorse EE if the activities associated with it are meaningful, a dilemma between acknowledging the significance of the activities and the resulting workload (or limited time available), to indifference or even negative attitudes towards the value of EE activities. Table 7 summarizes the convergence of survey and interview findings.

In addition to acknowledging the value of EE, some teachers in type III schools (schools F and H) tended to over-emphasize the workload factor. One teacher stated that school F, having 24 classes, had a heavier workload than other schools with a larger number of classes; there was only one secretary, so teachers had to share the secretarial duties. Another teacher in school F also mentioned their workload:

> We are very busy . . . very busy . . . everyday . . . . I have to mark the exercise books . . . In recess, I usually have to help some pupils. Sometimes I am not free even in recess (Teacher 2, School F, interview).

One school H teacher commented that teachers had extra duties, such as stamping pupils' handbooks and newly purchased library books with the school's name. He also observed that school H held more tests each year than other primary schools.

In other words, the interview data appear to support the survey finding that perceived non-monetary cost-benefit is a significant factor in explaining teachers' receptivity and that schools with high receptivity scores tend to have higher scores on perceived non-monetary cost-benefits from EE than schools with low receptivity scores (table 2).

Perceived school and perceived other support, and issues of concern

Type I schools generally had relatively higher aggregate scores for perceived school support than other schools. These results supported the interview findings that the principals, a critical mass of some supporters, and their fellow-teachers, made serious efforts to promote EE. However, across all the case-study schools, the survey results identified low scores in perceived 'other support'. The interview results, in general, supported this finding, and showed that, in the eyes of the teachers and principals, the assistance from the Education Department and environmental organizations in promoting EE at schools was inadequate.

In addition, one item from the survey, 'the introduction of EE will lead to less time being available for the teaching of the subject syllabus', had a relatively high mean (4.64). Another item, 'the principle of forming positive environmental attitudes included in the *Guidelines* matches my knowledge and skills in teaching environmental education', had a relatively low mean (4.16) across the schools. The interview results, in convergence with the survey findings, showed that many teachers from different schools felt that EE promotion might hinder their normal teaching and that they might not have adequate knowledge about EE and the ways of implementing the curriculum change.

Dominance of organizational factors in the interview findings

The interview findings also initiated some 'new lines of thinking through attention to surprises and paradoxes' (Miles and Huberman 1994: 41). During the interviews, I was surprised by the extent to which organizational factors preoccupied teachers' talk about their receptivity to EE as a curriculum change. The term 'organization' refers to the management of activities that create favourable human and resource conditions to encourage the implementation of EE activities. The organization of EE encompasses the following aspects:

- timing and scale of programmes and activities;
- distribution of workload or division of labour;
- procedural clarity and planning; and
- appointing a co-ordinator or setting up a committee.

The interview results showed that teachers from type I and II schools tended to consider the timing of EE programmes and activities seriously. Some thought the activities should not be introduced abruptly; some teachers in primary schools with either morning or afternoon sessions preferred the activities to be organized on Saturdays, in assemblies, or after the examination periods. In addition, many teachers across different schools thought their colleagues would tend to support EE programmes or activities which were 'phased or short-lived': 'If the principal is inclined to sacrifice the weekly assembly or some activities, the teachers are generally willing to assist' (Teacher 7, School H, interview). Some teachers across different schools were concerned about the division of labour and workload:

> It depends on how the school administration distributes the work to the teachers. It should not be concentrated on just one or two teachers [who are] responsible for the work but all the teachers in the school should share the work (Teacher 1, School D, interview).

Most teachers are influenced by the 'practicality ethic' (Doyle and Ponder 1977) and one component of the this ethic is instrumentality or procedural clarity, i. e. the 'hows' of implementation (Fullan 1991: 128, Lee and Wong 1996):

> If the programme has been well-planned, carries with it a manifest goal and will have a meaningful effect, the teachers will be quite co-operative (Teacher 5, School C, interview).

In type I schools, a co-ordinator and a group of teachers promoting the curriculum change was the dominant organizational and planning pattern. In type II and III schools, some teachers also perceived the need to have a co-ordinator or a committee to oversee EE planning: a teacher in school D suggested that the establishment of an EE committee was the best way to promote EE in the school (Teacher 6, interview); a teacher in school F also suggested that the promotion of large-scale activities could be encouraged if 'a team' could be in charge of EE (Teacher 3, interview). The principal in school F remarked that one of the main obstacles to the adoption of EE in the school was the appointment of a person-in-charge to steer the promotion. In the case of school H, the principal remarked that:

> if . . . a committee is set up to steer the entire programming, the teachers will just be required to do some collection work and won't be required to hold time-wasting meetings. They will then not be unsupportive.

Discussion

The results in this study show that, in general, Hong Kong teachers revealed positive attitudes towards the *Guidelines* and behavioural intentions towards promoting EE. Teachers perceived that the *Guidelines* were practical and the non-monetary benefit of implementing the *Guidelines* greater than the cost. Teachers, however, thought that school support for EE implementation was moderate and that support from outside agencies, such as government departments and environmental organizations, was inadequate. The survey data also suggested that the variables from the model proposed by Waugh and his colleagues (Waugh and Punch 1985, 1987, Waugh and Godfrey 1993), i.e. perceived non-monetary cost-benefit of implementing the *Guidelines*, perceived practicality, perceived school support and other support, issues of concern are predictors for the teachers' behavioural intentions towards promoting EE.

These results confirm that model of receptivity to change devised by Waugh and his colleagues is applicable to the context of EE in primary schools in Hong Kong as well as in Western countries. Perceived non-monetary cost-benefit was found to be the most significant predictor in the regression analyses. The interview data from the case studies, which in general converged with the survey results, also illustrate the importance of perceived non-monetary cost-benefit of promoting EE. This is consistent with Doyle and Ponder's (1977) and Brown and McIntyre's (1993) claims that the cost component of the 'practicality ethic' is influential on teachers' receptivity to, and adoption of, curriculum change.

To improve teachers' receptivity to EE and their willingness to support it, principals and senior teachers need to explain the benefits and value of EE, particularly in terms of benefits such as better pupil learning, heightened environmental awareness on the part of pupils, increased satisfaction in teaching, and improvement in school and community environments. This is especially important, as effective EE implementation is not linked to a teacher's promotion prospects. As teachers usually place higher priorities on covering the syllabuses and have worries about excessive workloads, it is essential that administrators provide sufficient resources, as well as clear and practical procedures, for teaching EE through the various subjects. In addition, it is desirable that some strategies and mechanisms be set up in schools to alleviate teachers' fears and concerns about introducing EE. Regular staff meetings, internal circulars and meetings with external agents (such as EE and environmental monitoring professionals) may be helpful in cultivating teachers' positive attitudes towards change. The support of senior staff is crucial; senior staff can help to communicate the benefits of change, provide informal advice, and alleviate teachers' concerns. Teachers' receptivity to EE can be further enhanced if principals play a leading and supporting role by taking the initiatives in adoption and providing adequate support for change.

The primary function of the principal in schools is not to identify needs for others, but to involve the staff in identifying self-perceived needs and formulate a policy for staff development within which an innovation may take place. Lieberman and Miller (1992: 26) assert that 'the major task is to get people involved in *their* definition of the problem, *their* view of a meaningful activity'. Day *et al.* (1985: 110) suggest a list of questions the principal needs to pose in order to move towards successful leadership for change:

(1) What help and support do individuals need from me? Can this be made available?
(2) What help and support does the teacher/group of teachers need from (a) inside (b) outside the school? Can this be made available?
(3) What are the priorities for action? (What is appropriate at this time, at this stage of development?)
(4) Is the activity practical in terms of time, energy and resources?
(5) What will be the teachers' gain from the activity? (Doyle and Ponder's (1977) 'pragmatic sceptics' will want to know what's in it for them and their pupils, and what the 'cost' will be.)

However, such staff or professional development should be critical, inquiry-based, participatory, and practice-based, which entails both a critique of the EE values that underpin existing EE activities and organizational practices and a collaborative engagement in political struggles (such as struggle for more time and resources for EE) between the external agents and the teachers (Robottom 1987b). In addition, teachers can be empowered through emancipatory or critical action research, which helps improves teachers' willingness to take risks and experiment and change existing constraints through new practices and critical reflection (Grundy 1987).

The results of this study suggest that teachers face severe constraints in terms of 'universal tensions' in primary education, such as teaching the subjects in the limited amount of time in the school day, and organizational barriers (Lieberman and Miller 1992). The organization of EE activities, for example, was a major factor affecting teachers' receptivity. This is congruent with the literature on EE implementation pinpointing the peripheral status of EE in the school curriculum (Gough 1997), the existence of logistical barriers (Ham and Sewing 1987/1988), and the implementation of EE as a purely technical concern (Stevenson 1987, 1993). In particular, teachers were concerned over the scheduling, the scale, the duration, as well as the procedural clarity of the activities. These concerns are basically in line with the existing school change literature emphasizing effective planning of activities (Bishop 1986). The concerns of teachers about scheduling and organization of EE activities also provide empirical support for Hungerford and Peyton's (1986) claim that scheduling or time considerations, as well as programme co-ordination, are critical concomitant variables for implementing EE curricula.

The emphasis of non-monetary cost-benefit and effective planning of activities, if considered together with teachers' concerns about the issues of heavy workload and time for teaching, point to the importance of addressing teachers' work in any study of teachers' receptivity to curriculum change. This echoes the findings from a recent case study of site-based reform in an elementary school in the USA, in which Gitlin and Margonis (1995: 403) claim that:

until fundamental injustices in the character of teachers' work are addressed, meaningful reform is unlikely . . . . [R]eformers might be better off focusing on the preconditions for reform: giving teachers the authority and time they need to teach in ways they find educationally defensible . . . . [T]eachers' workloads should be decreased to allow time for planning, curriculum development, and innovative pedagogy.

Teachers' low receptivity or resistance may not be just a matter of inadequate school and outside assistance, but also may be related to the lack of change in teachers' workload.

An often-posed question on studies of teachers' receptivity to change is the relationship between receptivity and usage. In this study, in terms of the number of EE activities organized, there were no remarkable differences between type II and type III schools. It is notable, however, that none of type II and III schools had the same degree of EE development as type I schools. Furthermore, only the type I schools and type II school C revealed relatively high levels of institutional commitment in promoting EE: in school C there was support from a core group of teachers and organization of an environmental protection month. This suggests that high receptivity seems to be a necessary but not a sufficient condition for relatively high level of usage and institutional commitment at the school level.

The results of this study suggest that a general model of teacher receptivity to EE includes the following independent variables:

- perceived non-monetary cost-benefit (a main variable in the model of Waugh and his colleagues (Waugh and Punch 1985, 1987, Waugh and Godfrey 1993));
- perceived practicality of the *Guidelines* (a main variable in the model of Waugh and his colleagues);
- perceived school support (a main variable in the model of Waugh and his colleagues);
- perceived other support (a main variable in the model of Waugh and his colleagues);
- issues of concern (a main variable in the model of Waugh and his colleagues);
- timing and scale of EE programmes;
- distribution of workload or division of labour;
- procedural clarity and planning; and
- appointment of a co-ordinator or setting up a committee.

Although this study shows that the model of Waugh and colleagues is applicable in the context of primary schools in Hong Kong, the combined use of quantitative and qualitative methods in this study uncovers the component of 'organization of activities' that seems to be relatively neglected in the study of teachers' receptivity in the existing literature. Issues of timing and scale of EE programmes and activities, distribution of workload, procedural clarity and planning of activities and appointing a co-ordinator or setting up a committee emerge from the interviews that complement the survey findings. Future studies may include these components for testing the model's generalizability in other settings and studying the relationship between teachers' receptivity to, and their implementation of, educational change.

Acknowledgements

I would like to thank Wong Hin-wah for his guidance, and Noel Gough, Russell Waugh, Allan Walker and other referees for their valuable comments for improving this paper.

References

BISHOP, G. (1986) *Innovation in Education* (London: Macmillan).

BRIDGES, E. M. and REYNOLDS, L. B. (1968) Teacher receptivity to change. *Administrator's Notebook*, 16 (6), 1–4.

BROWN, S. and MCINTYRE, D. (1982) Influences upon teachers' attitudes to different types of innovation: a study of Scottish Integrated Science. *Curriculum Inquiry*, 12 (1), 35–51.

BROWN, S. and MCINTYRE, D. (1993) *Making Sense of Teaching* (Buckingham, UK: Open University Press).

CARLESS, D. R. (1997) Managing systemic curriculum change: a critical analysis of Hong Kong's Target-Oriented Curriculum initiative. *International Review of Education*, 43 (4), 349–366.

CURRICULUM DEVELOPMENT COUNCIL (1992) *Guidelines on Environmental Education in Schools* (Hong Kong: Education Department).

DAY, C., JOHNSTON, D. and WHITAKER, P. (1985) *Managing Primary Schools: A Professional Development Approach* (London: Harper & Row).

DOYLE, W. and PONDER, G. A. (1977) The practicality ethic in teacher decision-making. *Interchange*, 8 (3), 1–12.

FIEN, J. (1993) *Education for the Environment: Critical Curriculum Theorising and Environmental Education* (Geelong, Australia: Deakin University Press).

FLEMING, M. (1992) Teachers' Receptivity to Teaching Models. Doctoral dissertation, University of Arizona.

FULLAN, M. with STIEGELBAUER, S. (1991) *The New Meaning of Educational Change*, 2nd edn (New York: Teachers College Press).

GAYFORD, C. (1996) Environmental education in schools: an alternative framework. *Canadian Journal of Environmental Education*, 1, 104–120.

GIACQUINTA, J. B. (1975) Status, risk, and receptivity to innovations in complex organizations: a study of the responses of four groups of educators to the proposed introduction of sex education in elementary schools. *Sociology of Education*, 48 (1), 38–58.

GITLIN, A. and MARGONIS, F. (1995) The political aspect of reform: teacher resistance as good sense. *American Journal of Education*, 103 (4), 377–405.

GOUGH, A. (1997) *Education and the Environment: Policy, Trends and the Problems of Marginalisation* (Melbourne, Australia: Australian Council for Educational Research).

GOUGH, N. (1989) From epistemology to ecopolitics: renewing a paradigm for curriculum. *Journal of Curriculum Studies*, 21 (3), 225–241.

GOUGH, N. (1992) *Blueprints for Greening Schools* (Melbourne, Australia: Gould League).

GREENALL GOUGH, A. (1990). Red and green: two case studies in learning through ecopolitical action. *Curriculum Perspectives*, 10 (2), 60–65.

GRUNDY, S. (1987) *Curriculum: Product or Praxis?* (London: Falmer).

HAM, S. H. and SEWING, D. R. (1987/1988) Barriers to environmental education. *Journal of Environmental Education*, 19 (2), 17–24.

HUNGERFORD, H. R. and PEYTON, R. B. (1986) *Procedures for Developing an Environmental Education Curriculum: A Discussion Guide for UNESCO Training Seminars on Environmental Education* (Paris: UNESCO).

JEPHCOTE, M. and WILLIAMS, M. (1994) Teacher receptivity to the introduction of economic and industrial understanding. *Economics and Business Education*, 2 (4), 163–167.

KATZ, E. H., DALTON, S. and GIACQUINTA, J. B. (1994) Status risk taking and receptivity of Home Economics teachers to a statewide curriculum innovation. *Home Economics Research Journal*, 22 (4), 401–421.

KELLEHER, R. R. (1981) The Relationship of Three Parts of Field Experience Programs to Student Teacher Open-mindedness, Attitudes Towards Education, and Receptivity to Innovation. Doctoral dissertation, Michigan State Univerisity, East Lansing, MI.

LeCOMPTE, M. D. and PREISSLE, J. (1993) *Ethnography and Qualitative Design in Educational Research*, 2nd edn (San Diego, CA: Academic Press).

LEE, C. K. and WONG, H. W. (1996) *Curriculum: Paradigms, Perspectives and Design*, 2nd edn (Hong Kong: Chinese University Press).

LEE, C. K. J. (1993) Geography teaching in England and Hong Kong: contributions towards environmental education. *International Research in Geographical and Environmental Education*, 2 (1), 25–40.

LEE, J. C. K. (1997) Environmental education in schools in Hong Kong. *Environmental Education Research*, 3 (3), 359–371.

LEE, J. C. K. and DIMMOCK, C. (1998) Curriculum management in secondary schools during political transition: a Hong Kong perspective. *Curriculum Studies*, 6 (1), 5–28.

LEE, J. C. K. and GERBER, R. (1996) The lived experience of curriculum change: a Hong Kong perspective. *Curriculum and Teaching*, 11 (1), 35–47.

LIEBERMAN, A. and MILLER, L. (1992) *Teachers—Their World and Their Work: Implications for School Improvement* (New York: Teachers College Press).

LLEWELLYN PANEL (1982) *A Perspective on Education in Hong Kong: Report by a Visiting Panel* (Hong Kong: Hong Kong Government).

MILES, M. B. and HUBERMAN, A. M. (1994) *Qualitative Data Analysis: An Expanded Sourcebook*, 2nd edn (Thousand Oaks, CA: Sage).

MORRIS, P. (1995) *Curriculum Development in Hong Kong*, 2nd edn (Hong Kong: Faculty of Education, University of Hong Kong).

MORRIS, P. (1996) *The Hong Kong School Curriculum: Development, Issues and Policies*, 2nd edn (Hong Kong: Hong Kong University Press).

MORRIS, P. and MARSH, C. (1991) Patterns and dilemmas. In C. Marsh and P. Morris (eds), *Curriculum Development in East Asia* (London: Falmer), 255–271.

NG, T. L. G. (ed.) (1991) *New Environmental Paradigm Survey 1991* (Hong Kong: Hong Kong Environment Centre).

RICHARDSON, V. (1991) How and why teachers change. In S. C. Conley and B. S. Cooper (eds), *The School as a Work Environment: Implications for Reform* (Boston, MA: Allyn and Bacon), 66–87.

ROBOTTOM, I. (1987a) Contestation and consensus in environmental education. *Curriculum Perspectives*, 7 (1), 23–27.

ROBOTTOM, I. (1987b) Towards inquiry-based professional development in environmental education. In I. Robottom (ed.), *Environmental Education: Practice and Possibility* (Geelong, Australia: Deakin University Press), 83–119.

STEVENSON, R. B. (1987) Schooling and environmental education: contradictions in purpose and practice. In I. Robottom (ed.), *Environmental Education: Practice and Possibility* (Geelong, Australia: Deakin University Press), 69–82.

STEVENSON, R. B. (1993) Becoming compatible: curriculum and environmental thought. *Journal of Environmental Education*, 24 (2), 4–9.

STIMPSON, P. G. (1997) Environmental challenge and curricular responses in Hong Kong. *Environmental Education Research*, 3 (3), 345–357.

WAUGH, R. and GODFREY, J. (1993) Teacher receptivity to system-wide change in the implementation stage. *British Educational Research Journal*, 19 (5), 565–578.

WAUGH, R. F. and PUNCH, K. F. (1985) Teacher receptivity to system-wide change. *British Educational Research Journal*, 11 (2), 113–121.

WAUGH, R. F. and PUNCH, K. F. (1987) Teacher receptivity to system-wide change in the implementation stage. *Review of Educational Research*, 57 (3), 237–254.

Appendix: 'Receptivity to change' instrument

Part I. Attitude towards the Guidelines *on Environmental Education in Schools (4.42, 0.74)*

Teachers were asked to respond to 10 adjective pairs as a seven-category semantic differential with the Guidelines as the referent (mean and standard deviation for the respondents in parentheses).

The adjective pairs are:

(1) satisfactory/unsatisfactory (4.10, 1.21);
(2) valuable/invaluable (4.69, 1.25);
(3) wise/foolish (4.18, 1.21);
(4) permissive/restrictive (4.46, 1.09);
(5) intelligent/absurd (4.86, 1.11);
(6) realistic/idealistic (4.21, 1.29);
(7) effective/ineffective (4.22, 1.18);
(8) necessary/unnecessary (4.74, 1.13); and
(9) uncomplicated/complicated (4.28, 1.22).

Part II.

Teachers were asked to respond to items in parts II–VII as follows:

- 'disagree very strongly' (1),
- 'disagree strongly' (2),
- 'disagree' (3),
- 'neutral' (4),
- 'agree' (5),
- 'agree strongly' (6), and
- 'agree very strongly' (7).

Perceived non-monetary cost-benefit of the *Guidelines* to the teacher (4.66, 0.89):

(1) In weighing up the balance between the work generated for me by the *Guidelines* and my satisfaction with teaching, I think the *Guidelines* are worthwhile (4.58, 1.12).
(2) In weighing up the balance between the work generated for me by the *Guidelines* and my improvement in environmental knowledge, I think the *Guidelines* are worthwhile (4.79, 1.09).
(3) In weighing up the balance between the work generated for me by the *Guidelines* and improvement in classroom learning, I think the *Guidelines* are worthwhile (4.90, 1.06).
(4) In weighing up the balance between the work generated for me by the *Guidelines* and the increased commitment towards improving

environmental quality by the students, I think the *Guidelines* are worthwhile (5.02, 1.07).

(5) In weighing up the balance between the work generated for me by the *Guidelines* and praise by my school principal, I think the *Guidelines* are worthwhile (4.96, 1.07).

(6) In weighing up the balance between the work generated for me by the *Guidelines* and better environmental quality in schools and local communities, I think the *Guidelines* are worthwhile (4.13, 1.26).

(7) In weighing up the balance between the work generated for me by the *Guidelines* and improvement in my professional status as a teacher, I think the *Guidelines* are worthwhile (4.24, 1.29).

Part III. Perceived practicality of the Guidelines *(4.63, 0.86)*

(1) The principle of experiential learning integral to EE suggested by the *Guidelines* suits my classroom teaching style (4.78, 1.13).

(2) The EE principle of a balanced viewpoint maintained for environmental issues reflects my educational philosophy (4.75, 1.06).

(3) The principle of encouraging the individual pupil's contribution and participation suggested by the *Guidelines* is likely to be realized in my classroom context (4.69, 1.13).

(4) The principles of implementing EE through both the formal and informal curriculum are appropriate to meeting the needs of the pupils in my school (4.75, 1.04).

(5) The principle of forming positive environmental attitudes included in the *Guidelines* matches my knowledge and skills in teaching EE (4.16, 1.31).

Part IV. Issues of concern associated with implementing EE (4.18, 1.03)

(1) I am concerned that pupils have incorrect attitudes towards environmental issues (4.61, 1.32).

(2) I am concerned that General Studies, Social Studies, Health Education and Science receive more attention than other subjects (3.69, 1.47).

(3) I am concerned that the introduction of EE will result in lower academic performance among the students at this school (3.70, 1.45).

(4) I am concerned that the introduction of EE will lead to less time being available for the teaching of the subject syllabus (4.64, 1.44).

(5) The pupils' abilities are causing me concern in regard to the teaching of EE at this school (4.13, 1.45).

(6) Disciplinary problems are causing me concern in regard to the teaching of EE at this school (4.30, 1.55).

Part V. Perceived school support for teaching EE (4.10, 1.00)

(1) There are regular meetings at which I can raise my worries and doubts about the implementation of EE (4.56, 1.35).
(2) Whenever there are problems of implementing EE, there is a senior teacher whom I can ask for advice (4.45, 1.48).
(3) There is good support whenever I have problems, such as a shortage of books and equipment, related to EE (4.37, 1.39).
(4) There are regular school-based talks or training programmes at which I can learn how to teach EE (3.34, 1.61).
(5) The majority of teachers in this school support EE (3.97, 1.26).
(6) The principal encourages teachers to participate in training course related to EE (4.30, 1.32).
(7) At school meetings, the principal makes comments emphasizing the importance of introducing EE at this school (3.74, 1.43).

Part VI. Perceived other support for teaching EE in schools (3.30, 1.08)

(1) In my opinion, the government departments support the implementation of EE in my school (3.03, 1.29).
(2) In my opinion, the Curriculum Development Institute and the Education Department provide sufficient suggestions and assistance to help teachers acquire the methods of implementing EE in my school (3.04, 1.33).
(3) In my opinion, the environmental organizations in Hong Kong provide adequate support for promoting EE in my school (3.21, 1.29).
(4) In my opinion, the majority of parents in this school supports the implementation of EE in this school (3.85, 1.44).
(5) In my opinion, the local community organizations provide adequate environmental activities for pupil participation (3.36, 1.30).

Part VII. General behavioural intentions towards promoting EE (4.29, 1.07)

(1) In my behaviour and communications with other teachers, I will actively and openly support the introduction of EE at this school in the academic year 1995–1996 (4.39, 1.23).
(2) In my behaviour and communications with other teachers, I will praise the introduction of EE at this school in the academic year 1995–1996 (4.39, 1.21).
(3) In my behaviour and communications with other teachers, I will propose the introduction of EE at this school in the academic year 1995-1996 (4.32, 1.21).

(4) In my behaviour and communications with other teachers, I will tell them that the *Guidelines* are both flexible and feasible and hence supportable in the academic year 1995–1996 (4.24, 1.19).

(5) In my behaviour and communications with other teachers, I will advise them that the *Guidelines* can be adapted to the abilities of pupils at this school in the academic year 1995-1996 (4.24, 1.20).

Complementary curriculum: the work of ecologically minded teachers

CHRISTY M. MOROYE

Myriad international efforts exist to infuse and reform schools with ecological perspectives, but in the US those efforts remain largely on the fringes of schooling. The purpose of this study is to offer a perspective on this issue from inside schools. If one looks to the future success of environmental education, one must consider the work of teachers. To that end, this qualitative study explores the intentions and practices of three ecologically-minded teachers in traditional public high schools in the US. The research methodology used was eco-educational criticism, an arts-based inquiry method with an ecological lens. The teachers were interviewed about their intentions for their students, and then observed to see how those intentions were manifested in the classroom. A follow-up interview then synthesized the connection between their ecological beliefs and their general intentions for students.

Introduction

Public interest in global environmental issues has surged. From newspaper cover stories to political causes to sitcom story-lines, 'green' perspectives and conversations are becoming more commonplace. Both formal and non-formal education has, since the 1970s, been asked to respond to this growing concern (International Union for the Conservation of Nature and Natural Resources/UNESCO 1970, United Nations Conference on Environment and Development 1992), and, to that end, researchers, practitioners, government agencies, and communities have worked to implement environmental and ecological education models. However, these initiatives remain largely on the fringes of schooling, particularly in the US. The purpose of this study is not to elaborate on why environmental education remains on the 'outside', but rather to offer another perspective—from inside the schools themselves. That perspective comes from ecologically-minded teachers who work in traditional US public schools and who teach 'non-environmental' curricula,[1] that is, teachers who are not explicitly engaged in teaching about the environment or in environmental education programmes.

Environmental education is a collective, broad term encompassing many facets of earth-inclusive education. 'Traditional' environmental education has roots in nature study, conservation education, and outdoor education, and is often found in supplementary programmes and activities that occur in addition to the 'regular' curriculum (Heimlich 2002). A more recent movement has emerged toward 'ecological education' (see Orr 1992, Jardine 2000), which Smith and Williams (1999: 3) define as 'an emphasis on the inescapable embeddedness of human beings in natural systems'. Other models include place-based education (Sobel 2004, Noddings 2005, Smith 2007), eco-justice education (Bowers 2001, Martusewicz 2005), education for sustainability (Sterling 2001), and education for sustainable development (Jickling and Wals 2007), to name a few. Jickling and Wals (2007) point out that this last model, education for sustainable development, while somewhat contested, 'has become widely seen as a new and improved version of environmental education, most visibly at the national policy level of many countries' (p. 4), although such policies remain absent in the US. While myriad models exist, Gruenewald and Manteaw (2007: 173) note that environmental education continues to be 'marginalized, misunderstood as mainly about science, and in many places totally neglected'.

There may be many reasons for environmental education's neglect or 'failure' (Blumstein and Saylan 2007), but certainly, if we look to the future success of environmental education in any of the above models, we must consider the work of teachers. To that end, many researchers have investigated a variety of aspects of the roles of teachers in environmental education. Cutter-Mackenzie and Smith (2001) looked at teachers' environmental knowledge or 'eco-literacy', and their related beliefs about the importance of attitudes toward, rather than knowledge about, the environment. Robertson and Krugly-Smolska (1997) report on three sources of the 'gap' between environmental education theory and practice: (1) 'the practical', in terms of variables such as time, materials, and schedules, (2) 'the conceptual', referring to 'conflicting ideas and resources that (make it difficult) for teachers to understand what the task of environmental education really is'; and (3) 'teacher responsibility', referring to the idea that 'teachers are not completely certain that they are permitted to do many of the things that are necessary to accomplish the lofty social and political goals of environmental education' (p. 316). Other studies (Dillon and Gayford 1997, Cotton 2006a, b) discuss teachers' beliefs and actions related to controversial environmental issues in the curricula. These and other studies illustrate that environmental education is no easy task for teachers.

While other studies, such as the ones described above, have focused on teachers in sanctioned environmental education settings, I focus on teachers in traditional US public schools who happen to be ecologically-minded, but whose curricular responsibilities do not necessarily include environmental topics. I selected teachers in social studies and English/language arts for two reasons. First, social studies and language arts are largely unexplored environmental education territory (Heimlich 2002). Secondly, while environmental science and technology may play an important role in mediating the environmental crises we face, many suggest that cultural values play at least an equal part (see, e.g. Bowers 1993, Blumstein and Saylan 2007,

Gruenewald and Manteaw 2007). Subject areas like English/language arts and social studies, which contribute to transmitting and transforming cultural values, may have an important role in environmental and ecological education reform.[2]

By studying the intentions and actions of ecologically-minded teachers in public schools, I was able to discern themes that emerged naturally as a result of teachers' strongly held beliefs. One such theme is a new term I argue for as an addition to the curricular lexicon, the *complementary curriculum*. This is not an attempt to redefine curriculum—it already has many definitions (see Connelly *et al.* 2008); instead, it is an attempt to call attention to a particular type of curriculum and, by so doing, offer the potential for expanding ecological perspectives in schools. I start, therefore, with the broad definition of curriculum offered by He *et al.* (2008: 223):

> Curriculum for us is a dynamic interplay between experiences of students, teachers, parents, administrators, policy-makers, and other stakeholders; content knowledge and pedagogical premises and practices; and cultural, linguistic, sociopolitical, and geographical contexts.

Within this definition the complementary curriculum is situated in the kinds of experiences teachers provide for students, as well as in the 'pedagogical premises and practices' that result from the teachers' beliefs.

In his discussion of the 'curriculum shadow', Uhrmacher (1997) argues for the use of a variety of terms to specify different curricula. He distinguishes, for example, the shadow curriculum and the null and hidden curricula. The *shadow curriculum* identifies a 'disdained' or neglected curriculum that could in fact improve the pedagogy at hand (Uhrmacher 1997). As an example Uhrmacher points to a social studies teacher who, in the name of order and efficiency, lectured on the US Constitution rather than encouraging discussion, which could be considered a more democratic means of learning.

The *null curriculum* (Flinders *et al.* 1986, Eisner 2002) describes what is missing. It includes intellectual processes and subject matter (Eisner 2002), as well as affect (Flinders *et al.* 1986). The null curriculum might include singular topics or perspectives as well as entire fields of study.[3] The *hidden curriculum* identifies the norms of schooling. Thus, Jackson (1968) distinguishes the official curriculum from the associated skills required to master it, skills such as putting forth effort, completing homework, and understanding and operating within institutional norms. Together these and other 'unofficial' aspects of what is taught in schools constitute the hidden curriculum.

Of the three terms discussed here, the *complementary curriculum* is most closely associated with the hidden curriculum. However, there are at least two key differences between the two. First, the hidden curriculum has its origins in something more ominous, or at the very least more negative; that is, in Jackson's original definition, it referred to the processes of schooling that were not explicitly taught but were required for success. In contrast, the complementary curriculum is an addition that may enhance or hinder the school experience, and students are not required to master any related skills. The second difference between the hidden and the complementary curriculum is the source. The hidden curriculum emerges from a variety of places, such as the school structure, the bell schedule, furniture, administrative decisions, textbooks, paint colours, etc. The complementary curriculum has one source: the teacher.

These (and other) terms, Uhrmacher (1997) argues, help curricularists make distinctions that may otherwise go unnoticed. This is, I believe, the case with complementary curriculum, which I describe as the embedded and often unconscious expression of a teacher's beliefs. In the study described here, focused upon ecological beliefs, it may include the teacher's use of examples, personal stories, vocabulary, and pedagogical practices that relate to or emerge from ecological ideas, even though the curriculum does not necessarily include information *about* an earth-based idea like watershed or ecosystem health. Adding this term to our curricular lexicon, I argue, brings to light pathways to understanding and improving curriculum and instruction, particularly from an ecological standpoint.

Method of inquiry

The study was designed to respond to two questions:

- What are the intentions of ecologically minded teachers? and
- How are those intentions realized (or not realized) in a teacher's practice?

In order to describe and interpret the potentially subtle manifestations of the participants' beliefs and intentions, I used the methods of educational connoisseurship and criticism (Flinders 1996, Eisner 1998, 2002, see also Barone 2000, Uhrmacher and Matthews 2005).[4]

This study has a particular focus on ecological themes, and, while educational criticism is a broad term defining the research methodology, *eco-educational criticism* is the term I use to specify the particular ecological lens through which I filtered my observations and interpretations. By 'ecological' I mean situations, ideas, and issues that address the inescapable embeddedness between and among humans and the natural environment including but not limited to issues of relationship (Smith and Williams 1999), care (Noddings 2005), decision-making (Heimlich 2002), and sustainability and global equity (Smith and Williams 1999). I was specifically seeking to understand how ecological concepts and themes emerged in non-ecological[5] contexts.

In this paper I provide educational criticisms in the form of vignettes with the intention to bring to light the manifestations of teachers' ecological beliefs in the classroom. In a previous study (Moroye 2005) I also used eco-educational criticism to describe teachers who did not necessarily hold to ecological beliefs, but whose practices could be described by ecological themes. In future studies this method could be used to draw forth additional ecological themes, as well as to analyse a variety of educational contexts and models for their ecological implications.

Two large US public high schools, Seneca Lake High School[6] (SLHS) and Highline High School (HHS),[7] served as the sites for my research. The three participants discussed here are US public high school teachers; two of the three teach English, and one teaches social studies. I first conducted an individual formal interview using a protocol in which the questions referred to the teachers' intentions, their ecological beliefs, and their educational practice in general. Next, I observed each teacher for 3–6 weeks. I concluded with a follow-up interview, which often synthesized the connection between the teachers' ecological beliefs and their practices. Working with one teacher at a time afforded me the opportunity to immerse myself in their work and to better understand the architecture of their practice. I then wrote accounts of each teacher that included the four aspects of an educational criticism: description, interpretation, evaluation, and thematics (Eisner 2002). Portions of those criticisms are included here in the form of vignettes.

Findings

As stated above, two questions guided this study: What are the intentions of ecologically minded teachers? How are those intentions realized in a teachers practice? As Eisner (1988) points out, the intentional dimension of schooling is important because intentions 'influence the kind of opportunities students will have to develop their minds... and intentions tell the young what adults think is important for them to learn; they convey our values' (p. 25). While Eisner was speaking about the school's intentions, the idea works for teachers as well. Intentions guide, among other things, curricular choices, emphases, and omissions. Here I look at the intentions of individuals with common values that were not directly related to schooling and I explore how, if at all, their practice was affected by these values. It is important to note that I asked teachers about their ecological beliefs, as well as their intentions for their students. I was seeking to understand the teachers' ways of connecting the two.

To that end, I interviewed each participant both prior to and after conducting classroom observations. One purpose of the interview was to understand the teachers' intentions for their students and whether or not they thought their ecological beliefs were linked to those intentions. Mr Rye, the first participant, explained the connection in this way:

> I can't walk in and give daily lessons on drilling in the [Arctic National Wildlife] Refuge, and I can't walk in and talk on a daily basis about treatment of animals or of the natural world. But I can talk about [students'] treatment of other human beings, their view of their own lives, and the values and principles upon which they base their own lives.

As Mr Rye points out, his ecological beliefs are somewhat at odds with his teaching. As an English teacher he is not charged with the role of teaching environmental education. However, including ecological ideas in the classroom is important to him, so he chooses to infuse his practice with a broader principle that, for him, is connected to an ecological ethic. That principle is *integrity*:

> I think that, at the core of environmental issues is personal integrity, [which guides whether] we exploit something or choose not to exploit something. And

what I want to do with my students on a daily basis is to have them examine and, hopefully, develop their integrity.

Mr Rye also alludes to his own sense of integrity and that he tries to live his personal and professional lives in such a way that they are in alignment with his beliefs. He does so, however, with awareness that he does not want to alienate his students. 'I try not to project myself as an environmentalist as much as just a human being who loves nature and who considers [the environment in making decisions].' Furthermore, he wants his students to live 'authentic' lives: 'My deep concern is about the type of lives these guys are going to live, and are they going to live lives that are individual and interesting and somehow sacred, or... lives that are frighteningly generic?'

Mr Rye appears sensitive to either the real or perceived limits imposed upon him by the formal curriculum, as well as by the potential negative reactions of his students. Therefore, he discusses his intentions for his students in broad terms with 'integrity' and 'authenticity' at the heart of his goals for them. So how do these ideals play out in practice? Consider the following vignette and notice how his beliefs are woven into the lesson:

> 'I want this project to rock!' Mr Rye shouts in a pep talk to his senior [i.e. Year 12] English class. He is preparing them to write their autobiographies as their final senior paper. This class is considered 'remedial' for students performing below grade level, and many in the class are staffed in special education.
>
> 'You need AT LEAST four sheets of paper. Not very environmental, I know.' Mr Rye roars at his students, 'HOORAY! You don't have to write essays!' A student asks if they will have assignments that tell them what to write about. 'You are prophetic! We're gonna break it down—b-b-b break it down!' Mr Rye and the class erupt in laughter at his failed attempt at rap music.
>
> Mr Rye then begins to explain the first writing assignment. 'FOOD in 2005 is fascinating! Why am I asking you to write about food? This isn't health class. But studies show that food is the single most determining factor about how long you will live and the quality of your life.' Mr Rye explains that writing about food is really writing about their lives. He talks about the history of humankind and how it is easy to predict what people would eat based upon where they lived. 'What would people in Colorado eat? Buffalo, corn, wheat, potatoes, carrots. They didn't go to Whole Foods to pick up sushi. If you weren't able to import everything you wanted, you lived with what the land gave you.
>
> 'In our era it is unparalleled! You can choose to be a vegan and still have variety. You can choose to be a vegetarian. In this day and age it is fascinating to explore individuality because you have so much choice! You can go to a 7–11 [i.e. a convenience store] and get lunch. Now you can even get stuffed sausages—kinda scary! It's a crazy world. In 10 minutes from SLHS you can get Thai and Chinese.' He continues noting that within minutes of their school students can taste the world.
>
> Mr Rye gives students 8 minutes to write about food as he buzzes about from student to student helping them brainstorm and encouraging their writing. 'Have some fun. Be spontaneous! Believe it or not, the power of life is in the details. If you want to stay on the surface with this project, I can't stop you. But this is your life and it's so much more interesting than that!' Mr Rye cheers as he hands out skinny slips of blue paper that say the following:

Life Signifiers: Uncovering the Reality of You

You and...

1. Food—what you eat and why where you eat; what you cook yourself; what your parents cook for you; guilty pleasures—stuff you eat but know you should not eat; what you will not eat and why; your typical day:... your food philosophy: what food means to you.

A student asks, 'Can I just list my allergies?'

'Yes! What a great feature!' he says again. Mr Rye puts a few strong student examples up on the overhead and discusses how interesting they are. One example deals with a student's Jewish religion and culture and their implications on the food she eats. As each student shares his or her responses, Mr Rye calls each by name, affirms his or her answer, and finds humour in almost every statement.

Remember that Mr Rye has two overarching intentions for his students: integrity and authenticity. These intentions come to life in several ways. The writing prompt itself values self-awareness, which for Mr Rye is connected to integrity and authenticity. So in that regard, his intentions are manifested in the explicit or stated curriculum. However, we may also see a more complex force, Mr Rye's beliefs, permeating the lesson.

First, the written handout details the first of several writing prompts for the students' autobiographies. The handout is a thin slip of paper that signifies reduced paper consumption. Secondly, Mr Rye remarks on the number of sheets of paper (four) students will need saying, 'Not very environmental, I know'. Thirdly, Mr Rye's elaboration on the history of food indicates his own understanding of the relationship between food and human existence, which did not always include a quick stop at a convenience store for a hot dog.

Separately, these three examples may not mean much. However, taken together they form a subtle curriculum. That subtle curriculum is the manifestation of Mr Rye's ecological beliefs. Throughout my observations of all participants, I noticed that their beliefs often emerged in understated ways, such as in the examples they used, personal stories about their lives, certain emphases, and even in their common vocabulary. While they were often not explicitly 'teaching' an ecological concept or idea, they were simply showing that their ecological beliefs are just below the surface, that they are part of who they are and how they teach. Because their beliefs are not separate from their practice, are not compartmentalized into a different section of their lives, are integral to who they are in the classroom, I refer to this type of subtle curriculum as the *complementary curriculum*.

A second and related vignette further illustrates the complementary curriculum in Mr Rye's practice. His ecological beliefs again emerge in his explanation of the written curriculum. In particular, Mr Rye asks students to consider the history of humans' need for drinking liquids, and he takes them through a brief story contrasting the use of local resources to the present-day beverage industry. Another teacher could simply ask students to think about their favourite beverages; Mr Rye offers a more ecological perspective in which he urges students to think about what humans really need, not just what they desire.

After the students list their favourite foods and other food quirks, Mr Rye launches into the next topic—beverages. The next writing prompt he distributes, which is again on a small slip of blue paper, prompts students to explore the drinks they consume.

'What was "drink" for the history of humankind? WATER! Wine if you were lucky enough to live near grapes. Milk if you were lucky enough to have a willing cow. But check out a 7–11 [convenience store]! What drink options do you have? Five varieties of Slurpees, Gatorade—like 20 varieties, Powerade, Energy drinks—at least 10 of those, bottled water, sparkling water—what *is* that? How do they make it sparkle? Iced tea, soda—which doesn't quench your thirst—juice, and so on! And how do they get things to taste like that? This is the only culture in which we drink more liquids *other* than water, and we pay more money for bottled water even though [tap water in the US] is cleaner than water in almost any other country—even in toilets it's cleaner! Now we have flavoured water—no—it's *INFUSED,* not just flavoured!

'I want you to see how completely foreign this is to humankind—drink has never been a factor of individuality before. Maybe you choose different drink for different reasons—your concern for your health, your concern for the environment. That is what makes *you* interesting!'

As class time draws to a close, Mr Rye prepares them for the next day by discussing 12 signifiers of individuality. 'This is how we measure and show and understand individuality. The next signifier is clothing—you'll find this interesting at SLHS. We see clothes and they say something. For example, look at girls with tie dyes. Does she love the earth? Does she love animals? Did she have a paint explosion? Your clothing is a great measure of who you are, at least in this country. Did you know that the average world citizen owns FIVE items of clothing—TODAY! So tomorrow we will talk about your clothing and you.

In the previous two vignettes, we might apply several different curricular terms, each revealing something different about this teacher's practice. We could analyse the formal or written curriculum, which is exhibited in Mr Rye's writing activity, and determine if such an activity were useful to his students and perhaps to others. We could also comment on the null curricula, what is missing, and note that perhaps Mr Rye did not place enough emphasis on editing or grammar. Selecting from a variety of terms provides us with a starting point for analysis and potential improvement. Additional terms such as complementary curriculum may provide additional and useful points of analysis.

In the second scenario, the complementary curriculum is expressed in Mr Rye's explanation of the assignment. He emphasizes to students that beverages have not always come from refrigerated coolers at convenience stores. He draws the connection between what the land could provide and what humans could consume. He notes for students that not only are they able to get drinks from around the world regardless of local agricultural limits, but also that the beverages now available have an air of absurdity about them. In a sense, he points out how far away from 'natural' the beverage industry has strayed. However, Mr Rye doesn't simply point out the state of this industry; he connects it to student choice. He is helping them to see that they do have choices that express their individuality, and that those choices say something about how they live in the world. He does not condemn them for drinking 'infused' water, but points to a perspective they may want to consider, and that perspective requires that they consider the origins of the products they consume. This consideration is a new paradigm for many students (and adults) comfortable with their present consumption patterns. Then at the close of class, Mr Rye tells them that the average world citizen owns only five pieces of clothing. He again includes a broad global perspective, albeit brief, that students may consider as they write their own autobiographies.

Is the complementary ecological curriculum here valuable? From an ecological perspective, we might wonder if Mr Rye's comments in the first vignette about the use of paper will have any meaning to students. Does merely mentioning the environmental insensitivity of using too much paper result in environmental stewardship in his students? Probably not. Perhaps Mr Rye's comments merely show his students that environmental ideas are on his mind, and that may lead them to ask him questions about the environment later. Mr Rye notes that while his students don't often bring this up, he is 'deeply gratified by the fact that it does occur'. However, perhaps his discussion of food and drink provides his students with a different perspective, one which allows them a window into a kind of ecological thinking, one that considers the origins of the products we consume. Considering consumption patterns is a key cultural component to addressing the ecological crisis, so in this regard, the complementary curriculum supplements the formal curriculum with a much needed focus on connections between consumption and production.

However, some environmental education scholars might question whether Mr Rye's attention to individuality is actually counterproductive to certain ecological ideas (see Bowers 1993). A more ecological perspective might focus more on the balance of individual and community needs (see Bowers 2003). This critique points to a difficult issue when discussing complementary curriculum; it may lead to an evaluation of teachers' personally held beliefs. This difficulty is compounded by Mr Rye's sensitivity to his students. He says, 'I'm aware… as a teacher not to alienate some of my kids because if they see me "an environmentalist" will they tune out [other lessons]?' Mr Rye therefore chooses to focus on self-awareness and personal integrity instead of other potential ecological ideas. These two areas of focus also serve as a proxy for explicit ecological perspectives in Ms Snow's practice.

While Mr Rye's ecological beliefs as expressed through the complementary curriculum are apparent in the way he explains and elaborates upon the explicit curriculum, it is much more behind the scenes for Ms Snow, an English teacher at Highline High School. Outside of school Ms Snow is a Native American minister,[8] and therefore she talks about spiritual beliefs in connection with ecological principles. She explains her ecological beliefs and related intentions for her students in this way:

> The core of my ecological beliefs has to do with relationship… relationship with self, relationship with others, [and] respect for self and respect for others… It also has to do with taking responsibility. We take responsibility for how we conduct ourselves in relationship to how we use resources on the earth, for instance. And because I believe that we must act with the spirit of integrity

to preserve those resources for seven generations on down, then I think that learning things about the self and individuation and alchemy and the archetypes and all of those things really *is* in deep alignment [with my ecological beliefs].

Ms Snow feels constrained by the requirements of the courses she teaches; the English curriculum does not allow for reading environmental writing and in-depth discussion of issues. As Robertson and Krugly-Smolska (1997) point out, teachers—even in sanctioned environmental education settings—share similar concerns about what they are 'allowed' to do because they feel limited by what is expected in the formal curriculum. Instead, through studying texts like *Demian*, Ms Snow is addressing ecological ideas as she defines them. '*Demian*, Jungian psychology, [the] search for self and individualization, and being true to an inner voice... [all] have to do with relationships. Relationship with self, relationship with others.'

Ms Snow's discussion of her beliefs and intentions for students has a similar ring to that of Mr Rye. Each seeks to develop self-awareness and integrity in students. Ms Snow's intentions are apparent in the following vignette as she guides her students to think about what makes them unique and how they will share that uniqueness with the world. We also see how she addresses each student with care and respect, which facilitates thoughtful discussion in the class.

> The humming overhead reads, 'Most men lead lives of quiet desperation and go to the grave with the song still in them'. Respond to this famous quote by Henry David Thoreau. Do you think that this is a true assessment?' Ms Snow looks out over her senior [i.e. year 12] Humanities seminar class... They lean over notebooks occasionally glancing up at the overhead to reread Thoreau's words.
>
> 'Looks like you all had a lot to say about this one', Ms Snow smiles. 'Let's pick up with *Zelig*[9] and connect the ideas', she suggests, referring to a Woody Allen film they had recently viewed. They discuss the fear of being seen for whom we truly are and the risk we take when we allow ourselves to be real with others. 'Let's keep building on this. I know you're more awake than I am.'
>
> 'I think a lot of people might do that because they are afraid of what society might brand them. Like Martin Luther King, Jr. He took a risk,' one female student offers.
>
> 'Do you think he died with a song still in him?' Ms Snow asks.
>
> 'No. He lived it', she responds.
>
> 'A lot of it has to do with fear. Like if you let your true self out', another student says.
>
> 'Yeah. Isn't it about taking risks?' Ms Snow asks as she sits down in a chair in the front of the room. 'What if you do sing your song and people don't accept it?'
>
> 'No one expects anything more than mediocrity', a third student says.
>
> 'I don't agree with that', replies another.
>
> 'Okay. Good. Let's come back to that. I want to hear what Tracy has to say.'
>
> Tracy says, 'I think society wants you to strive. They want you to be the best. WE have to run this world.'

'Okay!' Ms Snow praises. 'We are getting some great responses here. Let's hear from Stacy, then Sarah.'

'The simplest things can be made so hard. It's like they expect you to work at a fast-food restaurant. Especially minorities. It's like minorities are still looked down upon—since you're Native American, you're just going to be a drunk. So just go back to the reservation', Stacy, an African American girl says.

'Stacy's goin'!' Ms Snow cheers. 'Let's hear from Sarah.'

'I think fear of society is only half of it. People are lazy. They have that quiet desperation in themselves, but they don't do anything about it. They just watch TV.'

Ms Snow wraps up the conversation and then addresses the whole class:

> I want to ask you a question, but I don't want you to answer it. We are reading Socrates and watching *Pleasantville* to find out who you are in the world. The question I want to ask you is—what is your song and how will you sing it? You are about to walk across a bridge—many of you into higher education. I am going to show you something; it's called 'An Invitation' written by a white woman. You don't have to be trapped in that moment of quiet desperation. Those moments can make us fight to sing that song. You are going to write a senior credo. You will like it!

Students read 'The Invitation' and consider it silently. The first stanza reads, 'It doesn't interest me what you do for a living. I want to know what you ache for, and if you dare to dream of meeting your heart's longing.' Then, in silence, Ms Snow puts on the video of *Pleasantville*, a story of a teenager who wants to break out of his black-and-white sit-com world.

While the natural world and consumption patterns do not filter into this discussion as they did in Mr Rye's classroom, Ms Snow's stated intentions, which emanate from her ecological beliefs, include helping students examine their lives in order to take responsibility for their relationships. While it is apparent that the above vignette is in alignment with Ms Snow's intentions for students, it also shows that to Ms Snow, as well as to Mr Rye, self-awareness is a building block of integrity, and one who has consciously developed integrity, they believe, will be more likely to consider ecological perspectives. They do not include explicit ecological curriculum, but instead focus on what they consider to be a related ecological principle.

In contrast, the third participant, Mrs Avila, does tend to include more explicit ecological ideas, and she, like Mr Rye, does so through her elaboration of the written or stated curriculum. Mrs Avila's beliefs lead her to cover some subjects in more depth and with a particular perspective; for her it is a matter of emphasis. However, unlike Mr Rye, Mrs Avila feels very comfortable infusing her ecological beliefs:

> In geography, we talk about population, which is a pretty common topic, ... [but] I feel totally comfortable deciding ... to talk about not just where does population grow, where does it shrink and why, but also the impact of population growth, depending on whether it is a society that is resource-intense ... I feel totally comfortable choosing to introduce the kids to that.

The following vignette illustrates this ecological emphasis as well as an extended, spontaneous discussion with her students about recycling. Notice the stated agenda and what actually occurs. Although lengthy, the vignette does illustrate a real situation in which the teacher uses questioning to guide the students' understanding away from a common line of thinking that ecological responsibility is inconvenient toward a more connected way of thinking about personal choice and action.

Mrs Avila's 9th grade World Geography students are greeted by her friendly demeanour and an overhead that has the Geography Agenda with the Colorado state geography standards for the day:

6.1. Students know how to apply geography to understand the past.
6.2. Students know how to apply geography to understand the present and plan for the future.

Today's activities

1. Complete presentations.
2. Discuss population's impact.
3. What causes population to grow or shrink?
4. Population pyramids in Lab A.

The starter has a picture of a population pyramid, which looks like an isosceles triangle with horizontal stripes. The starter tells them that this is a population pyramid and asks students to explain what it might mean. 'Guessing is okay!' Mrs Avila tells them.

Mrs Avila takes responses, and one student surmises that those at the bottom of the pyramid don't have a lot of money. 'Good thinking!' Mrs Avila responds. 'Ian, what did you put?'

'Nothing', Ian replies.

'What *will* you be writing down?' Mrs Avila asks again.

'I think maybe the bars show age.'

'Terrific thinking!' Mrs Avila praises. They then move on to student presentations. 'Who's the environment group?'

The group of four students makes their way to the front, and they discuss how we need clean air and water to live. They say that we as humans take more for ourselves, leaving little for other species, and they give specific examples about deforestation.

'Pause there,' Mrs Avila interjects. 'What was Brad talking about with BIODIVERSITY? What are we using up? Where are we getting 50% of our prescription drugs?'

'The Rainforest', a student in the audience answers.

'So biodiversity refers to plants and animals that exist. So do we benefit from biodiversity?'

'Yes, like with prescription drugs', another student responds.

'But what was Alice talking about? It's not only about us, is it?'

'No.'

'It is about the plants and animals—they become extinct! For example, let's think about eggshells. In order for them to be made of what they need—

calcium—birds eat snails; snails eat plants; and plants get calcium from soil. But why is calcium not in soil anymore?'

'ACID RAIN!' a student shouts.

'Acid rain caused by?'

'Burning of fossil fuels', the student responds.

Mrs Avila moves to stand near two talkative boys, but does not scold them. 'So when we get in our cars, do we say, "We're going to kill some birds today!"? No! But the unintended consequences are just as serious as the intended consequences.' As the discussion unfolds, Mrs Avila questions individual students about resource-use and -consumption patterns, and eventually turns to a discussion of waste.

'Where is this place called trash? Has anyone ever visited this place called trash?' Mrs Avila asks.

'You mean like a landfill?'

'Yeah. How long does the toothpaste tube stay there?'

'Forever?' one student guesses.

Mrs Avila prompts, 'How long are you planning to live? 100 years? I'm planning on 105, so you'll be taking care of me when I'm old. Will the tube be there when Armando is 100 years old?' Students shrug, and some say no, some yes. 'The tube I use is metal—it's recyclable.'

'What kind do you use?'

'Tom's of Maine.' [i.e. brand name]

'Oh. That organic stuff.'

'Yeah. So how long does it take for the toothpaste tube to dissolve? Thousands of years! Students gasp. 'The vast majority of my furniture is used—from the 1930s and 1950s. I make a conscious effort to recycle and to buy things that can be recycled.'

'Why?' a student asks.

'Because it's not just about me. I think about you guys when you are 105. I want you to have a planet worth living on. What we're talking about with global warming—300 scientists, the top in their field—say the earth's temperature is rising a couple degrees. Glaciers that have been in Greenland for thousands of years are melting. Penguins and polar bears are dying because they can't get their food. So guys, are these [population pyramids] just about the number of people growing?'

'No', several students respond.

'NO. It's about what?'

'How we use our resources and create junk and stuff,' one student says.

Mrs Avila then addresses a student with a plastic Coke bottle. 'Evan, what will you do with that Coke bottle?'

'Throw it away,' Evan says.

'Why? Why won't you recycle it?' Mrs. Avila asks.

'No recycle bins.'

'Okay! Why at HHS do we not have many recycle bins? There is an area of the school with recycle bins, but students can't go there. Is that a problem? Shana is drinking juice—and we're glad because juice is far better than Coke, no offence, Evan. But when you're done, will you ask me to recycle it for you?'

'No. You should have a recycle bin in here', Shana says.

'So it's up to me?' Mrs Avila asks.

'We should have a recycle day. All the students who get in trouble should pick up trash and recycling,' another student offers.

'They should have recycle bins,' another student says.

'Who is this "they"? Do *you* care?'

'It's more of a habit,' Shana says.

'How could we get you to change that habit? Do you all agree that if more recycle bins were available, you would recycle?' About eight students raise their hands to say yes.

'But we have to overcome laziness!' Shana says.

'Who needs to organize this movement?' Mrs Avila asks Shana.

'Everybody. Students.'

'Why students? Would some people listen to *you*? *You* personally? Armando, would you be willing to work with other students to increase the number of recycling bins?'

'Maybe.'

'What would make you more likely?'

'To know that students will use them,' Armando says.

'Did you know we used to have a recycling club?' Mrs Avila asks.

'No!' many students respond, shocked.

'It faded away because students stopped coming. Mr Hepner might be willing to do this again, but could this be student-driven?'

'Yes,' many respond.

'What would need to happen?'

'Talk to Ms Wright,' Shana says referring to the activities director.

'Is anyone willing?'

'Yes! I will!' Shana volunteers.

'Is this a big change in the scheme of things?'

'No. We are only one school', a student says.

'But maybe it will encourage other schools!' Shana offers.

Class is ending, and Mrs Avila encourages students to think about their conversation today. 'Who will follow through?' she asks as they leave. Several students stay after the bell to talk further with her about the recycling club and various other ideas.

The written curriculum as evidenced by the agenda does not accurately reflect what actually occurred in the classroom.[10] While Mrs Avila certainly did 'discuss population's impact', she did so in a way that elicited thinking in her students that, for some, led to action and for others to increased overall engagement in the class. I asked Mrs Avila in our second interview if anyone had followed up on offering more recycling in the building.[11]

> They haven't had action yet, but they are still talking about it. And, actually Shana, one of the girls who volunteered, is talking to me more in class now and even turned in some late work... She was certainly not doing well [before this lesson], but I am hoping that she is feeling a little more tied in.

I asked Mrs Avila why she thinks that the lesson resonated in particular with Shana:

> I have some ideas that maybe it was because I totally trusted that she would do it and that I was very enthusiastic when she volunteered. I am hoping that she at least sees that I do believe in her. I am not sure that she believes in herself a whole bunch.

The complementary curriculum in Mrs Avila's case is not only expressed in the stories and examples of her own life, but also in the types of thinking she elicited in students through a series of questions and statements in the impromptu discussion about recycling. To elicit that thinking, she employed a pedagogical technique of questioning which is similar to strategies discussed by Cotton (2006b) in her study of three geography teachers in the UK. Cotton identified three strategies teachers use to discuss controversial environmental education topics: 'Strategy 1: Eliciting students' personal views...; Strategy 2: Enabling students to discuss their own views...; and Strategy 3: Challenging students' views' (p. 227).

Cotton's study and this study are similar in that both identify 'real', not 'ideal' practices. However, the contexts are different in that all three of Cotton's participants were actively engaged in teaching environmental issues as part of the formal curriculum. Still, the strategies discussed (and in particular Strategies 1 and 3) are evident in Mrs Avila's practice and offer another example of this pedagogy at work.

Furthermore, Mrs Avila appears more focused on uncovering the origins of students' individual behaviours. She spends a lot of time eliciting students' rationales for their own behaviour. ('Evan, what will you do with that bottle when you are done with it?'). This fourth strategy of considering the rationale for one's own behaviour could be considered useful in contexts in which teachers are focused on action, or in which the focus is on habits of mind that affect behaviour ('Why won't you recycle it?'). While the teachers in Cotton's study were more engaged in debating complex and abstract issues (such as the governance of Antarctica), Mrs Avila and her students were dealing with seemingly simple and concrete behavior—recycling. Discussing this immediate and daily behaviour highlighted the locality and immediacy of personal choice for students. This strategy, or pedagogical practice, emerged from deeply-held beliefs and the lifestyle of Mrs Avila, and it took place in the context of the caring classroom community she consciously orchestrated. It may be difficult for other teachers to emulate, but in this case, the pedagogical practice that characterizes the complementary curriculum in Mrs Avila's work led some students to reflect upon their own behaviour and to ultimately reorganize the Environment Club at Highline.

Implications for teaching: toward an 'environmentally-sustainable pedagogy'

Mrs Avila's pedagogical choices help her to guide students to a more ecological frame of mind; she does so by expanding upon the formal social-studies curriculum. However, many ecological curriculum theorists suggest that environmentally-sustainable education should be characterized by a transdisciplinary curriculum (Van Kannel-Ray 2006). This kind of curriculum requires a communal effort and, I would argue, a whole-school reform effort.[12] The participants in the present study, however, did not have the benefit of working within whole-school curriculum framework, or even with like-minded others. Indeed, each teacher worked alone and in a single discipline. Therefore, to ask whether or not they are realizing a new model of ecological education is neither fair nor appropriate, but we may perhaps glean some aspects of what *environmentally-sustainable pedagogy* could look like:

> environmentally sustainable pedagogy as a theory of teaching can inform how to hold the individual and the community in relationship... It can offer a new identity to teachers as teaching with a moral imperative, as helping students to become more responsibly embedded in the natural world. (Van Kannel-Ray 2006: 122)

She suggests that pedagogical practices should emerge from the overarching ecological principles of 'intergenerational responsibility', 'organic perception', and 'sustainable outcomes' (p. 117). Each teacher from the present study contributes to a vision of these pedagogical practices through either intergenerational responsibility or organic perception (the present study is limited in understanding the effects on sustainable outcomes).

Intergenerational responsibility deals with balancing the individual's needs with the needs of the past and the future. Mr Rye begins to help weave this tale of balance in his writing exercises with students. He urges them to write in detail about their own individuality, but couches that uniqueness and related consumption in a broader perspective so as to avoid seeing 'the individual as the epicentre of the universe' (Bowers 1995: 7). Furthermore, this type of focus seems to be in line with Bonnett's (2002) discussion of education for sustainability as a frame of mind which seeks to 'reconnect people with their origins and what sustains them *and* to develop their love of themselves' (p. 271). Reminding them that until recently water was the predominant drink for humankind, and that also until recently humankind relied upon local food sources, Mr Rye brings a deeper awareness of the connections between humans and their environments to his students and highlights students' understandings of their own choices. Mr Rye does not, however, ask students to change their behaviours or to even consider the environmental or social ramifications of their choices. On the other hand, Mrs Avila does urge students to consider the effects of their choices, particularly the ways they handle trash and recycling. Her efforts seem particularly fruitful in that the Recycling Club gained renewed membership and activity.

Organic perception is an indication of an individual's perceived connection with the natural world (Van Kannel-Ray 2006). Seeing oneself as connected, or as Ms Snow puts it 'in relationship', limits our tendencies to exploit others, both human and other-than-human. Therefore, Ms Snow's work may also make a contribution to environmentally-sustainable pedagogy in her cultivation of a caring community. Not only does Ms Snow have a deep commitment to fostering relationships with her students, she facilitates students' relationships with each other through encouragement, creating space for students to have their voices heard, and by making it safe for them to discuss different ideas with each other, even in a very diverse setting. This is done in the context of individual purpose and a discussion of each student's 'song'. The learning community becomes a place that fosters organic perception.

Complementary curriculum, the embedded and often unconscious expression of one's beliefs, is the manifestation of a teacher's wholeness or completeness, of his or her integrity.[13] In his essay 'The heart of a teacher: identity and integrity in teaching', Palmer (1997) discusses the importance of teachers' awareness and development of identity and integrity in teaching. By identity Palmer means 'an evolving nexus where all the forces that constitute my life converge in the mystery of self... Identity is a moving intersection of the inner and outer forces that make me who I am' (p. 17). By integrity Palmer means 'whatever wholeness I am able to find within that nexus as its vectors form and re-form the pattern of my life. Integrity requires that I discern what is integral to my selfhood, what fits and what does not' (p. 17). For Palmer, a teacher's identity and integrity—not technique and method—are what make them great teachers:

> My ability to connect with my students, and to connect them with the subject, depends less on the methods I use than on the degree to which I know and trust my selfhood—and am willing to make it available and vulnerable in the service of learning. (p. 16)

Complementary curriculum is the expression of this identity and integrity, of what Palmer (1997: 16) calls the 'integral and undivided self'. As illustrated in the vignettes presented above, this expression might emerge in a variety of planned or spontaneous ways, often dependent upon the particular moment and context as orchestrated by the teacher. This is what makes complementary curriculum different from the myriad of other terms in our curricular lexicon: the source of complementary curriculum comes uniquely from the teacher and her personal passions and beliefs.

While the focus of this study is on the expression of ecological beliefs and therefore complementary *ecological* curriculum, this idea might be applied to other beliefs or passions, such as an artistic sensibility or commitment to social justice. In order to explore and understand the complementary curriculum of such beliefs, the researcher would need to first interview the teacher so that she may articulate her beliefs and passions. Next, the researcher would observe the teacher's work to see how if at all the beliefs are infused in practice. For example, these passions might be expressed through the use of music or stories of artistic encounters, or through a biographical study of social activists, or first-hand accounts of participating in social change. It is important to note that the teachers' beliefs may emerge intentionally or unintentionally, consciously or not. Therefore, a follow-up interview with the teacher can foster a discussion of the teacher's intentions and beliefs with the researcher's observations. The researcher is then better able to evaluate how the expression of that teacher's beliefs—the complementary curriculum—influences pedagogy, curriculum, assessment, class structure, or other dimensions of schooling.

In addition to conducting a follow-up interview, sharing the educational criticisms (or observations) with the teachers may illuminate for them previously unseen connections between their beliefs and practice. Such was the case in the present study, and after I shared the educational criticisms with each teacher, I was struck by their responses. Ms Snow writes:

> I learned about how our internal belief systems shape the teaching process. Before I understood the nature of [your] study, I could not accurately articulate why I had sometimes been very happy and other times very unhappy with teaching. Now I understand the very necessary and intrinsic core of how our ecological belief systems and (for me, at least) a corresponding spiritual belief system shapes the art of our relationships with our wonderful students. (Personal communication, 8 September 2006)

While this study looks particularly at ecological beliefs, having a similar dialogue with teachers about their particular beliefs and then illustrating for them how those beliefs come to light in their practice may lead to a more developed sense of their teaching integrity, and further research could also explore how the complementary curriculum affects students directly.

Implications for ecological teacher education

Because of the skills, beliefs, and knowledge required to implement environmental education curricula, many point to the importance of ecological perspectives in teacher education programmes (Tilbury 1996, Oulton and Scott 1997, Corcoran 1999b). Some teacher educators have investigated the lives of ecologically-minded teachers and what factors caused them to become ecologically aware. Corcoran (1999b) details the process of writing an environmental autobiography, through which he guides his undergraduate pre-service teachers. Corcoran affirms the belief that environmental education in teacher education is the 'priority of priorities' (Tilbury 1996, cited in Corcoran 1999b: 179).

Corcoran says that environmental autobiographies can help us identify what makes humans want to live sustainably, an issue at the heart of environmental education. Corcoran says, 'A desire to protect the natural world arises from a deep sense of affinity with the land and nonhuman beings' (p.179). He terms this 'biophilia', or a love for other living beings, which Corcoran believes is 'central to our nature as humans' (p. 180). This is where he begins with environmental educators—with this innate sense of connection explored through environmental autobiography.

Corcoran (1999a) also completed a study of environmental educators in which he sought to understand the significant childhood life-experiences that led environmental educators to feel a strong connection with the natural world. Mirroring a previous study in the UK by Palmer (1993), he surveyed 510 US teachers about their experiences in nature as children. The narratives have recurring themes such as parents and grandparents as environmental educators and role models; fear of the effects of environmental problems; world-view, faith, and spirituality; childhood time outside; and hope (Corcoran 1999a: 211–217). Corcoran believes that teachers who have had these significant life experiences will provide similar opportunities for their students to develop their own affinity for the natural world. The present study, in combination with those discussed and cited above, builds evidence that attention to the ecological beliefs of pre-service and in-service teachers may play an important role in the expansion of environmental and ecological education, whatever form they may take.

Complementary ecological curriculum also may have import for students. In the case of ecological education, Corcoran (1999a) notes that many who hold ecological beliefs trace the origins of those beliefs to a role model they had in childhood. Perhaps ecologically-minded teachers may become one of those role models as they demonstrate to students through the complementary curriculum that their ecological beliefs are just below the surface and guide their decisions and ways of being. It illustrates to students that ecological issues and ideas are connected to a variety of aspects of our lives, and that they are integral in the minds of the ecologically-minded teachers. These issues and ideas comprise parts of the teachers' identities, and they inform aspects of personal and global decisions. Complementary ecological curriculum reinforces the notion that the environment and ecological issues are not separate or supplemental; they are part and parcel of our everyday lives. Smith (2004) notes a similar phenomenon in his study of the Environmental Middle School. Teachers did not 'check their ideals at the door. They instead brought those ideals into every dimension of their work' (p. 77). Both studies indicate that teachers' ecological beliefs inform their practice, and therefore what students may experience.

Conclusion

In his discussion of educational criticism, Eisner (2002) considers whether or not we can generalize from such research. While criticism cannot predict outcomes, it can, Eisner argues, create 'forms of anticipation by functioning as a kind of road map for the future' (p. 243):

> Once having found that such and such exists in a classroom, we learn to anticipate it in other classrooms that we visit. Through our experience we build up a repertoire of anticipatory images that makes our search patterns more efficient. (p. 243)

This is the case, I believe, with complementary curriculum, ecological or otherwise. As critics, teacher educators, curricularists, and researchers, we can enter a classroom anticipating various expressions of teachers' personal beliefs. This recognition adds a layer to our understanding and evaluation of what is happening in a classroom, or to what could or should be happening. In this way, identifying, understanding, and evaluating the complementary curriculum is not only useful to teachers themselves, but also to those who aim to support teachers and schools in their efforts, particularly those important and difficult efforts to 'green' our schools.

Acknowledgement

I am grateful to Bruce Uhrmacher of the University of Denver and Peter Hlebowitsh of the University of Iowa for their thoughtful and constructive feedback on drafts of this paper.

Notes

1. By 'non-environmental' I simply mean educational contexts and models that are not explicitly focused on teaching environmental themes and ideas, such as a traditional school or an English classroom focused on the Western canon. Certainly all contexts can be considered ecological, although Orr (1992: 90) has said that 'all education is environmental education,'. In other words, it is impossible to separate humans and our constructed worlds from the planet on which we live.
2. It is important to note that many environmental education reforms call for integration of disciplines (see Orr 1992, Smith and Williams 1999, Jardine 2000). While this may indeed be an appropriate and necessary recommendation, the current reality of public schooling is that most US secondary schools are structured with disciplinary separation.
3. The current war in Iraq (Flinders 2006), some religious concepts, and in some cases evolution are all examples of what is not taught in US schools.
4. Eisner (1998) developed educational connoisseurship and criticism (henceforth called educational criticism) as method of qualitative inquiry intended to improve education. Connoisseurship is the art of appreciation and criticism the art of disclosure (Eisner 2002). Therefore, connoisseurship requires that the researcher have enough educational knowledge to be able to observe the subtleties and intricacies of the educational setting. The criticism, then, illuminates the connoisseur's perspective with the aim of educational improvement in mind.
5. See note 1.
6. The campus of SLHS boasts a collegiate setting with four separate buildings, three cafeterias, a variety of outdoor spaces to congregate, and extensive sports facilities. The school is situated on 80 acres adjacent to a large state park, and several of its classrooms overlook the reservoir. Students have a generous amount of autonomy. Of the 3700 students, approximately 86% are White, 2% are African American, 7% are Asian, and 5% are Hispanic.
7. HHS lies on 32 acres near a large public park and wetlands refuge. The single, more traditional high-school building has been recently remodelled to include an Academic Success Centre, a new athletic area, and refurbished entrances. Of the approximately 2000 students at Highline, 1% is Native American, 32% are African American, 6% are Asian, 16% are Hispanic, and 45% are White. Furthermore, students speak 52 home languages and come from 110 countries. Both schools have an average class size of about 25 students. SLHS and HHS participate in their district's large-scale curriculum implementation project in which all classes provide an opportunity to learn certain essential components in the core areas (English, mathematics, social studies, and science). Teachers are provided with extensive curriculum binders, but in most cases are not directed how to teach the essential core content. The formal curriculum is a compilation of the state of Colorado's standards as well as university-preparatory skills, and a major focus of the district is to improve performance on standardized state tests.

8. Ms Snow was trained by Native American teachers in various ceremonies for a number of years. For purposes of confidentiality, I have eliminated all other identifying details.
9. *Zelig* is the story of a man who transforms himself to be like those who surround him in order to gain approval.
10. Eisner (2002: 32–34) described that which actually happens in a classroom as the 'operationalized curriculum'.
11. After the conclusion of this study, HHS did resurrect the Environment Club. Many members came from Mrs Avila's class.
12. See, for example, the Portland Environmental Middle School (Smith 2004).
13. 'Complementary' literally means 'forming a complement, completing, perfecting' or 'of two (or more) things: mutually complementing or completing each other's deficiencies' (*Oxford English Dictionary* 1989). We might think of complementary angles, which when paired together make a right angle. We might also think of complementary colours, 'which, in combination, produce white or colourless light' (*Oxford English Dictionary* 1989).

References

Barone, T. (2000) *Aesthetics, Politics, and Educational Inquiry: Essays and Examples* (New York: Peter Lang).

Blumstein, D. T. and Saylan, C. (2007) The failure of environmental education (and how we can fix it). *PLoS Biology*, 5(5), e120. Available online at: http://www.pubmedcentral.nih.gov/articlerender.fcgi?artid=1847843, accessed 12 November 2008.

Bonnett, M. (2002) Education for sustainability as a frame of mind. *Environmental Education Research*, 8(1), 265–276.

Bowers, C. A. (1993) *Education, Cultural Myths, and the Ecological Crisis: Toward Deep Changes* (Albany, NY: State University of New York Press).

Bowers, C. A. (1995) *Educating for an Ecologically Sustainable Culture: Rethinking Moral Education, Creativity, Intelligence, and Other Modern Orthodoxies* (Albany, NY: State University of New York Press).

Bowers, C. A. (2001) *Educating for Eco-justice and Community* (Athens, GA: University of Georgia Press).

Bowers, C. A. (2003) *Mindful Conservatism: Rethinking the Ideological and Educational Basis of an Ecologically Sustainable Future* (Lanham, MD: Rowman & Littlefield).

Connelly, F. M., He, M. F. and Phillion, J. (2008) *The SAGE Handbook of Curriculum and Instruction* (Los Angeles, CA: Sage).

Corcoran, P. B. (1999a) Formative influences in the lives of environmental educators in the United States. *Environmental Education Research*, 5(2), 207–220.

Corcoran, P. B. (1999b) Environmental autobiography in undergraduate educational studies. In G. A. Smith and D. R. Williams (eds), *Ecological Education in Action: On Weaving Education, Culture, and the Environment* (Albany, NY: State University of New York Press), 179–188.

Cotton, D. R. E. (2006a) Implementing curriculum guidance on environmental education: the importance of teacher's beliefs. *Journal of Curriculum Studies*, 38(1), 67–83.

Cotton, D. R. E. (2006b) Teaching controversial environmental issues: neutrality and balance in the reality of the classroom. *Educational Research*, 48(2), 223–241.

Cutter-MacKenzie, A. and Smith, R. (2003) Ecological literacy: the 'missing paradigm' in environmental education (Part one). *Environmental Education Research*, 9(4), 497–524.

Dillon, P. J. and Gayford, C. G. (1997) A psychometric approach to investigating the environmental beliefs, intentions, and behaviours of pre-service teachers. *Environmental Education Research*, 3(3), 283–297.

Eisner, E. W. (1988) The ecology of school improvement. *Educational Leadership*, 45(5), 24–29.

Eisner, E. W. (1998) *The Enlightened Eye: Qualitative Inquiry and the Enhancement of Educational Practice* (Upper Saddle River, NJ: Merrill).

Eisner, E. W. (2002) *The Educational Imagination: On the Design and Evaluation of School Programs*, 3rd ed. (New York: Macmillan).

Flinders, D. J. (1996) Teaching for cultural literacy: a curriculum study. *Journal of Curriculum and Supervision*, 2(4), 351–366.

Flinders, D. J. (2006) We can and should teach the war in Iraq. *Education Digest*, 71(5), 8–12.

Flinders, D. J., Noddings, N. and Thornton, S. J. (1986) The null curriculum: its theoretical basis and practical implications. *Curriculum Inquiry*, 16(1), 33–42.

Gruenewald, D. A. and Manteaw, B. O. (2007) Oil and water still: how No Child Left Behind limits and distorts environmental education in US schools. *Environmental Education Research*, 13(2), 171–188.

He, M. F., Phillion, J., Chan, E. and Xu, S. (2008) Immigrant students' experience of curriculum. In F. M. Connelly, M. F. He and J. Phillion (eds), *The SAGE Handbook of Curriculum and Instruction* (Los Angeles, CA: Sage), 219–239.

Heimlich, J. E. (ed.) (2002) *Environmental Education: A Resource Handbook* (Bloomington, IN: Phi Delta Kappa Educational Foundation).

International Union for the Conservation of Nature and Natural Resources (IUCN)/ UNESCO (1970) *Final Report on an International Working Meeting on Environmental Education in the School Curriculum—International Union for the Conservation of Nature and Natural Resources in Cooperation with UNESCO as Part of UNESCO's International Education Year* (Gland, Switzerland: IUCN).

Jackson, P. W. (1968) *Life in Classrooms* (New York: Holt, Rinehart & Winston).

Jardine, D. W. (2000) '*Under the Tough Old Star': Ecopedagogical Essays* (Brandon, VT: Foundation for Educational Renewal).

Jickling, B. and Wals, A. (2007) Globalization and environmental education: looking beyond sustainable development. *Journal of Curriculum Studies*, 40(1), 1–21.

Martusewicz, R. (2005) Eros in the commons: educating for eco-ethical consciousness in a poetics of place. *Ethics, Place and Environment*, 8(3), 331–348.

Moroye, C. M. (2005) Common ground: an ecological perspective on teaching and learning. *Curriculum and Teaching Dialogue*, 7(1.2), 123–137.

Noddings, N. (ed.) (2005) *Educating Citizens for Global Awareness* (New York: Teachers College Press).

Orr, D. W. (1992) *Ecological Literacy: Education and the Transition to a Postmodern World* (Albany, NY: State University of New York Press).

Oulton, C. R. and Scott, W. A. H. (1997) Linking teacher education and environmental education: a European perspective. In P. J. Thompson (ed.), *Environmental Education for the 21st Century: International and Interdisciplinary Perspectives* (New York: Peter Lang), 45–57.

Oxford English Dictionary (1989)(Oxford: Oxford University Press).

Palmer, J. (1993) Development of concern for the environment and formative experiences of educators. *Journal of Environmental Education*, 24(3), 26–30.

Palmer, P. J. (1997) The heart of a teacher: identity and integrity in teaching. *Change*, 29(6), 15–21.

Robertson, C. L. and Krugly-Smolska, E. (1997) Gaps between advocated practices and teaching realities in environmental education. *Environmental Education Research*, 3(3), 311–326.

Smith, G. A. (2004) Cultivating care and connection: preparing the soil for a just and sustainable society. *Educational Studies*, 36(1), 73–92.

Smith, G. A. (2007) Place-based education: breaking through the constraining regularities of public school. *Environmental Education Research*, 13(2), 189–207.

Smith, G. A. and Williams, D. R. (1999) *Ecological Education in Action: On Weaving, Education, Culture, and the Environment* (Albany, NY: State University of New York Press).

Sobel, D. (2004) *Place-based Education: Connecting Classrooms & Communities* (Great Barrington, MA: The Orion Society).

Sterling, S. (2001) *Sustainable Education: Re-visioning Learning and Change* Schumacher Briefing, No 6 (Foxhole, UK: Green Books).

Tilbury, D. (1996) Environmentally educating teachers: the priority of priorities. *Connect*, 15(1), 1.

Uhrmacher, P. B. (1997) The curriculum shadow. *Curriculum Inquiry*, 27(3), 317–329.

Uhrmacher, P. B. and Matthews, J. (eds) (2005) *Intricate Pallette: Working the Ideas of Elliot Eisner* (Upper Sadle River, NJ: Pearson).

United Nations Conference on Environment and Development (earth Summit) (1992) *Agenda 21: Chapter 36: Promoting Education, Public Awareness and Training.* Available online at: http://www.un.org/esa/sustdev/documents/agenda21/english/agenda21chapter36.htm, accessed 3 November 2008.

Van Kannel-Ray, N. (2006) Guiding principles and emerging practices for environmentally sustainable education. *Curriculum and Teaching Dialogue,* 8(1/2), 113–123.

Conclusion

Curriculum, critique and crisis in environmental education

ALAN REID

Introduction

In this concluding reflection on 'Curriculum Challenges *for* and *from* Environmental Education', we briefly consider how events such as the World Environmental Education Congress and the work of the United Nations on education, environment and sustainability might be critiqued from both emergent and theoretically-driven considerations in curriculum studies, focusing on, pace Dunlop, what is found–and found wanting–in the curriculums of environmental education.

To begin, preparing such a collection is a strong reminder of the importance of remaining cognizant of 'where you have come from, for where you might go'. Thus from the standpoint of working both *within* and *with* a particular academic domain, any account of the past through to the most speculative of suggested futures of a field of study (e.g. Kopnina, 2014) must surely recognise the continuing need to name and discuss intellectual sympathies and the value of disciplined thought. Some of these are expressed bluntly as most evident in courses of study and syllabuses that do little more than further a 'politics of unsustainability' (Blühdorn & Welsh, 2008). Elsewhere, similar critiques of the shortcomings of status quo approaches to curriculum include the lessons that can be drawn from another closely-related form, citizenship education, particularly in inter- and trans-disciplinary contexts. For example, in the *Journal of Curriculum Studies*—and available in our online companion collection—Tupper and Cappello (2012) have studied Saskatchewan high school students' understandings of the 'good' citizen, and the ways citizenship education is 'discursively produced in officially mandated school curriculum'. Their findings suggest the reproduction of 'commonsense narratives of "good" citizenship, including socially sanctioned concern for the environment, a sense of nationalism and national pride, respect for relationships and a communal ethos, and the official discourse of multiculturalism' (p. 37).

Looking at what else might be found, Hawkey (2014) provides something of a counter perspective emphasising that while the quest for epistemic distinctiveness has its problems, it should not be jettisoned, particularly if curriculum is to address 'Big Ideas'. In relation to the construction of history curriculum, Hawkey argues if students and teachers are to better account for human *and* natural history, a porosity to 'curricular boundaries' must be exploited when confronting topics such as climate change and globalisation. In fact, these are phenomena that Morton (2013) has elsewhere labelled 'hyperobjects', to remind us of the presence of things in the Anthropocene that, 'are massively distributed in time and space relative to humans' (p. 1) and can be understood as 'genuine nonhuman objects that are not simply the products of a human gaze' (p. 199). More pointedly, following Hawkey, should curriculum in environmental education and education more broadly address not just 'objects' but 'hyperobjects'? Or are there more pressing concerns to find and pursue?

Drawing on Wendy Brown's account of criticism, crisis and critique, the collection can be read to suggest we need to find and pursue a clear alternative to what amounts to a prevailing 'education for *unsustainable* development'. It may also suggest that becoming 'worldly wise' is invariably more than simply a matter of curriculum rhetoric or flourish. Thus in wrapping this collection up, we now consider how the collection's questions of curriculum orientation, instrumentalism and extensionism might in themselves be critiqued, including in relation to what counts as 'mere critique,' 'indulgent critique,' or '(un)timely critique'. Because it is together, and self-critically, that we can further each other's, as much as our own, understandings of curriculum challenges *for* and *from* environmental education.

(Re)setting the scene

The World Environmental Education Congress (WEEC) is an international gathering of practitioners, academics and researchers, civil society members, policy-makers, industry representatives and other stakeholders interested in the intersection of education and environment across a range of settings (http://www.environmental-education.org/).

As an intentional response to the environmental crisis (as much as the perception of crisis in education on these matters), in 2015, this biennial event focused on 'how people and planet can develop together', with a program of contributions that addressed 11 key themes:

1. Taking children seriously in addressing Global Challenges
2. Reclaiming sense of place in the digital age
3. Environmental education and poverty reduction
4. Learning in vital coalitions for green cities
5. (Re) emerging concepts for environmental stewardship and sustainability
6. Mind the gap! Moving from awareness to action
7. Assessing environmental and sustainability education in times of accountability
8. Beyond the green economy: educating and learning for green jobs in a green society
9. New perspectives on research in environmental and sustainability education
10. Educational policy development for environment & sustainability
11. Education and learning for climate change.

From what we might take as a standard curriculum studies perspective on these themes, we note that many of the key concerns at WEEC were expected to address various staple topics within this and related curriculum journals, namely: identifying and prioritising core components, content, contexts and expected outcomes for curriculum, and asking how these relate to education in general, alongside those that constitute both a worthwhile and recognisably *environmental* education.

To illustrate, Theme 1, 'Taking children seriously in addressing Global Challenges', raises both logical and logistical questions of who actually needs to be doing the bulk of the associated learning. Coupled with this is, who is attributed primary (and then other levels of) responsibility in environmental education in furthering particular educational and environmental goals, alongside their associated outcomes? Is it children, for example, or youth, adults, lifelong learners, communities or institutions—or in fact, a particular configuration or assemblage of these, including over the lifespan of learners and their teachers in their various 'communities of practice' (see Covert, 1969; Jardine, 1994)?

Such questions re-emerge, albeit modified, in relation to Themes 4, 6, 8 and 11. For example, how might environmental education be best conceptualised and enabled to happen? Relatedly, what is the value of participation and possibly credentialising this activity, be that in schooling through to vocational education and training, or outside of schools all together? Next, how might this relate to social learning and public pedagogies in addressing environmental challenges, including the competencies that have been identified to address these well? And finally, is all this in ways congruent with, or distinct from other 'adjectival educations' and education however generally conceived and practiced, given some of the tensions these may inevitably create? (See, for example, Cotton, 2006, on the divergence between teachers' beliefs and the espoused intentions for curriculum materials; and Fahey, 2012, on curriculum directives, reform and alignments.)

The remaining themes at WEEC can also be shown to pick up on questions of how 'vital' the curriculum is in and to environmental education, and the promise of particular (and perhaps sometimes overlooked) features expected to work together underneath the canopy of environmental education. For instance, how might any erstwhile commitment to place-based pedagogy help counter an ever-encroaching 'digital age' (2) (in the Journal, see Brookes, 2002, on curriculum 'contingencies' and the 'struggles for territory in a field of practice without a field of study'; and Backman, 2011, on locating and classifying curriculum imperatives bio-regionally, culturally, and *subject*-ively)? Next, how might the themes previously associated with the UN's Millennium Development Goals—and with the Post-2015 agenda, those of the Sustainable Development Goals—delimit and anticipate particular curricular priorities (e.g. *poverty reduction* – 3, and, *environmental stewardship and sustainability* – 5) (see UNESCO, 2017)? In addition, how are action taking and agency both expected and circumscribed within a wide range of provision, in education and other sectors? And last but not least, what of the stalwarts of many such gatherings, expressed in the standard terms of examining the commonplace versus new horizons in curriculum-based knowledge and its assessment (7), research (9) and policy (10)? (See also, Bernstein, 1971; Hopmann, 2003; cf. Sauvé et al., 2005, and Jickling & Wals, 2008.)

Questioning the adjectival: or, reconsidering the metaxu of the human, humane and post-human

Digging into the programmes for WEEC, there can be a distinct sense that the 'more-than-human' world could be a point of difference as much as critique to many other curriculum-related conferences outwith the field of environmental and cognate areas of education. If therein lies some of its significance to curriculum theory and practice in general, if not the focus of curriculum scholarship about education and the environment then critique might focus on, for example, the extent to which environmental education

- generates deeper attention to biodiversity in education;
- engages the promise and prospects of providing wide-ranging outdoor education; or
- attends *thickly* (rather than cursorily) to the 'voices' and 'being' of others who share the living conditions and phenomenon of life on this planet, but who may until now have remained otherwise 'hushed' (in the collection, see Gough, 1989; cf. Llewellyn et al., 2010).

Yet even if we pursue such inquiries, attending to more of the history and contestation on such established or emerging interests, given its scope and purpose, we might expect a collection from this Journal to move beyond such considerations. Specifically, we must ask what the *Journal of Curriculum Studies* might have to offer those engaging with the UN and the biennial WEEC event and series, as well as to those equally (and otherwise) interested in the intersections of education, environment, and more recently, sustainability (Sumner, 2008; Clémençon, 2012)?

As introduced earlier on in this collection, the studies brought together here serve to illustrate how various contributors to the Journal have encountered, framed, combined, advanced and critiqued curriculum scholarship on education and environment over time. Works that have proven seminal and germinal in curriculum studies and environmental education (as well as that which has remained somewhat *marginal* or for that matter, overreached themselves), have also helped elucidate key aspects of the temporary to perennial challenges of curriculum thinking and scholarship, framed by their authors and 'conversation partners'.

Examining the problems and challenges their argumentation raises in terms of philosophical, practical and/or programmatic concerns, we might detect recurring themes (and possibly gaps) to the wider conversations such events as the World Environmental Education Congress hope to address. Most notably these are on how such work relates to the identity, grounding, trajectory, value and significance of environmental and sustainability education, for education in general, and curriculum studies in particular. Interlocking with these are matters of how such work in the Congress mostly speaks *from* and *to*—or perhaps, *might accidentally or tactfully ignore?*—some of the intra- and inter-generational concerns of curriculum scholars, practitioners and developers, down the years.

So what we hope becomes possible through advancing a critical reading of and response to the studies in this collection is a deeper appreciation of something of the richness and potency of the variety of theories, methods, cases and discourses that are engaged. Many of these extend considerations about education and environment beyond matters of the school-based curriculum towards wider critical questions of the 'internal dynamics' and 'external conditions' for instigating environmental and/or sustainability education. To elaborate:

Curriculum critique

First, we must recognise that *curriculum* is often engaged by contributors to this collection via a well-trodden position of *critique*, particularly when curriculum is treated primarily as an object of study. Typically, this proceeds from a disposition that seeks to criticise what is offered as curriculum, setting out why, for example: (a) there is neither due nor sufficient attention to environmental conditions and challenges in curriculum, and/or (b) that environmental education itself is actually inadequate on some ground or other (e.g. Bonnett, 2007; Bowers, 1990). To illustrate, Bowers (1990) premises his deliberations about the epistemological foundations, orientations and orthodoxies of an (environmental) educational field by targeting the inadequacies of those related to its use of educational technologies. For example:

> As formal education involves transmitting culture to the next generation, the question of whether the culture that is to become the basis of thought and behaviour, contributes to a further deterioration of critical life sustaining natural systems, should be basic to any discussion of curriculum policy and practice.
>
> (p. 72; cf. Bengtsson & Ostman, 2016)

In identifying such a theme for criticising the work of the field, we note that Wendy Brown (2005, pp. 5–6) outlines how critique is 'an old term that derives from the Greek *krisis*', where:

> The sifting and sorting entailed in Greek *krisis* focused on distinguishing the true from the false, the genuine from the spurious, the beautiful from the ugly, and the right from the wrong, distinctions that involved weighing pros and cons of particular arguments—that is, evaluating and eventually judging evidence, reasons, or reasoning. *Krisis* thus comes close to what we would today call deliberation, and its connotations are quite remote from either negativity or scholasticism. Since this practice also has a restorative aim in relation to the literal crisis provoking it, there could be no such thing as "mere critique," "indulgent critique," or even "untimely critique." Rather, the project of critique is to set the times right again by discerning and repairing a tear in justice through practices that are themselves exemplary of the justice that has been rent.

Brown also notes the subsequent separation of criticism and critique from a sense of crisis since those times, even as a notion of 'critical condition' (as used in medical and political discourse) maintains its currency and value (p. 7):

> A critical condition is thus a particular kind of call: an urgent call for knowledge, deliberation, judgement, and action to stave off catastrophe.

In light of this, we can also observe an equivalent sense of urgency both brewing within, and permeating throughout, the vast majority of introductory and concluding comments in the manuscripts included in the collection, as well as in justifying attention to environmental education in curriculum more broadly (see too, Sterling, 2010). However, what appears to be at stake is not simply the additive, subtractive or combinatory in curriculum as a response to crisis, but whether it *so nearly is*—or *ever can*—address pressing environmental considerations adequately, richly and generatively.

Curriculum as critique

So in light of such hesitations and qualifications, we might detect another more cautious goal in the exercise of critique, that emerges through attention to the rhetorical and compositional features of curriculum as identified in these studies (see Popkewitz, 2009). This is illustrated relatively easily by the argument that an environmental education that doesn't deliberately set out to develop a critical environmental sensibility amongst students and teachers simply isn't 'worthy of the name' (Bonnett's phrase, 2007). For example, in the collection, Gayford (1986) discusses the persistence of the effects of 'subject chauvinism' in policy, analysis and teachers' lives, including in their professional development. Similarly, Gough (1989) explores the rhetoric, epistemology and presumed superiority detectable in various mainstream to eco-curriculum orientations (e.g. 'Earth Education'—Van Matre, 1990, cf. Dahlbeck, 2014; Lotz-Sisitka, 2017). Equally, via means of a case study, Greenall Gough and Robottom (1993) discuss a coastal water pollution project that illustrates the values and challenges of interdisciplinary and internationalised, socially critical eco-curriculum activity and action. While as Jenkins (1994) argues, 'real-life impact' as a criterion for curriculum evaluation might transform both science and environmental education, whereas Zembylas (2005) begins to tease out, amongst other things, 'the radical implications of the goal of "teaching science for social justice".'

The senses of critique at work here then can be recast as another relatively straightforward question of investigating *what* and *who* is the curriculum relevant to, and *why*? From a specifically environmental sustainability perspective, this may well be engaged in terms of particular or relative interests related to 'economy', if not 'societal' or 'environmental' concerns, as raised by a standard 3-pillar model of sustainable development, and the Brundtland-informed understandings of the concept discussed in our introductions. But it might be more than this too, including to show that other calls and demands on curriculum might reframe dialogue and debate in new ways. As in the examples listed in the second section of this collection, on 'Accounting for Curriculum in Environmental Education', this might be by articulating and advocating clear attention to a 'less anthropocentric' philosophy and framing of pedagogy and practice, including the responsibilities these might entail (see also Davies, 2014).

In fact, when Bonnett (2007), 'explores the view that environmental education—indeed 'any' education—worthy of the name needs to bring a range of searching questions concerning nature to the attention of learners, and to encourage them to develop their own on-going responses to those questions' (p. 707), we might also recall that Foster (2011) examines various ecological and intergenerational 'horizons' and 'virtues', and how these might be inflected in the demands of active learning governed by the 'virtues of critical self-awareness, exploratory-creative commitment, and a robust tolerance for uncertainty' (p. 383). So as a second substantial line taken in the offering of critique, we can recognise that in inspecting and interrogating curriculum from a range of stakeholder, empirical, philosophical, sociological or policy-related perspectives, certain key areas will be subject to particular scrutiny. As with Foster, these centre on those connections authors (or environmental educationalists) claim or dispute as suggestive of worthwhile learning, teaching, curriculum and their evaluation.

Curriculum in crisis?

But what happens if we accept that environmental educators, educationalists and scholars are in fact formed in a curriculum vocabulary they did not choose? Scholarship on environmental education as well as in the Journal more broadly has been quick to link questions of critique to reflexively accounting for aspects of the 'good, bad and the ugly' in a range of settings and scholarly pursuits. These include critically evaluating the shifting vocabularies and grammars of preparatory and continuing professional development of teachers and educators (if not their priorities and lacunae), as key to change and innovation, alongside matters of resistance and complication in this area (e.g. Cotton, 2006; Gayford, 1986). A particular area of concern is the forces and currents of professionalism (including deprofessionalisation) of teachers in local as well as broader curriculum development and innovation across a range of scales and timeframes (e.g. Lee, 2000; Jickling & Wals, 2008), a matter we might also speculate as extending to scholars too? Overlaying much of this is a challenge also expressed in the aims and scope of this Journal. This serves as a wider expectation of work presented both at the World Environmental Education Congress and for curriculum scholars working in this area: is what is being produced and discussed as innovative papers *those that analyse the ways in which the social and institutional conditions of education and schooling contribute to shaping curriculum*?

On this, we can note that while some things get taken up, some simply shouldn't, in scholarship or by certain scholars (e.g. when a certain 'critique' is actually an act of cultural appropriation). Thus it may be with some tentativeness that we note that those studies developing a comparative perspective in this collection afford critical insights by, for example:

- drawing on concepts of *Bildung* and action competence to evaluate national and international efforts and programs in environmental and health education, and by arguing for interdisciplinary perspectives and 'genuine' rather than 'tokenistic' or 'moralistic' participation in addressing environmental issues through education; see:

o Jensen (2004), for a case study from Danish education;

o Backman (2010), on the need to distinguish participation in activities from learning in Sweden, and what that implies for 'learning progressions'; and

o Biesta (2012) on education as credentialisation, socialisation and subjectification, more broadly speaking, using the ideas of *paideia* and *Bildung* to explore what it means to become 'world-wise' not just 'symbol-wise' via curriculum more generally;

• using various forms of critical discourse analysis to show how environment is 'institutionalised' in official curricular texts (including guidelines and examination papers) to then examine the tensions and problematics of using 'education to tackle social inequality'; see:

o Ross (2007), on Scottish curricula texts; and

o Jardine (1994, p. 510) on how in North America, 'well-intentioned' ecopedagogies might, however, 'unintentionally work against the real, Earthly conditions under which pedagogy is actually possible';

• applying eco-educational criticism (as an arts-based inquiry method with an ecological lens) to explore the intentions and practices of ecologically-minded teachers; see:

o Moroye (2009), for example, on traditional public high schools in the US, and how these fit with efforts to infuse environmental education and reform schools in the West towards supporting 'complementary curriculum';

• developing socially critical approaches to schooling and environmental education, such as in Australia to provide 'students with a map of the existing culture and society and a map of what a better society might be like . . . [to create] an informed commitment to the improvement of society' by empowering 'students to participate in a democratic transformation of society' (Greenall Gough & Robottom, 1993); and

• reading environmental education curriculum priorities and practices through another subject lens (e.g. broadly, science, arts or humanities) to argue for a 'real world action orientation' to environmental education, and thus revitalising debates about the nature and purpose of schooling in an age confronted by environmental and other problems of global significance; see:

o Jenkins (1994, p. 607) on how this might also involve noting how teachers 'abandon existing and familiar practices in favour of strategies which involve engagement with issues which are . . . controversial, messy and have to be brought into focus only to lack a unique, or even (initially or eventually) an agreed, solution'.

On that final point, we note that Jenkins (1994, p. 606) also writes:

> environmental education involves more than the acquisition of knowledge about a range of environmental matters and is concerned with the generation and implementation of solutions to practical problems or needs. Given this, environmental education exposes with particular clarity the complex

interactions among social, economic, personal and other value positions associated with almost any environmental issue such as atmospheric pollution or acid rain. Interestingly, research offers little support for the idea that the ways in which people treat the environment can be modified simply by helping them acquire the relevant environmental knowledge or attitudes (Hungerford and Volk 1990 [sic]). What seems to be missing is the engagement of such knowledge and attitudes with the commitment to doing something about an environmental issue. Commonly, such an issue is initially of local, rather than national, international or global concern, although the local issue may be a particular example of a wider problem.

Thus in accounting for this situation and the fair expectation that curriculum should make sense of the age we live in, we trust that this collection and its various contributions and commentaries show there are both areas for and lines of critique that lead to obvious questions about curriculum traditions, scholarship and development (Biesta & Stams, 2001). To illustrate, while Jickling and Wals (2008) identified 'globalizing forces' that are altering environmental education significantly, now the UN Decade of Education for Sustainable Development is over (2005–2014), are we still witnessing various efforts stemming from UNESCO 'to convert environmental education into education for sustainable development' that raise 'issues arising from . . . anomalies in light of the nature and purposes of education'? Do we need new forms and foci for analysis and developments here, on aspects of receptivity to curriculum change and innovation, or reticence and reluctance given recent curriculum priorities and developments (Gough, 2013)? Might there be something akin to an 'isostatic readjustment' afoot in curriculum thinking about the environment as the Decade unwinds and environmental education traditions recover, as some hope (Jickling & Sterling, 2017)? Or are these beyond recovery, marginalised in the face of a dominant ESD, which itself is marginal in the face of a dominant 'education for *un*sustainable development'?

As mentioned earlier, trumping all this might well be the question of ensuring current and future generations of teachers as curriculum makers engage 'with the idea of the Anthropocene as a mechanism' that shifts 'teacher education practice towards a deeper entanglement with sustainability, starting from a [school] subject perspective' (Coles et al. 2017). If not, what else is required, or what else must be done (Luke, 2001)?

In conclusion, to restate some of the perennial questions of curriculum studies, this collection asks readers:

- which educative experiences should a curriculum foster and why?
- what should be the scope of a worthwhile curriculum and how should it be decided and organized?
- when should distinctive curricula be provided to different groups of students? and
- how should (a) curriculum be best enacted and evaluated?

We also restate that related questions and associated challenges of *orientation*, *instrumentality* and *extensionism* sketched in the editorial and introduction to this collection, that concern insider and outsider scholars of environmental education and associated curricula, raise associated challenges of:

(i) 'mainstream' versus 'alternative' curriculum orientations and movements;

(ii) the place and critique of 'instrumentalism' in curriculum policy and practice; and

(iii) how attending to a particular form of 'extensionism'—in essence, those attempts to stretch the curriculum to include another area of curriculum without removing or compromising attention to others—has come to shape key debates on the curriculum priorities and politics regarding environmental and related areas.

So in closing this collection, we hope it has shown the *Journal of Curriculum Studies* provides a 'safe' but also 'discomfortable' space to discuss contextualised and nuanced understandings of education in relation to environment and/or sustainability, where the very concepts themselves are treated as those which remain, to some extent, open and unresolved. In other words, it is by paying critical attention to important social, cultural, geographical, historical and pedagogical differences, that curriculum scholarship can contribute to both new and critical curriculum theories in this field, as much as it might critique and shatter inadequate forms of environmental education in and for these times, and education more generally.

References

Backman, Erik (2011) Friluftsliv: a contribution to equity and democracy in Swedish Physical Education? An analysis of codes in Swedish Physical Education curricula, *Journal of Curriculum Studies*, 43:2, 269–288.

Bengtsson, Stefan, & Östman, Leif (2016) Globalisation and education for sustainable development: exploring the global in motion, *Environmental Education Research*, 22:1, 1–20.

Bernstein, Basil (1971) On the classification and framing of educational knowledge. In: Michael F. D. Young (ed.), *Knowledge and Control: New Directions for the Sociology of Education* (pp. 47–69) (London: Collier Macmillan).

Biesta, Gert (2012) Becoming world-wise: an educational perspective on the rhetorical curriculum, *Journal of Curriculum Studies*, 44:6, 815–826.

Biesta, Gert J. J., & Stams, Geert J. J. M. (2001) Critical thinking and the question of critique: some lessons from deconstruction, *Studies in Philosophy and Education*, 20:1, 57–74.

Blühdorn, Ingolfur, & Welsh, Ian (eds.) (2008) *The Politics of Unsustainability: Eco-Politics in the Post-Environmental Era* (London: Routledge).

Bonnett, Michael (2007) Environmental education and the issue of nature, *Journal of Curriculum Studies*, 39:6, 707–721.

Bowers, Chet (1990) Educational computing and the ecological crisis: some questions about our curriculum priorities, *Journal of Curriculum Studies*, 22:1, 72–76.

Brookes, Andrew (2002) Lost in the Australian bush: outdoor education as curriculum, *Journal of Curriculum Studies*, 34:4, 405–425.

Brown, Wendy (2005) *Edgework: Critical Essays on Knowledge and Politics*. (Princeton: Princeton University Press). [From Chapter 1, Untimeliness and punctuality: critical theory in dark times, pp. 1–16]

Clémençon, Raymond (2012) Welcome to the Anthropocene: Rio+20 and the meaning of sustainable development, *The Journal of Environment & Development*, 21:3, 311–338.

Coles, Alf, Dillon, Justin, Gall, Marina, Hawkey, Kate, James, Jon, Kerr, David, Orchard, Janet, Tidmarsh, Celia & Wishart, Jocelyn (2017) Towards a teacher education for the Anthropocene. In: Peter Corcoran, Joseph Weakland & Arjen Wals (Eds.) *Envisioning Futures for Environmental and Sustainability Education* (pp. 73–85) (Wageningen: Wageningen Academic Publishers).

Cotton, Debby (2006) Implementing curriculum guidance on environmental education: the importance of teachers' beliefs, *Journal of Curriculum Studies*, 38:1, 67–83.

Covert, Douglas C. (1969) Toward a curriculum in environmental education, *Journal of Environmental Education*, 1:1, 11–12.

Dahlbeck, Johan (2014) Hope and fear in education for sustainable development. *Critical Studies in Education*, 55:2, 154–169.

Davies, Bronwyn (2014) Legitimation in post-critical, post-realist times, or whether legitimation? In: Alan D. Reid, E. Paul Hart, & Michael A. Peters (Eds.), *A Companion to Research in Education* (pp. 443–450) (Dordrecht: Springer Netherlands).

Fahey, Shireen J. (2012) Curriculum change and climate change: inside outside pressures in higher education, *Journal of Curriculum Studies*, 44:5, 703–722.

Foster, John (2011) Sustainability and the learning virtues, *Journal of Curriculum Studies*, 43:3, 383–402.

Gayford, Chris (1986) Environmental education and the secondary school curriculum, *Journal of Curriculum Studies*, 18:2, 147–157.

Gough, Noel (1989) From epistemology to ecopolitics: renewing a paradigm for curriculum, *Journal of Curriculum Studies*, 21:3, 225–241.

Gough, Noel (2013) Thinking globally in environmental education: a critical history. In: Robert Stevenson, Michael Brody, Justin Dillon, & Arjen Wals (Eds.), *International Handbook of Research on Environmental Education* (pp. 33–44) (Washington, DC: Routledge).

Greenall Gough, Annette (1991) Greening the future for education: changing curriculum content and school organization, *Journal of Curriculum Studies*, 23:6, 559–571.

Greenall Gough, Annette & Robottom, Ian (1993) Towards a socially critical environmental education: water quality studies in a coastal school, *Journal of Curriculum Studies*, 25:4, 301–316.

Hawkey, Kate (2014) A new look at big history, *Journal of Curriculum Studies*, 46:2, 163–179.

Hopmann, Stefan (2003) On the evaluation of curriculum reforms, *Journal of Curriculum Studies*, 35:4, 459–478.

Hungerford, Harold R., & Volk, Trudi L. (1990) Changing learning behaviour through environmental education, *Journal of Environmental Education*, 21:3: 8–12.

Jardine, David W. (1994) 'Littered with literacy': an ecopedagogical reflection on whole language, pedocentrism and the necessity of refusal, *Journal of Curriculum Studies*, 26:5, 509–524.

Jenkins, Edgar W. (1994) Public understanding of science and science education for action, *Journal of Curriculum Studies*, 26:6, 601–611.

Jensen, Bjarne Bruun (2004) Environmental and health education viewed from an action-oriented perspective: a case from Denmark, *Journal of Curriculum Studies*, 36:4, 405–425.

Jickling, B. & Sterling, S. (Eds.) (2017) *Post-Sustainability and Environmental Education: Remaking Education for the Future* (Palgrave Macmillan).

Jickling, Bob, & Wals, Arjen (2008) Globalization and environmental education: looking beyond sustainable development, *Journal of Curriculum Studies*, 40:1, 1–21.

Kopnina, Helen (2014) Future scenarios and environmental education, *The Journal of Environmental Education*, 45:4, 217–231.

Lee, John Chi-Kin (2000) Teacher receptivity to curriculum change in the implementation stage: the case of environmental education in Hong Kong, *Journal of Curriculum Studies*, 32:1, 95–115.

Lotz-Sisitka, Heila (2017) Decolonisation as future frame for environmental and sustainability education: embracing the commons with absence and emergence. In: Peter Corcoran, Joseph Weakland & Arjen Wals (Eds.) *Envisioning Futures for Environmental and Sustainability Education* (pp. 45–62) (Wageningen: Wageningen Academic Publishers).

Llewellyn, Kristina R. Cook, Sharon A., & Molina, Alison (2010) Civic learning: moving from the apolitical to the socially just, *Journal of Curriculum Studies*, 42:6, 791–812.

Luke, Tim W. (2001) Education, environment and sustainability: what are the issues, where to intervene, what must be done? *Educational Philosophy and Theory*, 33:2, 187–202.

Moroye, Christy (2009) Complementary curriculum: the work of ecologically minded teachers, *Journal of Curriculum Studies*, 41:6, 789–811.

Morton, Timothy (2013) *Hyperobjects: Philosophy and Ecology after the End of the World* (Minneapolis: University of Minnesota Press).

Popkewitz, Tom (2009) Curriculum study, curriculum history, and curriculum theory: the reason of reason, *Journal of Curriculum Studies*, 41:3, 301–319.

Pritchard, Tom (1968) Environmental education – its social relevance in North-West Europe. (Strasbourg: Council of Europe) (unpublished report CE/Nat (68) 67).

Ross, Hamish (2007) Environment in the curriculum: representation and development in the Scottish physical and social sciences, *Journal of Curriculum Studies*, 39:6, 659–677.

Sauvé, Lucie Brunelle, Renée, & Berryman, Tom (2005) Influence of the globalized and globalizing sustainable development framework on national policies related to environmental education, *Policy Futures in Education*, 3:3, 271–283.

Sterling, Stephen (2010) Living *in* the Earth: towards an education for our times, *Journal of Education for Sustainable Development*, 4:2, 213–218.

Sumner, Jennifer (2008) From academic imperialism to the civil commons: institutional possibilities for responding to the United Nations Decade of Education for Sustainable Development, *Interchange*, 39:1, 77–94.

Tupper, Jennifer A., & Cappello, Michael P. (2012) (Re)creating citizenship: Saskatchewan high school students' understandings of the 'good' citizen, *Journal of Curriculum Studies*, 44:1, 37–59.

UNESCO (2017) Education for Sustainable Development Goals: Learning Objectives (Paris: UNESCO). http://unescodoc.unesco.org/images/0024/002474/247444e.pdf

Van Matre, Steve (1990) *Earth Education... A New Beginning* (Warrenville, Ill.: The Institute for Earth Education).

Zembylas, Michalinos (2005) Science education: for citizenship and/or for social justice?, *Journal of Curriculum Studies*, 37:6, 709–722.

Index